BIOLOGY OF GROWTH FACTORS

Molecular Biology, Oncogenes, Signal Transduction, and Clinical Implications

ADVANCES IN EXPERIMENTAL MEDICINE AND BIOLOGY

Recent Volumes in this Series

Volume 226
MOLECULAR MECHANISM OF MUSCLE CONTRACTION
Edited by Haruo Sugi and Gerald H. Pollack

Volume 227
OXYGEN TRANSFER FROM ATMOSPHERE TO TISSUES
Edited by Norberto C. Gonzalez and M. Roger Fedde

Volume 228
THE MOLECULAR IMMUNOLOGY OF COMPLEX CARBOHYDRATES
Edited by Albert M. Wu and L. Garry Adams

Volume 229
LIPOXINS: Biosynthesis, Chemistry, and Biological Activities
Edited by Patrick Y-K Wong and Charles N. Serhan

Volume 230
CELL AND MOLECULAR BIOLOGY OF THE UTERUS
Edited by Wendell W. Leavitt

Volume 231
ADVANCES IN POST-TRANSLATIONAL MODIFICATIONS OF PROTEINS AND AGING
Edited by Vincenzo Zappia, Patrizia Galletti, Raffaele Porta, and Finn Wold

Volume 232
CELLULAR Ca^{2+} REGULATION
Edited by Douglas R. Pfeiffer, Jeanie B. McMillin, and Steve Little

Volume 233
CANCER METASTASIS
Edited by Giorgio Prodi, Lance A. Liotta, Pier-Luigi Lollini, Spiridione Garbisa, Sergio Gorini, and Kurt Hellmann

Volume 234
BIOLOGY OF GROWTH FACTORS: Molecular Biology, Oncogenes, Signal Transduction, and Clinical Implications
Edited by Jeffrey E. Kudlow, David H. MacLennan, Alan Bernstein, and Avrum I. Gotlieb

---

A Continuation Order Plan is available for this series. A continuation order will bring delivery of each new volume immediately upon publication. Volumes are billed only upon actual shipment. For further information please contact the publisher.

BIOLOGY OF GROWTH FACTORS

Molecular Biology, Oncogenes, Signal Transduction, and Clinical Implications

Edited by

Jeffrey E. Kudlow

Departments of Clinical Biochemistry and Medicine and
Banting and Best Diabetes Centre
University of Toronto and
Toronto General Hospital
Toronto, Ontario, Canada

David H. MacLennan

Banting and Best Department of Medical Research
University of Toronto
Toronto, Ontario, Canada

Alan Bernstein

Department of Medical Genetics and Medical Biophysics
University of Toronto and
Division of Molecular and Developmental Biology
Mount Sinai Hospital Research Institute
Toronto, Ontario, Canada

and

Avrum I. Gotlieb

Department of Pathology and
Banting and Best Diabetes Centre
University of Toronto and
Toronto General Hospital
Toronto, Ontario, Canada

SPRINGER SCIENCE+BUSINESS MEDIA, LLC

Library of Congress Cataloging in Publication Data

Symposium on Biology of Growth Factors: Molecular Biology, Oncogenes, Signal Transduction, and Clinical Implications (1987: Toronto, Ont.)
Biology of growth factors.

(Advances in experimental medicine and biology; v. 234)
"Proceedings of the Triennial Symposium on Biology of Growth Factors: Molecular Biology, Oncogenes, Signal Transduction, and Clinical Implication, held June 17–19, 1987, Toronto, Ontario, Canada"—T.p. verso.
Includes bibliographies and index.
1. Growth promoting substances—Physiological effect—Congresses. 2. Pathology, Molecular—Congresses. 3. Oncogenesis—Congresses. 4. Blood-vessels—Pathophysiology—Congresses. 5. Autoimmunity—Congresses. 6. Cells—Growth—Regulation—Congresses. 7. Cell proliferation—Congresses. I. Kudlow, Jeffrey E. II. Title. III. Series. [DNLM: 1. Cell Communication—congresses. 2. Cell Transformation, Neoplastic—congresses. 3. Growth Substances—Physiology—congresses. 4. Oncogenes—congresses. 5. Receptors, Endogenous Substances—physiology—congresses. W1 AD559 v.234 / QZ 202 S9797b 1987]
QP801.G74S96 1988 616.07′1 88-17945
ISBN 978-1-4757-1982-6 ISBN 978-1-4757-1980-2 (eBook)
DOI 10.1007/978-1-4757-1980-2

Proceedings of the Triennial Symposium of the Banting and Best Diabetes Centre on Biology of Growth Factors: Molecular Biology, Oncogenes, Signal Transduction, and Clinical Implications, held June 17–19, 1987, in Toronto, Ontario, Canada

Originally published by Plenum Press, New York in 1988
Softcover reprint of the hardcover 1st edition 1988

PREFACE

Growth factors are elaborated to control the growth of cells in such physiological processes as wound healing, tissue regeneration and the immune response. Abnormal production of these growth factors, their receptors or intracellular mediators of their action may lead to disease states including oncogenesis. This volume will focus on exciting developments in defining the precise molecular lesions that permit the conversion of controlled proliferative signals to neoplasia, on the possible involvement of growth factors in the development of blood vessel diseases as seen in diabetes and atherosclerosis, on the altered immune surveillance that leads to autoimmunity and on the fundamental mechanisms by which growth factors signal their target cells. We expect that the contents of this volume will help promote understanding of the role of these fundamental biological processes and their alterations in a wide variety of disease states and stimulate new investigations in this important area of biomedical research.

The Editors

CONTENTS

PERSPECTIVES ON THE BIOLOGY OF GROWTH FACTORS

Irving B. Fritz

Banting and Best Dept. of Med. Res.
University of Toronto
Toronto, Ontario, M5G 1L6

In this introduction to the "Symposium on the Biology of Growth Factors", I shall offer perspectives gained from insights provided by classical endocrinology.

Hormones (like insulin) are usually produced primarily by a single cell type (i.e. the β cells of the Islets of Langerhans) in response to specific signals (i.e. an elevation of glucose levels in the plasma). The hormones are transported, most often by a vascular route, to cells which have specific receptors, to which the hormones bind with high affinity. Occupied receptors undergo conformational changes which are recognized by a "transducer", and a cascade of reactions is thus initiated. Rates of particular multienzyme pathways (such as glycogen synthesis, glycolysis, lipolysis, etc.) are subsequently altered, in a manner characteristic of the phenotype of the cell being stimulated. These late events are mediated by modulation of the activity of an enzyme which catalyzes a rate-determining reaction in the multi-step sequence. A specific phosphorylation of an enzyme by a specific protein kinase is often implicated. The most complex example, perhaps, is that offered by the regulation of overall rates of glycolysis by the regulation of the levels of fructose 2,6-bisphosphate (Pilkis et. al., 1984). 6-Phospho-fructo-2-kinase, fructose 2,6-bisphosphatase is a bifunctional enzyme which catalyzes the formation of fructose 2,6-diphosphate from fructose--6-phosphate and ATP when in the non-phosphorylated form, and which catalyzes the hydrolysis of fructose 2,6-bisphosphate to fructose-6-phosphate and inorganic phosphate when the enzyme is in the phosphorylated form.

Classical systemic hormones, such as "Growth Hormone" (Somatotropin) elicit increased cell divisions in the intact organism, but addition of these hormones rarely stimulates isolated cells in culture to proliferate. Insulin offers an apparent exception, in that high concentrations can induce some cells in culture to enter mitosis. These effects are not direct ones, however, but instead are mediated by interactions of insulin with somatomedin C (IGF-I) receptors (for reviews, see Rechler and Nissley, 1985; King and Kahn, 1985). Other examples exist in which the stimulation by classical hormones of growth *in vivo* is mediated indirectly by enhancing the production of a *local* growth factor. Experiments reported by Salmon and Daughaday (1957) were probably the first to demonstrate this phenomenon. In pointing out the role of a hepatic "sulfation factor" in mediating the *in vivo* effects of somatotropin, the findings of Daughaday and his colleagues led to the discovery of the somatomedins, and to the knowledge that somatotropin acts on hepatic cells to increase somatomedin production (for review, see Daughaday, 1985). Similarly, agents which enhance ovarian follicular growth when administered *in vivo* do not have mitogenic effects *in vitro*, whereas local growth factors do elicit increased granulosa cell division (Hammond and English, 1987).

Classical endocrinologists had a relatively easy way to determine functions of hormones. They could satisfy some of Koch's postulates simply by ablating the primary source of the hormone (i.e. the pancreas or the cells in the case of insulin), and then determining the nature of components required for correction of the metabolic or growth impairments observed in the hormone-ectomized animal.

TABLE 1

COMPARISON OF CLASSICAL ENDOCRINOLOGY HORMONES WITH GROWTH CONTROL FACTORS

PROPERTY	CLASSICAL ENDOCRINOLOGY HORMONES	GROWTH CONTROL FACTORS
Origin of hormone	Single cell-type	Multiple cell-types
Control of production	Specific signal	Largely unknown
Transport of hormone	Vascular: humoral	Diffusion: local
Cell targets	Few cell types	Multiple cell types
Alter rates ?	++++	+
Modulation of growth ?	+	++++
Protein Kinases ?	Usually YES	Putatively YES

Alas, no such easy route is open to the current wave of cell and molecular biologists concerned with the functions of local growth control factors (Table 1). Comparison of the hormones of classical endocrinology with those of the more recently discovered growth control factors immediately reveals a host of differences as well as similarities. Most notably, various growth factors (e.g. somatomedins, EGF, PDGF and others) can be produced by many different cell types. If one wishes to ablate these local growth factors from the organism, it is necessary to master techniques more sophisticated than surgery or simple pharmacology. Fortunately, recombinant DNA technology makes it theoretically possible to prevent the expression of specific growth factors, with the use of anti-sense mRNA inserts.

Classical endocrinology has made many contributions to our understanding of the control of growth (Table 2) (for review, see Tepperman and Tepperman, 1987). The information, techniques and concepts developed in this area during the last half-century have served us well in defining problems and suitable approaches to them. For example, it has been well documented that occupancy of a specific receptor by its ligand can initiate a cascade of events, often associated with activation of adenylate cyclase activity. The role of G proteins, or transducin, in these processes is being well delineated. The second messenger concept, arising from the work of Sutherland, Rall and their colleagues, has been extended to include not only cAMP, which can regulate the activities of specific protein kinases, but also by components other than cAMP. Well-known examples include the influences of α,β-diglyceride on protein

TABLE 2

Contributions of Classical Endocrinology To Understanding of Growth Control Systems

1. **Catalogue of hormones**
2. **Discriminator (receptor occupancy) concept**
3. **Concept of transduction, and control of cyclase activity; role of G protein**
4. **Second messenger concept**
   a) cAMP
   b) Ca^{++}
5. **Controls of protein kinase activities**
6. **Role of phosphorylation of key proteins in the control of complex metabolic pathways**

kinase C activity; calcium ion dependent protein kinases; and tyrosine protein kinases which are turned on in response to EGF or insulin action. During the last few decades, classical biochemistry has joined forces with classical endocrinology to make spectacular gains in our knowledge and understanding of the control of metabolic processes. In the early 1940's, the Cori laboratory discovered that epinephrine elicits glycogen breakdown but not glycogen synthesis. Since then an enormous amount of information has accumulated which demonstrates the crucial role of phosphorylation of phosphorylase and glycogen synthase in these processes (for review, see Tepperman and Tepperman, 1987). Other protein kinases influence overall rates of glucose conversion to pyruvate, and rates of lipolysis by influencing activities of rate-determining enzymes. It is readily apparent that current investigations on the mechanisms of action of growth factors represent logical extrapolations from concepts emanating from classical endocrinology concerning the mechanisms of control of complex metabolic pathways. Continuity of the spectrum is obvious.

To underline this point, consider the statement that "the growth factor control system is like the classical endocrine system of the microenvironment" (Fig. 1). [I have quotation marks around this statement because I cribbed it from an assigned term paper in this year's course on Growth Factors. The graduate student (Matthew Moyle) who wrote this in his essay assures me that he in turn had cribbed it from my lectures, and gave me permission to use the phrase]. The reasoning behind the statement emphasizes similarities, differences, and interactions between investigations of classical endocrinology and investigations of growth control factors. Factors locally released by target cells in response to classical hormones may act as autocrine or

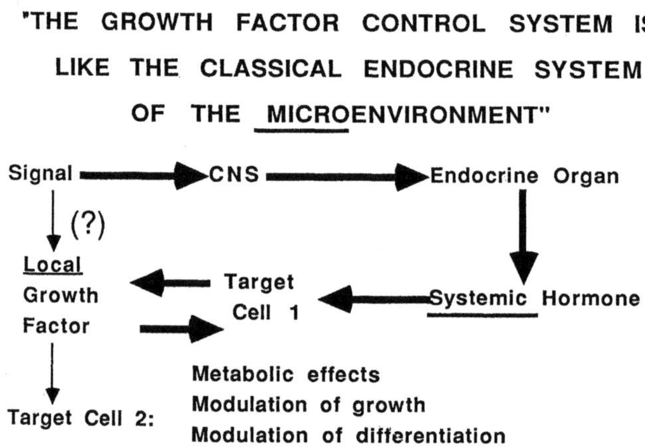

Fig. 1.

paracrine agents. Signals from the external environment (or changes in the internal environment, such as alterations in plasma levels of substrates or Ca^{++}) are detected by specific receptors. The occupancy of receptors in appropriate sensory cells (such as those in the CNS) initiates a cascade leading to changes in the production and release of traditional hormones by source cells (such as various hypothalamic or adenohypophysial cells). The "classical" hormones produced and secreted then act on specific target cells at a distance. Thus, TSH, for example, initiates the synthesis of triiodothyronine or thyroxin by the thyroid. However, such hormones may also trigger the formation of local growth factors. We have already mentioned the hepatic production of IGF-1 stimulated by somatotropin. In addition, somatomedin C production by ovarian granulosa cells is enhanced by estrogens and gonadotropins (Hsu and Hammond, 1987). Examples exist in which somatomedin C production by other cell types is stimulated by different hormones.

Alternatively, a growth factor producing cell could act autonomously, independent of regulation by classical hormones (for review, see Heldin and Westermark, 1984) (Fig. 2). In this scheme, control in normal cells is mediated by hormones or ligands (1) combining with receptors (2) to generate a cascade (3) which results somehow in stimulation of cell division (Fig. 2A). In other circumstances, however, the cell could generate its own growth factor (such as the PDGF derivative in v-sis oncogene-transformed cells), and thereby regulate its own growth with an autocrine factor (Fig. 2B, arrow 1). Similarly, a neighboring cell could be influenced by this factor. Alternatively, there could be a mutation in the receptor such that occupancy by the

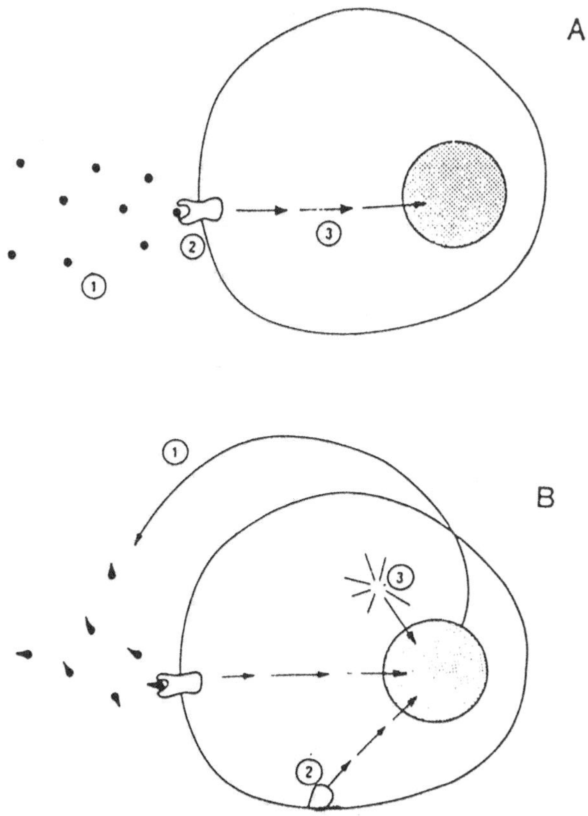

Fig. 2. Part A) Sites at which cell division can normally be regulated by hormones or growth factors (1); receptors (2); and by transduction mechanism leading to a metabolic cascade which stimulates DNA replication (3). In panel B, cell division could be autonomously regulated by secreting growth factors (autocrine regulators) (1); by having altered types pf receptors which did not require occupation by ligands to initiate nuclear division (2); or which had indigenously rapid rates of DNA replication (3). From Heldin and Westermark (1984).

usual ligand was not required to permit maximal generation of the cascade leading to nuclear division, as for example in the v-erb B oncogene-transformed cells in which a truncated EGF-like receptor exists (Fig. 2B, arrow 2). Finally, other intracellular events controlling the cell cycle could be totally independent of control by hormones or receptors (Fig. 2B, arrow 3).

How could such controls for rates of cell division have originated and evolved? In prokaryotes, the availability of substrates (or the lack thereof) is clearly a crucial factor in determining the rates of cell proliferation. However, even in the presence of optimal or excess substrates, mechanisms evolved to limit cell growth or to favor differentiation.

Consider the case in sponges. Proliferation and growth by non-sexual replication occurs under usual optimal conditions. Nevertheless, each archeocyte in the sponge retains the capacity to differentiate into a gonad in which germ cells undergo meiosis (for reviews, see Willmer, 1970; Fell, 1974; and Fritz, 1986) (Fig. 3). The generation of haploid cells and sexual reproduction is programmed in response to specific signals from the environment which are detected by specialized cells. What regulates the turn-off of mitosis and the turn-on of differentiation in these sponge cells? It seems possible that the production of local growth or differentiation factors may be under direct modulation by environmental stimuli. Under these conditions, with sponges, it is not possible to define "CNS"-type receptors, or a "classical" endocrine system. It seems more likely that a less specialized somatic cell would release a factor in response to a detectable change in the environment. This speculation provides the basis for the question mark in the arrow leading directly from "signal" to "local

DIFFERENTIATION OF SPONGE CELLS TO FORM SPERMARY FOLLICLE:
Potential Importance of Ratio of Modulation Factors to Growth Factors

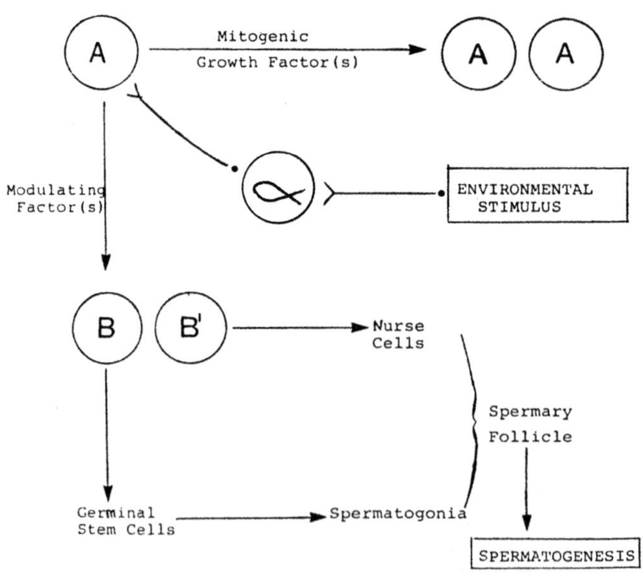

Fig. 3. For details, see the text, and review by Fritz (1986), from which this diagram is taken.

growth factor" in Fig. 1. In metazoan organisms, two sets of mechanisms, one systemic and the other local, could be operative in generating different levels of control of target cell responses. Parallels between levels of direct and indirect controls in the modulation of growth or differentiation in response to changes in the environment can be readily envisaged (Fig. 4) (for review, see Fritz, 1986). A shared final common pathway is postulated to be regulated by complex sets of checks and balances, which become increasingly numerous and subtle during evolution.

What mechanisms govern the control of differentiation of cells? How do these mechanisms differ from those controlling the proliferation of cells? What modulates possible transitions during the cell cycle to determine commitment? (Fig. 5). How does receptor transduction under one set of circumstances lead to cell division, and under a different set of circumstances lead to be altered functions in the same cell type? Approaches to these difficult questions pose a tremendous challenge. Answers will refine and extend the sorts of "simpler" questions posed previously by classical endocrinologists. Nevertheless, I hope that contemporary cell and molecular biologists will remain well aware of the magnitude of the splendid foundation we have inherited from our predecessors. Our heritage, bequeathed by prior practioners of the art, is indeed a rich one. It augurs well for continuing advances in knowledge, and perhaps even in wisdom, concerning the extraordinarily complex controls of growth and differentiation which have evolved.

By the nature of reductionist approaches, we begin with a detailed analysis of each growth factor, its chemistry, and its mechanisms of action. Hopefully, we shall progress to an understanding of mechanisms of interactions among growth factors in the control of cellular responses, and hence to a greater understanding of the biology of the intact organism.

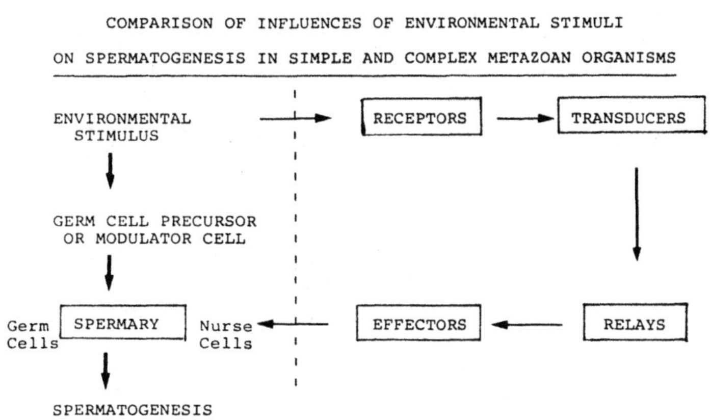

Fig. 4. For details, see the text, and review by Fritz (1986), from which this diagram is taken.

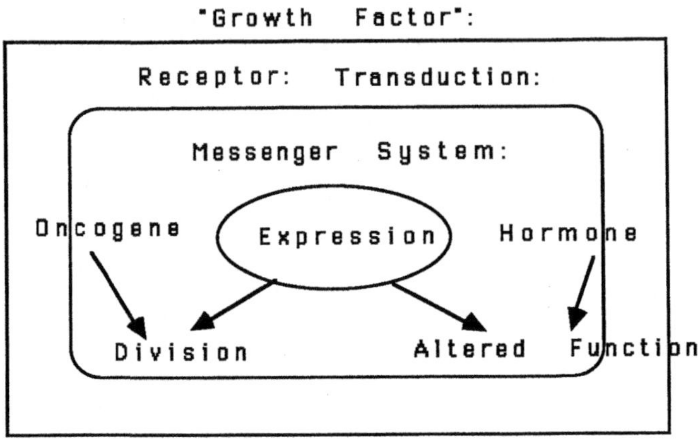

Fig. 5. For details, see the text, and review by Fritz (1986), from which this diagram is taken.

REFERENCES

Doughaday, W. H., 1985, Prolactin and Growth in Health and Disease, in: "Contemporary Endocrinology," S. H. Ingbar, ed., Plenum Publishing Corp., N.Y. Vol. 2:27.

Fell, P. E., 1974, Porifera, in: "Reproduction of Marine Invertebrates," A. C. Giese, and J. S. Pearse, eds., Academic Press, N.Y. Vol. 1:51.

Fritz, I. B., 1986, Reflections on the evolution of the regulation of spermatogenesis, Prog. Clin. Biol. Res., 226:371.

Hammond, J. M., and English H. F., 1987, Regulation of deoxyribonucleic acid synthesis in cultured porcine granulosa cells by growth factors and hormones, Endocrinol., 120:1039.

Heldin, D. H., and Westermark, B., 1984, Growth factors. Mechanisms of action and relations to oncogenes, Cell, 37:9.

Hsu, C-J., and Hammond, J. M., 1987, Gonadotropins and estradiol stimulate immunoreactive insulin-like growth factor-I production by porcine granulosa cells in vitro, Endocrinol., 120:198.

King, G. L., and Kahn, C. R., 1985, Effect of insulin on growth in vivo and cells in culture, in: "Control of Animal cell Proliferation," A. L. Boynton, and H. L. Leffert, eds., Academic Press Inc., N.Y. Vol. 1:204.

Pilkis, S. J., Regen, D. M., Stewart, B. H., Chrisman, T., Pilkis, J., Kountz, P., Pate, T., McGrane, M., El-Maghrabi, M. R., and Claus, T., 1984, Rat liver 6-phosphofructo 2-kinase/fructose 2.6-bisphosphatose: A unique bifunctional enzyme regulated by cyclic AMP-dependent phosphorylation, in: "Enzyme Regulation by Reversible Phosphorylation-Further Advances," P. Cohen, ed., Elsevier Science Publishers B.V., N.Y. Pp. 95-122.

Rechler, M. M., and Nissley, S. P., 1985, The nature and regulation of the receptors for insulin-like growth factors, Ann. Rev-Physiol., 47:425.

Salmon, W. D., and Daughaday, W. H., 1987, A hormonally controlled serum factor which stimulates sulfate incorporation by cartilage in vitro, J. Lab Clin. Med., 49:825.

Tepperman, J., and Tepperman, H. M. 1987, "Metabolic and Endocrine Physiology," Year Book Medical Publishers, Inc., N.Y., Fifth edition. Pp. 29-42.

Willmer, E. N., 1970, "Cytology and Evolution," Academic Press, N.Y., 2nd edition.

PLATELET-DERIVED GROWTH FACTOR -- ITS ROLE IN HEALTH AND DISEASE

Russell Ross and Elaine W. Raines

Department of Pathology
University of Washington
Seattle, WA

INTRODUCTION

It has been well known for many years that serum is required for the growth of most diploid cells in culture. Several attempts were made to determine the source of the growth activity present in serum. Partially purified fractions that contained mitogenic activity were isolated, but none of these demonstrated the cellular source of this proliferative activity. In 1974, it was observed that platelets were the source of the principal mitogenic activity present in whole blood serum and absent in cell-free plasma-derived serum.[1,2] These observations were made initially for smooth muscle cells[1] and subsequently for mouse embryo 3T3 cells.[2] At that time it was not clear whether there was more than one growth factor present in platelets. However, the major mitogen released by platelets, which is the subject of this report, has been termed the platelet-derived growth factor (PDGF).

Subsequently it has been demonstrated that platelets contain several other growth factors, including an epidermal growth factor-like (EGF-like) molecule,[3] transforming growth factor beta (TGFβ),[4] a mitogenic factor for vascular endothelial cells,[5,6] and potentially other factors yet to be characterized. Each of these growth factors appears capable of acting on a reasonably well-defined spectrum of cells and tissues. Specifically, the PDGF molecule (see below) has direct effects on most mesenchymally derived connective tissue-forming cells, including fibroblasts, osteoblasts, chondroblasts, smooth muscle cells, and glial cells. It appears, however, to have no direct effect on vascular endothelium or on epithelial cells, since neither of these two cell types contain receptors for PDGF.[7] In addition, PDGF may also have effects on lymphoid cells (see below); however, these effects appear to be indirectly mediated since lymphoid cells also have no receptors for PDGF.

THE MOLECULE

PDGF is a ubiquitous growth factor which can be formed by many different cell types (see below). Since the discovery of its presence in human platelets, it has been demonstrated that the molecules derived from other cell types may consist of different combinations of the same peptides secreted by activated platelets. PDGF from human platelets is a cationic glycoprotein (PI ≃ 9.8) with an Mr of ≃ 30,000.[8-11] On analytical silver-stained SDS polyacrylamide gels, 3 to 4 bands are characteristically observed ranging in molecular weight from 28,000 to 35,000. Human platelet-derived PDGF consists largely of a heterodimer of two peptides of 16,000 and 14,000 molecular weight, termed A and B chains, respectively.[12] Most other cells (see below) appear to synthesize and secrete either A-A or B-B homodimers, or both, rather than an A-B heterodimer like the platelet (unpublished observations). Human platelets also contain variable amounts of A-A or B-B homodimers (unpublished observations); however, the relative quantities of each and the conditions under which they occur remain to be determined. In contrast to human platelets, porcine platelets appear to contain principally B-B homodimers of PDGF.[13] A-A homodimers from the osteosarcoma cell line U-2OS have been isolated and characterized,[14] and in all cases these homodimers appear to be mitogenic.

PDGF contains up to 16 cysteine residues and is extensively disulfide bonded through both intra-chain and inter-chain disulfide bonds. Amino acid sequence analysis demonstrates a large amount of homology between the A and B chains (56%).[15,16] The A-chain gene is located on chromosome 7 in humans and that the B-chain gene is found on chromosome 22.

Comparison of the amino acid sequence of the B-chain of PDGF with the other known sequences demonstrates a striking (96%) homology between the B-chain of human PDGF and the protein product of an oncogene (v-sis) derived from the simian sarcoma retrovirus, termed $p28^{sis}$.[15,16] This molecule (p^{28-sis}) contains additional N- and C-terminal sequence distinct from PDGF, but undergoes apparent processing steps to yield molecules virtually identical to the B-chain of PDGF.[17,18] To specifically examine the biological activity of the B-chain of PDGF, it was expressed in yeast by Kelly et al.[19] They were able to express a B-B homodimer that retained all of the biological activity observed in human purified PDGF. Similarly, A-A homodimers have been purified and sequenced from the U-2OS osteosarcoma line.[14] The A-chain is distinct from the B-chain, but 56% homologous at the level of the mature proteins.[20] There have been numerous reviews that have provided greater details of the molecule and its relation to the v-sis oncogene.[21-24]

THE PDGF RECEPTOR

Receptors for PDGF on responsive cells appear to have a molecular weight of 170,000-180,000. A number of observations support the concept that the binding site for ^{125}I-PDGF is a functional receptor responsible for induction of its biological activity.[7] This receptor was first identified using affinity cross-linking techniques with disuccinamide reagents which coupled ^{125}I-PDGF bound to cell membrane preparations and identification of a PDGF-receptor complex in the membrane preparations that could not be competed for by factors other than PDGF.[25] Hart et al.[26] have developed a monoclonal antibody against the human fibroblast receptor. They showed that there are several lower molecular weight precursor forms within the cell and that the mature cell surface form of the receptor consists of a glycoprotein of approximately 180,000.

When PDGF binds to its receptor, it induces phosphorylation of the receptor principally on tyrosine moieties.[27-29] The mouse PDGF receptor has been purified, sequenced, and cloned by Yarden et al.[30] by taking advantage of this phosphorylation on tyrosine residues. The phosphorylation is due to activation of a tyrosine kinase located in the receptor molecule in a region split into two components separated by approximately 100 amino acids on the cytosolic side of the receptor.[30] Similarly, several other receptors contain tyrosine kinase activity and are autophosphorylated upon ligand binding, including the EGF receptor and the insulin receptor. This ability to catalyze transfer of phosphate groups from ATP to tyrosine on proteins is rare among protein kinases, and it has been estimated that less than 0.03% of cellular protein-bound phosphate is on tyrosine.[31] It has been suggested that tyrosine-induced phosphorylation is related to the process of the generation of a mitogenic signal in cells when PDGF binds to its receptor. However, this has not yet been proven to be the case.

Rapid Clearance of PDGF In Vivo

Plasma contains undetectable levels of PDGF using either receptor competition binding assays or an ELISA based upon the development of a polyclonal monospecific anti-PDGF IgG. When PDGF is injected intravenously into normal baboons, it is cleared with a $t_{\frac{1}{2}}$ of less than 2 minutes.[32] As a result of these observations, it is probable that PDGF is incapable of acting as a systemic hormone, but rather acts locally on connective tissue cells at sites where it may be secreted in sufficient concentration to elicit a response. Thus the specific tissue localization of cells capable of secreting PDGF may be critical in determining local concentrations of PDGF and resulting PDGF activity.

Modulation of PDGF Formation and Activity

Both heterodimers and homodimers of the A- and B-chains of PDGF are active as detailed above. Although homologous, the A- and B-chains are distinct gene products and may be regulated differentially by specific factors. For example, Leof et al.[33] observed that AKR-2B cells express mRNA for PDGF B-chain and secrete PDGF-like molecules in response to TGFβ. Other conditions that appear to induce gene expression include cell transformation,[20] culturing on plastic,[34] treatment with phorbol esters,[35] transfection and growth selection of cultured cells (Sakariassen et al., unpublished observations), and extreme growth conditions such as hypoxia (Sakariassen et al., unpublished observations). From these studies it appears that the genes encoding PDGF A and PDGF B are regulated independently.

PDGF may also exist in inactive forms within cells. For example, it has been demonstrated that when cultured endothelial cells are stimulated with thrombin, they secrete PDGF initially in the absence of protein synthesis, presumably from an inactive intracellular precursor that is somehow activated prior to or during the secretory process.[36]

PDGF localization may also be affected by the fact that it can bind to matrix constituents such as collagen.[37] PDGF will also bind to heparin,[8] although it does not bind as tightly as does basic fibroblast growth factor.[38] PDGF will also bind to plasma proteins such as alpha$_2$-macroglobulin, which can inhibit binding of PDGF to its receptor if the alpha$_2$-macroglobulin is present in sufficient concentration.[39,40] Thus interaction with various proteins, including matrix proteins, may protect PDGF against degradation, may facilitate its clearance from plasma, or may enhance its activity at local sites within the connective tissue.

RESPONSES STIMULATED BY PDGF

Phospholipase Activation

When PDGF binds to its receptor, it virtually immediately induces activation of phospholipase C-type activity,[41] as well as phospholipase A_2-type activity.[42] Such phospholipase activation results in release of free arachidonic acid and in the formation of diacylglycerol. This may also lead to the formation of inositol tris-phosphate, which may result in C-kinase activation.

The release of free arachidonate provides substrates for the formation of eicosanoids, including PGE_2 and PGI_2.[41] The increased levels of these prostaglandins could result in increased bone resorption[43] as well as vasodilation[44] and anti-platelet activity.[45] At a somewhat later time period, PDGF also leads to the synthesis of new cyclooxygenase by the cells, providing a second peak of prostaglandin formation.[46] These activities are enhanced in the presence of either very low density or low density lipoproteins, as well.[47] In addition, PDGF induces increased numbers of LDL receptors on cells that normally contain such receptors, and thus may modulate in important ways both fatty acid metabolism and lipid uptake, including cholesterol metabolism.[48,49]

Connective Tissue Matrix Alterations

PDGF can stimulate both collagen synthesis and its degradation. Narayanan and Page[50] observed that PDGF will stimulate the formation of type V collagen by gingival fibroblasts and also may regulate the synthesis of types III and IV collagen. In addition to forming collagen, PDGF can induce collagenase activity upon exposure to human dermal fibroblasts.[51,52] Collagenase secretion begins approximately 8-10 hours after PDGF binding. Since PDGF is a potent mitogen, the remodeling of connective tissue in association with cell proliferation induced by this molecule may be important in those disease entities where this occurs.

Chemotaxis

PDGF induces directed migration of fibroblasts and smooth muscle cells, and the dose that provides maximal chemotaxis is equivalent to that which provides maximal mitogenesis.[53,54] PDGF has also been shown to be chemotactic for monocytes and polymorphonuclear neutrophilic leukocytes,[55] neither of which respond mitogenically to PDGF but are stimulated to release specific granule contents when they are exposed to this mitogen.[56,57] The chemotactic property of PDGF may be a critical activity of this peptide, since the capacity to attract cells in a concentration gradient of PDGF may be important in retaining the cells at a given site and in inducing a local mitogenic response.

Indirect Effects of PDGF on Growth

PDGF appears to affect the growth of some cells that have no detectable receptors for PDGF, apparently via an indirect mechanism. Picogram amounts of PDGF will partially restore the proliferative response of antigen-specific T-cells cultured in medium containing reduced serum.[58] PDGF is not mitogenic for such T-cells in the absence of antigen and does not act synergistically with T-cell growth factor, nor does it modulate T-cell surface molecules. The effect of PDGF can be induced by preincubating antigen-presenting cells (E-rosette-negative cells from peripheral blood, principally B-cells and monocytes) with PDGF. When these cells are incubated with PDGF, their major histocompatibility complex class II antigens appear to be increased on their surface. It is not clear whether the enhancement of T-cell growth by PDGF is due solely to the increased expression of MHC class II antigens on these accessory E-rosette-negative cells, nor is it known which of the cells mediates this effect.

PDGF can also enhance erythropoiesis in an indirect fashion. Enhancement of erythropoiesis requires the presence of accessory mesenchymal cells such as marrow fibroblasts. Thus erythroid colony cells cultured in the presence of plasma-derived serum are enhanced to multiply in a dose-dependent manner by increasing concentrations of PDGF.[59,60] This enhancement requires co-culture with either fibroblasts or smooth muscle cells.

PDGF also increases the production and release of IGF-1 by fibroblasts and smooth muscle cells, which may also enhance the mitogenic response of cells to PDGF.[61]

Vasoconstriction

PDGF has been characterized as one of the most potent vasoconstrictive molecules that has yet been identified. Berk et al.[62] observed vasoconstriction upon exposure to PDGF in a concentration-dependent manner using strips of rat aorta, and demonstrated that PDGF was more potent than angiotensin-II in this activity. PDGF also induces an increase in cytosolic-free calcium in cultured rat aortic smooth muscle cells, and such calcium release may participate not only in this vasoconstrictive activity, but in the proliferative activity that ultimately may occur.

CELLS CAPABLE OF SECRETING PDGF

Platelets

Blood platelets, the product of the megakaryocyte, were shown to contain PDGF in a class of granule termed the alpha granule.[63-65] Platelets contain very poorly developed protein synthetic apparatus, and are induced to release PDGF-like materials upon exposure to thrombin, collagen, or a variety of substances that induce the "platelet release reaction." There is approximately 0.06 ng PDGF per 10^6 human platelets (approximately 1200 molecules per platelet)[32] and PDGF accounts for approximately 50% of the mitogenic activity derived from platelets present in human serum.[66]

Studies of megakaryocyte lysates demonstrate that they contain significant amounts of growth factor activity for bone marrow fibroblasts, and contain a 30,000 Mr form as well as higher Mr forms of PDGF (Raines, Rabellino and Ross, unpublished observations). Lysates obtained from megakaryocyte-enriched marrow have not been shown to contain inactive precursors of PDGF.

Mononuclear Phagocytes

It was determined in 1975 that mononuclear phagocytes, originally obtained from guinea pig peritoneum, can release a growth factor in culture upon stimulation to be phagocytic; this growth factor was originally termed macrophage-derived growth factor.[67] Activation of these cells with opsinized particles and induction of phagocytosis would induce such a response. Glenn and Ross[68] showed that human peripheral blood monocytes did not secrete mitogenic material, whereas after they were activated to become macrophages, they did so. Subsequently, part of the mitogenic activity was found to be due to PDGF.[69-71] Activated macrophages were also shown to secrete factors that were mitogenic for endothelial cells[72] and for epithelial cells (Madtes et al., unpublished observations). Since the monocyte does not appear to store mitogenic material,[68] the formation of growth factors may depend upon synthesis of new proteins by these cells. Shimokado et al.[69] and Martinet et al.[70] demonstrated that freshly isolated human peripheral blood monocytes do not express the gene for the B-chain of PDGF in accordance with their lack of secretion of this material. When the cells are activated, both B-chain and A-chain are induced (Sakariassen et al., unpublished observations) and PDGF-like material is secreted by these cells. Since macrophages secrete a broad spectrum of growth factors for epithelium, connective tissue mesenchyme, and endothelium, determination of control of growth factor gene expression by these cells will be important in understanding the role of the monocyte/macrophage in the different inflammatory responses that result in fibrous tissue proliferation.

Endothelial Cells

Vascular endothelial cells in culture release growth-promoting activity for cells, including PDGF-like material.[73,74] When human umbilical vein endothelial cells or bovine aortic endothelial cells are scraped from the intact artery, the levels of PDGF B-chain by Northern analysis are very low. When similar cells are grown in culture, however, B-chain mRNA is induced 10- to 83-fold, and secretion of PDGF-like material occurs.[75,76] Cultured human umbilical vein endothelial cells can also be induced to secrete higher amounts of PDGF by coagulation factors, including activated thrombin and activated factor X.[36,77] In addition, a macrophage product that appears to be formed in reasonably large quantities by activated macrophages, tumor necrosis factor (TNF, cachectin), also will induce expression of PDGF mRNA in vascular endothelium followed by secretion of PDGF (Sakariassen et al., unpublished observations). Thus various forms of "injury" to endothelium and interactions of vascular endothelial cells at sites of blood coagulation or at sites of inflammation with activated macrophages may result in PDGF formation by arterial endothelial cells.

Smooth Muscle Cells

Smooth muscle cells from rat aorta have been shown, under particular circumstances, to be able to form PDGF-like molecules. Newborn rat pup aortic smooth muscle cells secrete PDGF-like material, whereas adult (3 months) cells do not.[78] Similarly, smooth muscle cells obtained from an intimal proliferative lesion induced in the rat carotid artery by balloon catheter de-endothelialization will also produce a PDGF-like molecule in cell culture.[79] It is not yet clear whether the capacity to form PDGF-like molecules by smooth muscle cells is a phenomenon that is unique to the rat or whether this occurs in other species, including the human. If gene expression and secretion can be induced in human arterial smooth muscle cells, it will be important to determine the factors responsible for this activity.

Transformed Cells

A large number of neoplastically transformed cells secrete PDGF-like molecules. In simian sarcoma virus (SSV) transformed cells, it would be expected that the gene for the B-chain, which is transferred into the cells by the retrovirus, would express PDGF-like activity.[18,80,81] It has also been observed that a large number of cells that are transformed by other agents, such as cells transformed by other retroviruses, by different DNA viruses, spontaneous transformants that arise in culture, as well as cells obtained from a series of different human tumors, can secrete PDGF-like molecules.[80] Under these circumstances, it must be assumed that the normal cellular gene for PDGF is expressed in these cells as the result of some aspect of the process of cell transformation, leading to secretion of PDGF by the tumor cells.

THE ROLE OF PDGF IN BIOLOGY AND IN PATHOLOGY

Thus PDGF is a relatively ubiquitous growth factor that can be formed by many normal as well as tumor cells, and when formed in increasing amounts could result in fibroproliferative responses in tissues, where it could lead to or be involved in the development of disease.

PDGF undoubtedly plays an important role in the process of normal wound repair. Since PDGF can be initially deposited in wounds by platelets and subsequently can be formed by activated macrophages, its presence in healing wounds could be a prime determinant of fibroblast proliferation, connective tissue formation, and thus of wound repair. Platelets represent an initial evanescent response in wounds. Monocytes, the source of the macrophages in wounds, enter into the wound relatively early and remain for varying periods of time depending upon the size of the wound. Since both platelets and monocyte/macrophages can secrete elements that can induce not only connective tissue proliferation but epithelial and vascular endothelial proliferative responses, they probably represent the critical sources of the growth factors responsible for the process of normal wound repair.

Embryogenesis may also be critically affected at particular phases by PDGF. The cytotrophoblast of the human placenta can form PDGF,[82] and it is conceivable that many other embryonic cells may form and secrete PDGF during specific phases of development. This remains to be more extensively investigated.

Atherosclerosis, the principal cause of death in the United States and western Europe, consists of a proliferation of principally smooth muscle cells within the innermost layer of the artery wall, the intima. A series of observations of the cellular interactions that lead to atherosclerosis demonstrate that monocyte/macrophages are one of the earliest cells involved in this process, and in several circumstances may play a critical role in the genesis of the important proliferative vascular lesion.[83-85] Other cellular sources of PDGF that may be involved at different phases of lesion development include activated endothelial cells, platelets, and medial smooth muscle cells themselves (see above).

Since many neoplastic cells have been shown to form platelet-derived growth factor-like molecules, and since many of these cells contain receptors for PDGF, this molecule also may play a role in the neoplastic response itself, stimulating some of the neoplastically transformed cells, possibly in an autocrine fashion. In addition, many neoplastic growths *in vivo* induce a desmoplastic or connective tissue proliferative response in the immediate vicinity of the tumor. Since many tumors express and secrete PDGF, it is possible that in those circumstances where a desmoplastic response occurs, resulting in increased stroma and cells around the tumor, the response may be due to PDGF released by the tumor cells.

Finally, myelofibrosis may also be, in part at least, a result of PDGF secretion. In chronic myelogenous leukemia, myelofibrosis is common and these transformed cells could form PDGF relatively easily, as has been demonstrated in culture. In addition, ineffective megakaryocytopoiesis, resulting in a high proportion of developing megakaryocytes that die and are lysed within the marrow, may also lead to abnormal release of PDGF within the marrow environment and thus to a fibroproliferative response in the marrow.

Thus it can be seen that this relatively ubiquitous growth factor may play critical roles in many aspects of normal biology as well as in numerous pathologic responses ranging from cardiovascular disease (atherosclerosis) to different forms of neoplastic transformation.

REFERENCES

1. R. Ross, J. A. Glomset, B. Kariya, and L. Harker. A platelet-dependent serum factor that stimulates the proliferation of arterial smooth muscle cells in vitro. Proc. Natl. Acad. Sci. USA 71:1207 (1974).
2. N. Kohler and A. Lipton. Platelets as a source of fibroblast growth-promoting activity. Exp. Cell Res. 87:297 (1974).
3. Y. Oka and D. N. Orth. Human plasma epidermal growth factor/β-urogastrone is associated with blood platelets. J. Clin. Invest. 72:249 (1983).
4. R. K. Assoian, A. Komoriya, C.A. Meyers, D. M. Miller and M. B. Sporn. Transforming growth factor-beta in human platelets. J. Biol. Chem. 258:7155 (1983).
5. G. L. King and S. Buchwald. Characterization and partial purification of an endothelial cell growth factor from human platelets. J. Clin. Invest. 73:392 (1984).
6. K. Miyazono, T. Okabe, A. Urabe, T. Fumimaro and C. -H. Heldin. Purification and properties of an endothelial cell growth factor from human platelets. J. Biol. Chem. 262:4098 (1987).
7. D. F. Bowen-Pope, R. A. Seifert and R. Ross. The platelet-derived growth factor receptor, in: "Control of Animal Cell Proliferation," A. L. Boynton and H. L. Leffert, eds., Academic Press, New York, p. 281 (1985).
8. E. W. Raines and R. Ross. Platelet-derived growth factor. I. High yield purification and evidence for multiple forms. J. Biol. Chem. 257:5154 (1982).
9. C. -H. Heldin, B. Westermark and A. Wasteson. Platelet-derived growth factor: isolation by a large-scale procedure and anlysis of subunit composition. Biochem. J. 193:907 (1981).
10. H. N. Antoniades. Human platelet-derived growth factor (PDGF): purification of PDGF-I and PDGF-II and separation of their reduced subunits. Proc. Natl. Acad. Sci. USA 78:7314 (1981).
11. T. F. Deuel and J. S. Huang. Platelet-derived growth factor: purification, properties, and biological activities. Prog. Hematol. 13:201 (1983)
12. A. Johnsson, C. -H. Heldin, A. Wasteson, B. Westermark, T. F. Deuel, J. S. Huang, P. H. Seeburg, A. Gray, A. Ullrich, G. Scrace, P. Stroobant and M. D. Waterfield. The c-sis gene encodes a precursor of the B chain of platelet-derived growth factor. EMBO J. 3:921 (1984).
13. P. Stroobant and M. D. Waterfield. Purification and properties of porcine platelet-derived growth factor. EMBO J. 3:2963 (1984).
14. C. -H. Heldin, A. Johnsson, S. Wennergren, C. Wernstadt, C. Betsholtz and B. Westermark. A human osteosarcoma cell line secretes a growth factor structurally related to a homodimer of PDGF A-chains. Nature 319:511 (1986).
15. R. F. Doolittle, M. W. Hunkapiller, L. E. Hood, S. G. Devare, K. C. Robbins, S. A. Aaronson and H. N. Antoniades. Simian sarcoma virus onc gene, v-sis, is derived from the gene (or genes) encoding a platelet-derived growth factor. Science 221:275 (1983).
16. M. D. Waterfield, G. T. Scrace, N. Whittle, P. Stroobant, A. Johnsson, A. Wasteson, B. Westermark, C. -H. Heldin, J. S. Huang and T. F. Deuel. Platelet-derived growth factor is structurally

related to the putative transforming protein p28sis of simian sarcoma virus. Nature 304:35 (1983).
17. K. C. Robbins, H. N. Antoniades, S. G. Devare, M. W. Hunkapiller and S. A. Aaronson. Structural and immunological similarities between simian sarcoma virus gene product(s) and human platelet-derived growth factor. Nature 305:605 (1983).
18. A. Johnsson, C. Betsholtz, K. von der Helm, C. -H. Heldin and B. Westermark. Platelet-derived growth factor agonist activity of a secreted form of the v-sis oncogene product. Proc. Natl. Acad. Sci. USA 82:1721 (1985).
19. J. D. Kelly, E. W. Raines, R. Ross and M. J. Murray. The B chain of PDGF alone is sufficient for mitogenesis. EMBO J. 4:3399 (1985).
20. C. Betsholtz, A. Johnsson, C. -H. Heldin, B. Westermark, P. Lind, M. S. Urdea, R. Eddy, T. B. Shows, K. Philpott, A. L. Mellor, T. J. Knott and J. Scott. cDNA sequence and chromosomal localization of human platelet-derived growth factor A-chain and its expression in tumour cell lines. Nature 320:695 (1986).
21. T. F. Deuel and J. S. Huang. Roles of growth factor activities in oncogenesis. Blood 64:951 (1984).
22. R. Ross, E. W. Raines and D. F. Bowen-Pope. The biology of platelet-derived growth factor. Cell 46:155 (1986).
23. C. D. Stiles. The molecular biology of platelet-derived growth factor. Cell 33:653 (1983).
24. C. -H. Heldin and B. Westermark. Growth factors: mechanism of action and relation to oncogenes. Cell 37:9 (1984).
25. K. Glenn, D. Bowen-Pope and R. Ross. Platelet-derived growth factor. III. Identification of a platelet-derived growth factor receptor by affinity labeling. J. Biol. Chem. 257:5172 (1982).
26. C. E. Hart, R. A. Seifert, R. Ross and D. F. Bowen-Pope. Synthesis, phosphorylation and degradation of multiple forms of the platelet-derived growth factor receptor studied using a monoclonal antibody. J. Biol. Chem., in press.
27. L. J. Pike, D. Bowen-Pope, R. Ross and E. G. Krebs. Characterization of platelet-derived growth factor-stimulated phosphorylation in cell membranes. J. Biol. Chem. 258:9383 (1983).
28. A. R. Frackelton Jr., P. M. Tremble and L. T. Williams. Evidence for the platelet-derived growth factor-stimulated tyrosine phosphorylation of the platelet-derived growth factor receptor in vivo. J. Biol. Chem. 259:7909 (1984).
29. C. -H. Heldin, B. Ek and L. Ronnstrand. Characterization of the receptor for platelet-derived growth factor on human fibroblasts. Demonstration of an intimate relationship with a 185,000 dalton substrate for the PDGF-stimulated kinase. J. Biol. Chem. 258:10054 (1983).
30. Y. Yarden, J. A. Escobedo, W. -J. Kuang, T. L. Yang-Feng, T. O. Daniel, P. M. Tremble, E. Y. Chen, M. E. Ando, R. N. Harkins, U. Francke, V. A. Fried, A. Ullrich and L. T. Williams. Structure of the receptor for platelet-derived growth factor helps define a family of closely related growth factor receptors. Nature 323:226 (1986).
31. J. A. Cooper, D. F. Bowen-Pope, E. Raines, R. Ross and T. Hunter. Similar effects of platelet-derived growth factor and epidermal growth factor on phosphorylation of tyrosine in cellular proteins. Cell 31:263 (1982).
32. D. F. Bowen-Pope, T. W. Malpass, D. M. Foster and R. Ross. Platelet-derived growth factor in vivo: levels, activity, and rate of clearance. Blood 64:458 (1984).
33. E. B. Leof, J. A. Proper, A. S. Goustin, G. D. Shipley, P. E. DiCorleto and H. L. Moses. Induction of c-sis mRNA and activity similar to platelet-derived growth factor by transforming growth

factor β: a proposed model for indirect mitogenesis involving autocrine activity. Proc. Natl. Acad. Sci. USA 83:2453 (1986).
34. T. B. Barrett, C. M. Gajdusek, S. M. Schwartz, J. K. McDougall and E. P. Benditt. Expression of the sis gene by endothelial cells in culture and in vivo. Proc. Natl. Acad. Sci. USA 81:6772 (1984).
35. T. Papayannopoulou, E. Raines, S. Collins, B. Nakamoto, M. Tweeddale and R. Ross. Constitutive and inducible secretion of platelet-derived growth factor analogs by human leukemic cell lines coexpressing erythroid and megakaryocytic markers. J. Clin. Invest. 79:859 (1987).
36. J. M. Harlan, P. J. Thompson, R. Ross and D. F. Bowen-Pope. α-thrombin induces release of PDGF-like molecule(s) by cultured human endothelial cells. J. Cell Biol. 103:1129 (1986).
37. J. C. Smith, J. P. Singh, J. S. Lillquist, D. S. Goon and C. D. Stiles. Growth factors adherent to cell substrate are mitogenically active in situ. Nature 296:154 (1982).
38. Y. Shing, J. Folkman, R. Sullivan, C. Butterfield, J. Murray and M. Klagsbrun. Heparin affinity: purification of a tumor-derived capillary endothelial cell growth factor. Science 223:1296 (1984).
39. E. W. Raines, D. F. Bowen-Pope and R. Ross. Plasma binding proteins for platelet-derived growth factor that inhibit its binding to cell-surface receptors. Proc. Natl. Acad. Sci. USA 81:3424 (1984).
40. J. S. Huang, S. S. Huang and T. F. Deuel. Specific covalent binding of platelet-derived growth factor to human plasma alpha-2-macroglobulin. Proc. Natl. Acad. Sci. USA 81:342 (1984)
41. A. J. R. Habenicht, J. A. Glomset, W. C. King, C. Nist, C. D. Mitchell and R. Ross. Early changes in phosphatidylinositol and arachidonic acid metabolism in quiescent Swiss 3T3 cells stimulated to divide by platelet-derived growth factor. J. Biol. Chem. 256:12329 (1981).
42. W. T. Shier. Serum stimulation of phospholipase A_2 and prostaglandin release in 3T3 cells is associated with platelet-derived growth-promoting activity. Proc. Natl. Acad. Sci. USA 77:137 (1980).
43. A. H. Tashjian Jr., E. F. Voelkel, M. Lazzaro, F. R. Singer, A. B. Roberts, R. Derynck, M. E. Winkler and L. Levine. α and β human transforming growth factors stimulate prostaglandin production and bone resorption in cultured mouse calvaria. Proc. Natl. Acad. Sci. USA 82:4535 (1985).
44. S. Moncada and J. R. Vane. Arachidonic acid metabolites and the interactions between platelets and blood-vessel walls. N. Engl. J. Med. 300:1142 (1979).
45. B. Samuelsson, M. Goldyne, E. Granstrom, M. Hamberg, S. Hammarstrom, and C. Malmsten. Prostaglandins and thromboxanes. Ann. Rev. Biochem. 47:997 (1978).
46. A. J. R. Habenicht, M. Goerig, J. Grulich, D. Rothe, R. Gronwald, U. Loth, G. Schettler, B. Kommerell and R. Ross. Human platelet-derived growth factor stimulates prostaglandin synthesis by activation and by rapid de novo synthesis of cyclooxygenase. J. Clin. Invest. 75:1381 (1985).
47. A. J. R. Habenicht, H. A. Dresel, M. Goerig, J. -A. Weber, M. Stoehr, J. A. Glomset, R. Ross and G. Schettler. Low-density lipoprotein receptor-dependent prostaglandin synthesis in Swiss 3T3 cells stimulated by platelet-derived growth factor. Proc. Natl. Acad. Sci. USA 83:1344 (1986).
48. A. Chait, R. Ross, J. J. Albers and E. L. Bierman. Platelet-derived growth factor stimulates activity of low density lipoprotein receptors. Proc. Natl. Acad. Sci. USA 77:4084 (1980).

49. L. D. Witte, J. A. Cornicelli, R. W. Miller and D. S. Goodman. Effects of platelet-derived and endothelial cell-derived growth factors on the low density lipoprotein receptor pathway in cultured human fibroblasts. J. Biol. Chem. 257:5392 (1982).
50. S. A. Narayanan and R. C. Page. Biosynthesis and regulation of type V collagen in diploid human fibroblasts. J. Biol. Chem. 258:11694 (1983).
51. C. C. Chua, D. E. Geiman, G. H. Keller and R. L. Ladda. Induction of collagenase secretion in human fibroblast cultures by growth promoting factors. J. Biol. Chem. 260:5213 (1985).
52. E. A. Bauer, T. W. Cooper, J. S. Huang, J. Altman and T. F. Deuel. Stimulation of in vitro human skin collagenase expression by platelet-derived growth factor. Proc. Natl. Acad. Sci. USA 82:4132 (1985).
53. G. R. Grotendorst, T. Chang, H. E. J. Seppa, H. K. Kleinman and G. R. Martin. Platelet-derived growth factor is a chemoattractant for vascular smooth muscle cells. J. Cell. Physiol. 113:261 (1982).
54. H. Seppa, G. Grotendorst, S. Seppa, E. Schiffmann and G. R. Martin. Platelet-derived growth factor is chemotactic for fibroblasts. J. Cell Biol. 92:584 (1982).
55. T. F. Deuel, R. M. Senior, J. S. Huang and G. L. Griffin. Chemotaxis of monocytes and neutrophils to platelet-derived growth factor. J. Clin. Invest. 69:1046 (1982).
56. L. T. Williams, H. N. Antoniades and E. J. Goetzl. Platelet-derived growth factor stimulates mouse 3T3 cell mitogenesis and leukocyte chemotaxis through different structural determinants. J. Clin. Invest. 72:1759 (1983).
57. D. Y. Tzeng, T. F. Deuel, J. S. Huang, R. M. Senior, L. A. Boxer and R. L. Baehner. Platelet-derived growth factor promotes polymorphonuclear leukocyte activation. Blood 64:1123 (1984).
58. R. B. Acres, J. R. Lam and M. Feldman. Effects of platelet-derived growth factor and epidermal growth factor on antigen-induced proliferation of human T-cell lines. Immunology 54:9 (1985).
59. N. Dainiak, G. Davies, M. Kalmanti, J. Lawler and V. Kulkarni. Platelet-derived growth factor promotes the proliferation of erythropoietic progenitor cells in vitro. J. Clin. Invest. 71:1206 (1983).
60. F. Delwiche, E. Raines, J. Powell, R. Ross and J. Adamson. Platelet-derived growth factor enhances in vitro erythropoiesis via stimulation of mesenchymal cells. J. Clin. Invest. 76:137 (1985).
61. D. R. Clemmons and J. J. Van Wyk. Evidence for a functional role of endogenously produced somatomedin-like peptides in the regulation of DNA synthesis in cultured human fibroblasts and porcine smooth muscle cells. J. Clin. Invest. 75:1914, 1985.
62. B. C. Berk, R. W. Alexander, T. A. Brock, M. A. Gimbrone and C. R. Webb. Vasoconstriction: a new activity for platelet-derived growth factor. Science 232:87 (1986).
63. L. D. Witte, K. L. Kaplan, H. L. Nossel, B. A. Lages, H. J. Weiss and D. S. Goodman. Studies of the release from human platelets of the growth factor for cultured human arterial smooth muscle cells. Circ. Res. 42:402 (1978).
64. D. R. Kaplan, M. J. Broekman, A. Chernoff, G. R. Lesznik and M. Drillings. Platelet alpha-granule proteins: studies on release and subcellular localization. Blood 53:604 (1979).
65. J. M. Gerrard, D. R. Phillips, G. H. R. Rao, E. F. Plow, D. A. Walz, R. Ross, L. A. Harker and J. G. White. Biochemical studies of two patients with the gray platelet syndrome. J. Clin. Invest. 66:102 (1980).
66. C. -H. Heldin, B. Westermark and A. Wasteson. Demonstration of an

antibody against platelet-derived growth factor. Exp. Cell Res. 136:255 (1981).
67. S. J. Leibovich and R. Ross. A macrophage-dependent factor that stimulates the proliferation of fibroblasts in vitro. Am. J. Pathol. 84:501 (1976).
68. K. Glenn and R. Ross. Human monocyte-derived growth factor(s) for mesenchymal cells: activation of secretion by endotoxin and concanavalin A. Cell 25:603 (1981).
69. K. Shimokado, E. W. Raines, D. K. Madtes, T. B. Barrett, E. P. Benditt and R. Ross. A significant part of macrophage-derived growth factor consists of at least two forms of PDGF. Cell 43:277 (1985).
70. Y. Martinet, P. B. Bitterman, J. -F. Mornex, G. R. Grotendorst, G. R. Martin and R. G. Crystal. Activated human monocytes express the c-sis proto-oncogene and release a mediator showing PDGF-like activity. Nature 319:158 (1986).
71. J. -F. Mornex, Y. Martinet, K. Yamauchi, P. B. Bitterman, G. R. Grotendorst, A. Chytil-Weir, G. R. Martin and R. G. Crystal. Spontaneous expression of the c-sis gene and release of a platelet-derived growth factorlike molecule by human alveolar macrophages. J. Clin. Invest. 78:61 (1986).
72. B. M. Martin, M. A. Gimbrone Jr., E. R. Unanue and R. S. Cotran. Stimulation of nonlymphoid mesenchymal cell proliferation by a macrophage-derived growth factor. J. Immunol. 126:1510 (1981).
73. P. E. DiCorleto and D. F. Bowen-Pope. Cultured endothelial cells produce a platelet-derived growth factor-like protein. Proc. Natl. Acad. Sci. USA 80:1919 (1983).
74. P. E. DiCorleto. Cultured endothelial cells produce multiple growth factors for connective tissue cells. Exp. Cell Res. 153:167 (1984).
75. T. B. Barrett, C. M. Gajdusek, S. M. Schwartz, J. K. McDougall and E. P. Benditt. Expression of the sis gene by endothelial cells in culture and in vivo. Proc. Natl. Acad. Sci. USA 81:6772 (1984).
76. T. Collins, D. Ginsburg, J. M. Boss, S. H. Orkin and J. S. Pober. Cultured human endothelial cell express platelet-derived growth factor chain 2: cDNA cloning and structural analysis. Nature 316:748 (1985).
77. C. Gajdusek, S. Carbon, R. Ross, P. Nawroth and D. Stern. Activation of coagulation releases endothelial cell mitogens. J. Cell Biol. 103:419 (1986).
78. R. A. Seifert, S. M. Schwartz and D. F. Bowen-Pope. Developmentally regulated production of platelet-derived growth factor-like molecules. Nature 311:669 (1984).
79. L. N. Walker, D. F. Bowen-Pope, R. Ross and M. A. Reidy. Production of PDGF-like molecules by cultured arterial smooth muscle cells accompanies proliferation after arterial injury. Proc. Natl. Acad. Sci. USA 83:7311 (1986).
80. D. F. Bowen-Pope, A. Vogel and R. Ross. Production of platelet-derived growth factor-like molecules and reduced expression of platelet-derived growth factor receptors accompany transformation by a wide spectrum of agents. Proc. Natl. Acad. Sci. USA 81:2396 (1984).
81. J. S. Huang, S. S. Huang and T. F. Deuel. Transforming protein of simian sarcoma virus stimulates autocrine growth of SSV-transformed cells through PDGF cell-surface receptors. Cell 39:79 (1984).
82. A. S. Goustin, C. Betsholtz, S. Pfeifer-Ohlsson, H. Persson, J. Rydnert, M. Bywater, G. Holmgren, C. -H. Heldin, B. Westermark and R. Ohlsson. Coexpression of the sis and myc proto-oncogenes in developing human placenta suggests autocrine control of

trophoblast growth. Cell 41:301 (1985).
83. A. Faggiotto, R. Ross and L. Harker. Studies of hypercholesterolemia in the nonhuman primate. I. Changes that lead to fatty streak formation. Arteriosclerosis 4:323 (1984).
84. R. Ross. The pathogenesis of atherosclerosis: an update. N. Engl. J. Med. 314:488 (1986).
85. R. G. Gerrity. The role of the monocyte in atherogenesis. I. Transition of blood-borne monocytes into foam cells in fatty lesions. Am. J. Pathol. 103:181 (1981).

MOLECULAR AND DEVELOPMENTAL BIOLOGY ASPECTS

OF FIBROBLAST GROWTH FACTOR

Denis Gospodarowicz

Cancer Research Institute
University of California Medical Center
San Francisco, California

INTRODUCTION

Basic and acidic fibroblast growth factors (FGFs) are two closely related peptides of M_r=16000 that act as potent mitogens and differentiation factors for a wide variety of mesoderm and neuroectoderm derived cells. Basic FGF (bFGF pI 9.6) was first identified by its ability to cause the proliferation and phenotypic transformation of BALB-C 3T3 fibroblasts (Gospodarowicz,1974;Gospodarowicz and Moran,1974), while acidic FGF (aFGF, pI 5.6) was first identified by its ability to cause the proliferation and delayed differentiation of myoblasts (Gospodarowicz et al.,1975). It was later rediscovered on the basis of its ability to stimulate endothelial cell proliferation (Lemmon et al.,1982; Maciag et al.,1978).

Over the last 4 years, both growth factors have been purified to homogeneity, their amino acid sequences determined, and their cDNA cloned and sequenced (reviewed in Gospodarowicz et al.,1986a,b; Baird et al., 1986; Abraham et al.,1987). This information has already had significant impact on our understanding of a variety of mitogenic activities isolated from diverse origin. It has become clear that growth factors isolated from ovary, adrenal, kidney, eye, brain, placenta, macrophages, prostate, cartilage, and various tumors, are structurally and biologically identical, or at least, very similar to bFGF or aFGF (reviewed in Gospodarowicz, et al.,1986a). Availability of the pure mitogens has led to the recognition of a wide spectrum of activities for the two mitogens (Gospodarowicz,1985), most notably their potential to act as angiogenic factors in vivo. Basic FGF or aFGF are multifunctional, since they can either stimulate proliferation and induce or delay differentiation. They stimulate other critical processes in cell function as well. So far, however, the molecular mechanisms of action of the FGF's is unknown. Nevertheless, so many new and varied functions have now been described for FGF's that, at present, one must consider these peptides to be of special importance for the control of cell growth and differentiation.

STRUCTURE OF ACIDIC AND BASIC FGF AND THEIR GENE

Basic FGF has been purified from most mesoderm- or neuroectoderm-derived tissues or cells which have in common a strong angiogenic potential (Table I, Gospodarowicz, 1986a,b). Structural studies have shown that bFGF is a single chain peptide composed of 146 amino acids which can also exist in an NH_2-terminally truncated form missing the first 15 amino acids Gospodarowicz et al., 1985). The truncated form of bFGF is as potent as native bFGF, as demonstrated by radioreceptor binding and biological assays, indicating that the NH_2-terminal region of bFGF is neither involved in its binding to FGF cell surface receptors nor in its biological activity (Neufeld and Gospodarowicz, 1986). Related to bFGF is aFGF, which shares a 55% total sequence homology with bFGF (Esch et al., 1986). Acidic FGF is a 140-amino acid peptide which can also exist in an NH_2-terminally truncated form missing the first 6 amino acids: des. 1-6 aFGF. The high degree of homology between aFGF and bFGF suggests that they are derived from a single ancestral gene.

Evidence that a viral oncogene may code for a growth factor or part of a growth factor receptor has recently emerged from studies on the PDGF structure and that of the EGF receptor structure (reviewed in Sporn and Roberts, 1985). In the case of FGF, partial sequence homology with the predicted product of Int-2, a potential oncogene implicated in virally induced mammary cancer (Dickson et al, 1984) has been reported (Dickson and Peter, 1987).

Basic FGF seems to have been extremely well conserved through evolution. For example, bovine and human bFGF differ in only 2 of their 146 amino acids, giving an overall amino acid sequence homology of 98.7% (Abraham et al., 1986b). Avian bFGF and bovine bFGF have the same amino acid composition and avian bFGF cross-reacts on an equimolar basis with bovine bFGF in an RIA using rabbit anti-bFGF polyclonal antibodies. Thus, homologous epitopes are well conserved. Acidic FGF seems to be less well conserved, and the bovine form differs from the human by 11 residues out of a total of 146 (Abraham et al.,1987). Basic FGF and aFGF possess two potential binding domains for heparin, one being located near their NH_2-terminus, while the other is near the COOH-terminus. Both domains could be involved in the strong affinity of FGF toward heparin (reviewed in Gospodarowicz et al., 1986a,b; Baird et al., 1986).

Recently, the FGF genes have been cloned and complementary DNA sequences of both bFGF and aFGF have been synthesized. The genomic organization of the genes encoding bFGF and aFGF has been described (Abraham et al., 1986a,b; Mergia et al., 1987, Fig.1). The basic FGF gene encodes two exons widely separated by two introns: the first one separates codons 60 and 61, and the second separates codons 94 and 95. The aFGF gene has a similar organization, with 2 large introns located in identical positions in the coding sequence once basic and aFGF are properly aligned (Fig.1).

Restriction enzyme mapping of both genes indicates that their size is greater than 38 Kbp. Southern blot analysis of human genomic DNA has shown that there is only one bFGF and one aFGF gene. Therefore, all of the characterized or uncharacterized heparin binding endothelial cell mitogens related to bFGF or aFGF are the products of a single bFGF or aFGF gene (Abraham et al., 1986a;1987). The bFGF gene is localized on human chromosome 4, while that of aFGF is on chromosome 5 (reviewed in Abraham et al.,1987). This suggests that through a process of gene duplication and evolutionary divergence, bFGF and aFGF have become separate gene products.

In various cultured cells and tissues, the bFGF gene gives rise to two polyadenylated mRNA's of approximately 3.7 and 7.0 kB (reviewed in Gospodarowicz et al.,1987). The gene promoters have not yet been identified. The aFGF gene appears to encode a single mRNA species of approximately 4.8 kb (Jaye et al., 1986).

The primary translation product for either bFGF or aFGF is composed of 155 amino acids. Proteolytic cleavage from the precursor molecule of the first 9 (bFGF) or 15 residues (aFGF) would result in the generation of the mature proteins which can then be cleaved further in homologous positions to give the NH_2-truncated form of bFGF (des.1-15) or aFGF (des.1-6) (reviewed in Abraham et al.,1987; Gospodarowicz,1987) (Fig.1).

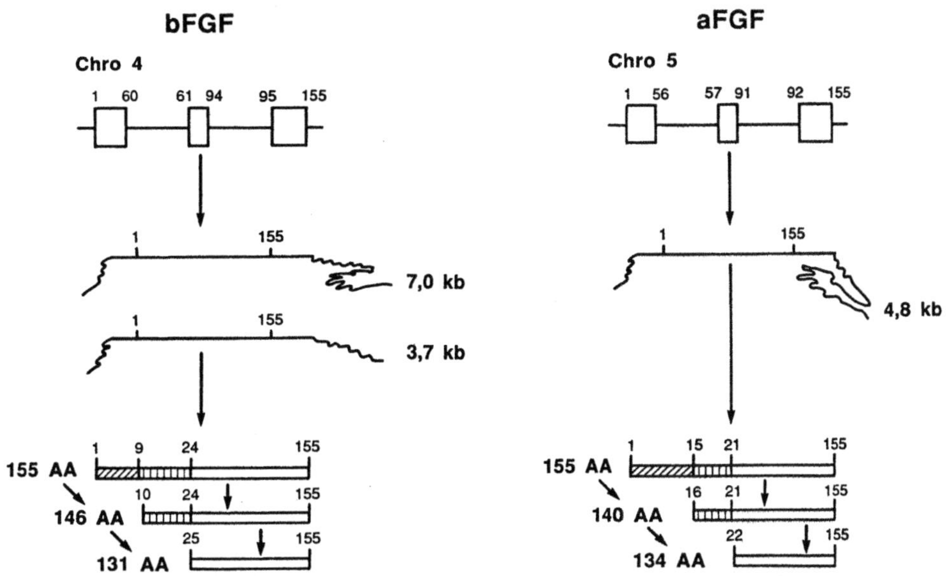

Figure 1 Genomic organization, mRNA transcripts, and cellular processing of bFGF and aFGF.

EXPRESSION OF BFGF AND AFGF IN TISSUES AND CULTURED CELLS

Tissues

So far, basic FGF has been purified from a wide variety of mesoderm- and neuroectoderm-derived normal or malignant tissues (Table 1)(reviewed in Gospodarowicz et al., 1986a,b).

Depending on the organ from which bFGF is isolated, the predominant forms are either the 146 or 131 amino acid long forms, with smaller amounts of the 155 amino acid long form being present. It is not known whether these various forms coexist in the tissues, or if they are artefactually created by specific proteases during FGF extraction and isolation (Ueno et al., 1986). So far, aFGF has only been found in brain, retina, bone matrix and osteosarcoma (Gospodarowicz et al.,1986a,b). In the case of aFGF, the 140 and 136 amino acid truncated aFGF forms are the predominant species, with a small amount of the 155 amino acid form being present.Since aFGF is 30- to 100-fold less potent than bFGF, it contributes to only 8% and 0.15% respectively of the total mitogenic activity present in crude brain or retinal extract, with the rest contributed by bFGF (Gospodarowicz et al.,1986b).

Table 1 Normal and neoplastic tissues or normal and transformed cells containing bFGF.

Normal or Transformed Tissue	Cultured normal diploid cells	Cultured tumor cells
Brain	Corneal endothelial cells	Y-1 Adrenal cortex cells
Retina	Capillary endothelial cells	Osteosarcoma U2OS
Pituitary	Pituitary cells	Ewing's sarcoma
Kidney	Ovarian granulosa cells	Rhabdomyosarcoma
Placenta	Adrenal cortex cells	Melanoma
Corpus Luteum	Lens epithelial cells	Hepatoma (Sk HP-1)
Adrenal Glands	Uterine epithelial cells	Retinoblastoma
Immune System	Myoblasts	
(Macrophage-Monocyte)	Retinal pigmented epithelial cells	
Prostate	Vascular smooth muscle cells	
Bone	Astrocytes	
	Osteoblasts	
Cartilage		
Chondrosarcoma		
Melanoma		

[a] So far, aFGF has only been detected in brain, retina, bone matrix, osteosarcoma osteoblasts, astrocytes, and fetal vascular smooth muscle cells.

Cultured cells

All organs that contain bFGF are heavily vascularized. This suggested that cells of the vascular system might synthesize bFGF. In fact, vascular endothelial cells express the bFGF gene, and they synthesize bioactive bFGF (Schweigerer et al.,1987a). In contrast, they do not express the aFGF gene or bioactive aFGF. Thus, bFGF could act as an autocrine growth factor for vascular endothelial cells. The synthesis of bFGF in vascular endothelial cells also provides an explanation for the rather ubiquitous distribution of bFGF. Basic FGF is also expressed in a wide variety of other normal diploid cells, all of which are sensitive to bFGF *in vitro* (Gospodarowicz et al.,1986b;1987) (Table 1). Various tumors derived from those cells also express bFGF (Table 1). This has led to the proposal that uncontrolled expression of bFGF could be involved in the development and in the progression of tumors (Schweigerer et al.,1987b; reviewed in Gospodarowicz et al.,1987).

The demonstration of the presence of bFGF in various normal diploid or tumor cells has also provided clues for its presence in the most unexpected locations. For example, it has recently been reported that mineralized matrix of osseous tissue harbors abundant mitogenic activity, represented in part by bFGF and aFGF (Hauschka et al.,1986b). The cellular sources of FGF present in bone matrix had not been defined. However, the demonstration of bFGF in cultured osteoblast (Gospodarowicz et al.,1987) suggests that bFGF could be produced by that cell type *in vivo*, and that it could act in an autocrine or paracrine manner. Basic

FGF is also produced in cells derived from retinoblastoma (Gospodarowicz et al.,1987), a tumor thought to be derived from photoreceptor cells. Thus, it is possible that photoreceptors are the source of bFGF within the normal retina. Although this localization of bFGF would be impossible to predict based on what was previously known of its biological activity, it has recently been demonstrated that bFGF is a component of the rod's outer segment, where it is strongly bound to rhodopsin (reviewed in Gospodarowicz et al.,1987). This suggests a possible role of bFGF in phototransduction. Finally, the localization of bFGF in cell types which have no function yet assigned to them provides new insights on their possible physiological functions. Pituitary glands are known to contain high concentrations of bFGF (Gospodarowicz,1974), but the cell types responsible for synthesis of pituitary bFGF was unknown. Recent studies have shown that the main cellular source of bFGF in pituitary are follicular cells which can contain as many as 5 X 10^5 bFGF molecules per cell (Ferrara et al.,1987). This cell type has no endocrine function and no physiological role yet assigned. It has, however, been suggested that follicular cells which are mainly concentrated in the pars tuberalis would provide support for the portal vessels. The presence of an angiogenic factor such as bFGF in follicular cells, could therefore relate to the development and maintenance of the differentiated state of the pars tuberalis microvasculature.

Basic FGF and aFGF have a very broad spectrum of activity, and other activities might be identified in the future. This would be compatible with their wide cross-species distribution, ranging from mammalian, avian, amphibian, and piscean, where both bFGF--and to a lesser extent, aFGF, seem to be well conserved (Gospodarowicz et al.,1987). Less is presently known about the degree of conservation of the FGF receptor across species since studies have been mostly concerned with cell types of mammalian origin.

THE FGF RECEPTORS

All cell types which respond to bFGF or aFGF bear specific FGF cell surface receptors. In the baby hamster kidney (BHK-21) cell line, the density of FGF cell surface receptors is 10- to 30-fold higher than that of normal diploid cells. In BHK-21 cells, bFGF binds to specific high-affinity cell surface receptors (k_d=0.27 nM; 1.2X10^5 binding sites per cell). This compares with a k_d for aFGF of 0.25 nM and 8.7X10^4 binding sites per BHK-21 cell. As expected from the high degree of structural homology between bFGF and aFGF, both mitogens bind to the same receptor. Basic FGF and aFGF do not bind to other growth factor receptors, nor do other growth factors bind to the FGF receptor (Gospodarowicz et al.,1986a,b). Cross-linking of bFGF or aFGF to the BHK-21 cell surface receptor indicates that qualitatively, both mitogens interact with the same two M_r 145 kD and 125 kD membrane components, which could differ by their degree of glycosylation. Quantitatively, bFGF appears to display a higher affinity than aFGF for the M_r=145 kD receptor species, while aFGF displays a higher affinity than bFGF for the M_r=125 kD receptor species (Neufeld and Gospodarowicz,1986). This could help in explaining the different biological potencies of bFGF versus aFGF (see below).

The FGF receptors of muscle cells, Swiss 3T3, lens epithelial cells, human umbilical vein endothelial cells, rhabdomyosarcoma, Ewing's sarcoma (SK-ES1), and PC-12 cells have also been characterized (Table 2). They all share similar molecular weight ranges, but might differ by their degree of glycosylation. However, in all cases both bFGF and aFGF interacts with the same receptor, with kD's ranging from 45 pM to 200 pM (Table 2) (reviewed in Gospodarowicz et al.,1987).

Table 2 Characterization of FGF receptors on various cell types.

Cell Types	MW(kD)	k_d	Receptors/cells
Muscle cells	165[ab]	11pM[ab]	$2X10^{3}$[ab]
Swiss 3T3	160[ab]	45pM[ab]	$6X10^{4}$[ab]
Murine capillary endothelial cells	150 and 130[a]	ND	ND
Human umbillical endothelial cells	130[ab]	ND	ND
PC-12	145[ab]	20pM[ab]	$3.6 X 10^{3}$[ab]
Rhabdomyosarcoma	125 and 145[ab]	ND	ND
SK-ES1 cells	150[b]	76pM[b]	$1.4 X 10^{4}$[b]
BHK-21 cells	125 and 145[ab]	270pM[ab]	$1 X 10^{5}$[ab]
Lens epithelial cells	145	53pM[b]	$2 X 10^{4}$[b]

ND = not determined
a = determined with aFGF
b = determined with bFGF
ab = determined with both basic and acidic FGF

BIOCHEMICAL RESPONSES INDUCED BY FGF

Although little information is available on the activities of aFGF, a variety of cellular responses has been studied with bFGF (Gospodarowicz,1985). Basic FGF increases membrane fluidity and ruffling activity, and these effects correlate with rapid changes in the dynamic structure of the actin cytoskeleton (Gospodarowicz,1985). Addition of bFGF to quiescent cultures of Swiss 3T3 cells induces within minutes diacylglycerol formation, protein kinase C activation and Ca^{2+} mobilization (reviewed in Gospodarowicz et al.,1986a,b).

Basic FGF also stimulates the synthesis of various specific proteins (reviewed in Gospodarowicz et al., 1986b;1987). The most dramatic, specific increases is of secreted rather than intracellular proteins. One of these secreted proteins, the "major excreted protein" (MEP), has recently been shown to be a thiol dependent cathepsin. FGF also stimulates the synthesis and release of a glycoprotein first named mitogen regulated protein (MRP). MRP is a member of the prolactin-growth hormone family and is identical to proliferin (reviewed in Gospodarowicz et al.,1987; Nilsen-Hamilton et al.,1986). It has been proposed that MRP-proliferin could act as an intercellular messenger, either as a hormone or a paracrine regulator. The bFGF-stimulated release of MEP and MRP is inhibited by TGF_B (reviewed in Nilsen-Hamilton et al.,1987). This correlates with the ability of TGF_B to modulate the biological activity of bFGF (see below).

Basic FGF regulates expression of cellular oncogenes, in

particular that of c-fos and c-myc. Both oncogenes appear to be involved in cell differentiation and cell proliferation,respectively (reviewed in Gospodarowicz,1985). The very rapid appearance of c-fos mRNA and protein after bFGF stimulation argues that c-fos activation affects gene expression by a direct mechanism, whereas activation of the c-myc may occur via indirect mechanisms. The sudden disappearance of c-fos mRNA also contrasts strikingly with the gradual decline of c-myc mRNA (reviewed in Gospodarowicz et al.,1986a,b;1987).

Like PDGF, bFGF can act as a competence factor for the normal diploid cells. Among the plasma components that are responsible for the progression of the cells committed to divide by exposure to bFGF are transferrin and high density lipoproteins (Gospodarowicz et al.,1982; Gospodarowicz, 1983).

IN VITRO BIOLOGICAL EFFECT OF FGF

Most of the biological studies with FGF have been done with the basic form. Only recently has the biological activity of aFGF started to be investigated. As expected from its high degree of structural homology with bFGF, aFGF has a mitogenic effect identical to that of bFGF, although it is, depending on the cell type, 30- to 100-fold less potent (Gospodarowicz, 1986a,b). The lesser potency of aFGF makes it a weak agonist of the basic form, and may reflect the qualitatively different interactions of aFGF versus bFGF with its cell surface receptor species (Gospodarowicz et al.,1986a,b). Basic FGF has both acute and longterm effects on the morphology and growth pattern of responsive cells. It increases their migratory activity (Gospodarowicz,1985), and it makes confluent cultures of BALB/c 3T3 cells look "transformed" in that it induces reduced cell-substratum adhesion, growth in crisscross pattern, and increased membrane ruffling (Gospodarowicz and Moran,1974). Basic FGF can also induce the growth in soft agar of non-transformed cells, and in that model it potentiates the effect of TGF_B (reviewed in Gospodarowicz et al.,1986b).

Basic FGF is a potent mitogen for mesoderm-derived cells (Gospodarowicz,1985) (Table 3), triggering cell proliferation with half-maximal and maximal effects at 1.5 and 10 pM, respectively. Basic FGF is mitogenic both for cells seeded at clonal density and for low-density cultures (Gospodarowicz,1979), and greatly reduces their average cellular doubling time. This is primarily due to a shortening of the G1 phase of the cell cycle (Gospodarowicz et al.,1978).

Basic FGF stabilizes the phenotypic expression of cultured cells. (Table 3) (Gospodarowicz, 1985). This property is particularly interesting, since it has made possible the long-term culturing of cell types that otherwise would lose their normal phenotype in culture when passaged repeatedly at low cell density (Gospodarowicz et al.,1978). This biological effect of bFGF has been studied exhaustively in endothelial cells derived from large vascular vessels or cornea that were cloned and maintained in the presence of bFGF and then deprived of it for various time periods (reviewed in Gospodarowicz et al.,1979; Gospodarowicz, 1985). This effect of bFGF on cell differentiation may be due to its ability to control the synthesis and deposition of various extracellular matrix (ECM) components that are known to affect cell surface polarity and gene expression. These include collagen, fibronectin, laminin, and proteoglycans (reviewed in Gospodarowicz and Greenburg,1981).

Basic FGF can induce capillary endothelial cells to invade a three-dimensional collagen matrix and to organize themselves to form characteristic tubules that resemble blood capillaries. Concomitantly, bFGF stimulates endothelial cells to produce a urokinase-type plasminogen activator (PA), a protease that has been implicated in the neovascular response. Thus bFGF can stimulate processes that are characteristic of angiogenesis in vivo, including endothelial cell migration, invasion, and production of plasminogen activator (Montesano et al., 1986).

Table 3 Cell types for which basic or acidic FGF is mitogenic, or affects differentiation.

Normal Diploid Cells	Basic FGF	Acidic FGF
Glial and astroglial cells	+(D)	+
Oligodendrocytes	+(D)	+
Trabecular meshwork cells	+	?
Endothelial cells from capillary, large vessel and endocardium	+(D)	+(D)
Corneal endothelial cells	+(D)	+(D)
Fibroblasts	+	+
Myoblasts	+(D)	+(D)
Vascular smooth muscle	+	+
Chondrocytes	+(D)	+(D)
Osteoblasts	+	+
Blastema cells	+	?
Adrenal cortex cells	+	+
Granulosa cells	+	+
Prostatic epithelial cells	+	+
Mesothelial cells	+	+
Neuronal cells	+	?
Established Cell Lines		
Rat fibroblast-1	+	+
Balb/c 3T3	+	+
Swiss 3T3	+	+
BHK-21	+	+
A-204 Rhabdomyosarcoma	+	?
PC-12	(D)	(D)

(D) = induces differentiation
\+ = positive effect on cell proliferation
? = effect on cell proliferation not determined

Basic FGF also acts as a differentiation factor for various cells including chondrocytes, sheep preadipocyte fibroblasts, astrocytes, oligodendrocytes. In nerve cells, bFGF can induce both neurite outgrowth and ornithine decarboxylase activity. Not all of the effects of bFGF on cell differentiation are stimulatory. For example, bFGF can delay the differentiation and fusion of normal diploid myoblasts (Gospodarowicz et al., 1976). In some established myoblast cell lines, basic and aFGF can decrease creatine phosphokinase expression (Lathrop et al.,1985).

FGF also significantly delays the ultimate senescence of cultured cells, such as that of granulosa cells, adrenal cortex cells,lens epithelial cells, or vascular endothelial cells. When maintained in the absence of bFGF, these cells have a limited lifespan, whereas in its presence, they can proliferate for more than 200 generations (reviewed in Gospodarowicz et al.,1978; Gospodarowicz, 1985).

AGENTS MODULATING THE BIOACTIVITY OF FGF

Heparin is one of the factors that can modulate the bioactivity of FGF. It potentiates the mitogenic activity of aFGF on BHK-21 or human umbilical endothelial cells by a hundred fold, thus rendering it as potent as bFGF. In PC-12 cells, heparin has similar effects on the neurotropic activity of aFGF (reviewed in Gospodarowicz et al.,1987).

It has been proposed that in vivo heparin could affect aFGF in much the same fashion as those effects reported in vitro. However, an assumption of complete correspondence may be misleading: all studies treating the potentiation of heparin/aFGF bioactivity have, to date, been performed using highly purified aFGF, which may have lost much of its own intrinsic bioactivity (normally exhibited in vivo) to acid treatment and other harsh protocols of the purification process (Gospodarowicz and Cheng,1986; Gimenez-Gallego et al., 19860. Therefore, when released by cells in vivo, and in its native state, aFGF could be as potent as bFGF.

Protamine sulfate can also modify the biological response to FGF. Protamine sulfate can act as an angiogenesis inhibitor in vivo, and it markedly inhibits the ability of bFGF or aFGF to stimulate the proliferation of vascular endothelial cells in vitro. The inhibition is reversible and the cells remain viable even after prolonged exposure to protamine sulfate. Protamine sulfate inhibits the mitogenic effects of both growth factors by preventing their interaction with FGF cell surface receptors (reviewed in Gospodarowicz et al.,1987). In contrast, protamine sulfate potentiates the mitogenic activity of EGF. This indicates that protamine sulfate can also act at cellular sites which are not associated with FGF receptors (Neufeld and Gospodarowicz,1987).

TGF_B, depending on the cell type upon which FGF acts, can either potentiate or inhibit its activity while having no activity of its own. This reflects the multifunctional properties of TGF_B which has been shown to both stimulate or inhibit cell proliferation or differentiation. TGF_B strongly potentiates the mitogenic activity of bFGF on osteoblasts. In contrast, TGF_B inhibits the mitogenic activity of bFGF and aFGF on vascular endothelial cells. The ability of TGF_B to inhibit the activation of PA as well as the release of other proteases, which are positively regulated by bFGF and which may be crucial for ECM degradation during angiogenesis, may be relevant to some of its inhibitory effects on vascular endothelial cell proliferation triggered by FGF. Since TGF_B and FGF can interact at the cellular level to modulate growth, it suggests that many of the biological activities of FGF observed in vitro and in vivo may be regulated by the presence of TGF_B and related proteins (i.e., inhibin) in the local cellular milieu (Baird and Durkin,1986; Frater-Schroder et al., 1986).

FGF AND THE EXTRACELLULAR MATRIX

In the early stage of embryonic development, different tissues composing an organ are formed as a result of strictly timed and spatially interrelated proliferative and differentiative events. This involves the interaction of cells with newly formed extracellular matrix (ECM), a process that promotes their proliferation and which stabilizes their newly acquired phenotype (Gospodarowicz et al.,1982). ECM components have been implicated in inductive tissue interaction, in somite chondrogenesis, the differentiation of corneal epithelium, salivary gland morphogenesis, tooth germ development, and kidney nephron formation (reviewed in Gospodarowicz,1985). It is possible that bFGF represents one of the ECM components involved in these events. Previous studies have shown that ECM produced by vascular or corneal endothelial cells can mimic all of the effects of bFGF, including those on cell proliferation and/or differentiation (Gospodarowicz and Tauber,1980; Gospodarowicz and Greenberg,1981; Gospodarowicz et al.,1982). Serum or plasma-derived growth factors, purified ECM components such as fibronectin, laminin, or various collagen types are not responsible for these effects, suggesting that FGF associated with ECM components might be the active factor. Indeed, bFGF has a high affinity for heparin. This glycosaminoglycan is closely related to heparan sulfate, which is produced in large quantities by both corneal and vascular endothelial cells, and is a structural component of their ECM (reviewed in Gospodarowicz et al.,1987).

The possibility therefore exists that FGF could be secreted by the cells in association with extracellular matrix components and become an integral part of the ECM. Indirect evidence for the integration of bFGF into an insoluble substrate such as the ECM can also be derived from the observation that neither media conditioned by capillary or corneal endothelial cells has a significant impact on their proliferation. In contrast, their own denuded ECM will induce them to rapidly proliferate and assume the proper phenotype once confluent. Thus bFGF, in contrast to other conventional growth factors such as TGF_B, EGF, and PDGF, may not be released in a soluble form (Gospodarowicz et al.,1986a,b). This is in agreement with the demonstration that neither the bFGF nor the aFGF gene seem to encode a conventional signal peptide sequence (Abrahams et al.,1986; Jay et al., 1986). Thus, both growth factors might be associated with ECM components and as such, be transported to the cell exterior, where they could interact with specific FGF cell surface receptors to induce their biological effects (Vlodavsky et al.,1987).

It is not clear how FGF might be liberated from the ECM. It is possible that this could occur following hydrolysis of ECM components. In that context, it is interesting to note that during morphogenesis, the areas with the greatest mitotic activity are located in areas where the hydrolysis of ECM occurs. Similarly, in the kidney, angiogenesis correlates well with the hydrolysis of the kidney mesenchymal stroma. In the adult, heparan sulfate present in ECM can be degraded by heparitinase, an enzyme released either by platelets, when they attach to the subendothelium, or by macrophages once they are activated (reviewed in Gospodarowicz et al.,1986b). This could ultimately lead to the solubilization of heparan sulfate/FGF complexes which would be biologically active, and could participate in various repair processes, including wound healing.

OTHER REGULATORY FUNCTIONS OF FGF

As in the case of EGF, which is known to modulate the synthesis and release by GH3 cells of prolactin and growth hormone, bFGF can increase the release of prolactin and decrease the basal level of growth hormone released in Ch_4C_1 cells (reviewed in Gospodarowicz et al.,1986b) or primary cultures of rat anterior pituitary cells (Baird et al.,1986).

In rat granulosa cells, which do not respond to bFGF by an increase in proliferative rate, bFGF inhibits the FSH-mediated induction of the LH receptor and reversibly attenuates the FSH-induced cell aromatase activity. This tends to suggest that FGF may play an inhibitory cytodifferentiative role in the ontogeny of the granulosa cells (reviewed in Baird et al.,1986; Gospodarowicz et al.,1987).

The positive correlation of plasminogen activator (PA) expression and the proliferation of solid tumors points to an instrumental role for this enzyme in neoplastic growth, a role that is likely to involve the degradation of matrix and basement membrane proteins by plasmin or plasmin-activated proteases. It has been shown that bFGF could increase PA production in certain mammary carcinoma cells without inducing their replication <u>in vitro</u>. Among the factors which could influence PA expression induced by bFGF is TGF_{B}. This growth factor has been shown to inhibit the production of secreted proteases in response to bFGF and increase the production of a PA inhibitor (reviewed in Nilsen-Hamilton et al.,1986;1987).

IN VIVO EFFFECTS OF FGF

In early embryonic development, the basic body plan arises because cells in different regions of the egg become programmed to follow different pathways (Slack,1983). During oogenesis, differences arise between the animal and vegetal halves of the eggs. Fertilization results in a subdivision of the vegetal half into a dorsal-vegetal and a ventro-vegetal region. Mesoderm is then induced from the animal hemisphere by signal(s) originating from the vegetal region of the egg (Nieuwkoop, 1969; Slack, 1983). This induction is an instructive phenomenon that suppresses epidermal differentiation of cells from the animal pole, and directs them instead to differentiate into mesodermal cells. Signal(s) originating from the dorsal-vegetal region lead to the formation of dorsal-type mesoderm, mostly consisting of notochord and somites, while signal(s) originating from the ventro-vegetal region lead to the formation of ventral-type mesoderm, consisting primarily of blood cells, mesenchyme, and mesothelium.

In recent studies Slack and his colleagues (1987) have investigated the possibility of bFGF mimicking the effect of the ventro-vegetal signal(s) responsible for the formation of ventral-type mesoderm. Tissue explants isolated from animal pole of stage 8 Xenopus blastulae normally differentiate into epidermis or undifferentiated epidermal cells. When similar explants are exposed to bFGF, cells differentiated instead into mesodermal structures. Between 2 to 30 ng bFGF/ml, the induction closely resembled ventral-type mesoderms formed by explants where ventrovegetal regions were combined with animal poles. They consisted of concentric arrangements of loose mesenchyme, mesothelium, and blood cells within an epidermal jacket. At higher bFGF concentrations (30 to 120 ng/ml) most of the explants contained significant amounts of muscle blocks (Slack et al., 1987). The inducing effect of bFGF seems to be highly specific, since it could not be mimicked by other growth factors such as TGF_{B} or $TGF\alpha$, TNF, Interferons α and γ, insulin, Interleukin-1α and β, G-CSF and GM-CSF (colony-stimulating factors).

Therefore, in early embryo, bFGF can act as a primordial differentiation factor, inducing the ectoderm to become mesoderm. This is in close agreement with previous <u>in vitro</u> studies which have shown that FGF had a transforming activity and could act as a morphogen, as well as a mitogen, on practically all mesoderm-derived cells studied to date. The FGF capability to induce mesoderm formation in early embryo is also of interest in view of its partial structural homology with the oncogene Int-2 (Dickson and Peters, 1987). This oncogene has been reported to be present only prior to day 7.5 of gestation in the mouse, and is most abundantly expressed in cells of the primitive endodermal lineage (Jakobovits et al., 1986).

In lower vertebrates (amphibians) bFGF can promote limb regeneration (Gospodarowicz and Mescher,1981). This tends to support the concept that bFGF could be involved in the neurotropic control of this process (reviewed in Gospodarowicz et al.,1986b). In addition to being a neurotrophic factor involved in limb regeneration, bFGF could also play a role in the early development of the nervous system. Basic FGF promotes both the survival and differentiation of nerve cells derived either from the hippocampal region or the cortex. Nerve cells have also been shown to contain bFGF (Pittman et al., 1986), and preliminary studies have demonstrated that neuronal cell populations derived from early embryonic brain do proliferate in response to bFGF, and later express cholinergic differentiation. All these effects point towards bFGF's possible role in CNS development. FGF could also have pronounced effects on the proliferation and differentiation of other brain cell populations, such as astrocytes and oligodendrocytes, by influencing their glial properties during normal development or subsequent to a specific pathogenic event (Gospodarowicz et al.,1987). Through its angiogenic properties bFGF could influence CNS development, since it has been reported that FGF could be responsible for capillary ingrowth into the brain (Risau, 1986).

Basic FGF, which has been detected in macrophages, could play a crucial role in the wound healing processes, following its release from the damaged cells. Interestingly, bFGF, unlike other growth factors such as EGF, PDGF or TGF_B, can stimulate both _in vitro_ as well as _in vivo_ the proliferation of all the cell types involved in wound healing (Gospodarowicz et al.,1986a,b). These include capillary endothelial cells, vascular smooth muscle, and fibroblasts, not withstanding other cell types which are involved in the wound healing of specialized territories, such as chondrocytes, myoblasts, etc. (Gospodarowicz, 1979;1985). Basic FGF also increases the formation of granulation tissue _in vivo_ (reviewed in Gospodarowicz,1985), and the synthetic function of fibroblasts and myoblasts. It also stimulates the rate of reepithelialization of the epidermis detached from dermis after blister induction. In other tissues such as cartilage, bFGF can promote chondrossification, and its presence in bone matrix could indicate that it could play an important role in the development and growth of osseous tissue (reviewed in Gospodarowicz et al.,1987).

Basic FGF acts as a potent angiogenic factor _in vivo_, as demonstrated by the the rabbit cornea, the chick chorioallantoic membrane (CAM) or the hamster cheek pouch assays. These observations are supported by the demonstration of bFGF as the major angiogenic agent in healthy vascularized tissues such as corpus luteum, adrenal gland, kidney, and retina (reviewed in Gospodarowicz et al.,1986a,b; Baird et al.,1986).

Recent studies have shown that bFGF is identical to the tumor angiogenic factor (Klagsbrun et al., 1986). Thus, bFGF might play an important role in tumor progression. By increasing capillary endothelial cell proliferation, and inducing the sprouting of new capillaries into solid tumors, bFGF might allow an increased supply of the tumors with O_2 and nutrients, and also the metastasis of the tumors. Basic FGF could also act at the level of the tumor cell itself. By increasing its PA level as well as by increasing the secreted levels of various proteases and collagenase, bFGF would facilitate the metastasic process and tumor invasion (reviewed in Gospodarowicz et al.,1987). It could also act as a mitogen for the tumor cells. The normal diploid cells listed in Table 3 are either uniquely dependent on bFGF or they require various growth factors including FGF, EGF, and PDGF in order to proliferate (reviewed in Gospodarowicz et al.,1978). Upon neoplastic transformation of cells that depend only on FGF, uncontrolled expression of bFGF in these cells could occur and make them divide in an uncontrolled manner. In the case of cells responding to multiple growth factors, uncontrolled expression of bFGF might make them independent of other exogenous growth factors during their growth phase.

CONCLUSION

The importance of FGF in the ontogeny of development and biopathology of mesenchymal tissues can no longer be ignored. The recent molecular characterization of FGF as well as the cloning and mapping of its genes, has lead to the general consensus that all the numerous heparin-binding growth factors were, in fact, represented by 2 single gene products: basic and acidic FGF. The further demonstration that both basic and acidic FGF have a high degree of structural homology and bind to the same receptor has lead to the conclusion that they had identical biological roles, although they differed in their specific activity.

Probably one of the most important questions to be resolved is the in vivo role of FGF. The high degree of structural conservation of bFGF through species as different as mammalian, avian and amphibian, as well as its presence in all vertebrates studied to date--including piscean--indicate that in vivo FGF could have a very primordial role. This is, in fact, what seems to be indicated by the studies of Slack et al. (1987), which have shown that bFGF can act as a primordial morphogen at one of the earliest embryonic stages, inducing the transformation of cells destined to be ectodermal,into mesenchymal cells. This is consistent with the in vitro properties of bFGF, which has been shown to act as mitogen as well as morphogen for all mesenchymal cells studied to date. It is also in agreement with the ability of bFGF to support the regeneration process in lower vertebrates.

One of the most popularized aspects of the in vivo biology of FGF is based on its angiogenic activity. Basic FGF has been shown to induce in early embryo the appearance of blood islands and, its effects on capillary endothelial cells in vitro or in vivo, as well as its distribution in various organs or tumors known to have angiogenic potential provide a common denominator for its widespread angiogenic activity. However, narrowing the experimental focus to FGF's angiogenic properties would produce an extremely limited view of its potential targets in vivo. The cell types listed in Table III clearly indicate in which organ an effect of FGF, either in tissue formation or repair process, should be investigated. In particular, the in vivo mitogenic and differentiation effects of bFGF on various cell types of the nervous system would be worth investigating, since it has been shown to control both the proliferation and differentiation of glial cells, oligodendrocytes, and astroglial cells, as well as acting as a triggering mechanism for the differentiation of nerve cells in vitro.

Although it is likely that bFGF could still control the proliferation and differentiation of mesenchymal cells later in the ontogeny of development, little is actually known about the implication of the activation of the FGF gene expression in neoplastic transformation. It has been speculated that activation of growth factor-controlling gene expressed early in embryogenesis and later repressed could lead to neoplastic tranformation (Sporn and Roberts, 1985). Transfection of normal diploid cells such as vascular endothelial cells with plasmids carrying FGF-cDNA, resulting in the constitutive expression of bFGF, could be a useful approach to testing that hypothesis. Cells expressing high bFGF levels could also be used for studies dealing with the locus of action of FGF (either intra- or extracellular locations), the transcellular transport of FGF and pathways of FGF release from the cells. Of equal importance will be studies dealing with mechanisms and factors controlling the expression of the FGF receptor, since this process would ultimately determine the time, as well as the cell type, which would be FGF-responsive. Such studies cannot be initiated until we have a better understanding of the molecular properties of the FGF receptor, and in particular, until we have cloned FGF receptor cDNA, which can later be used to study its expression.

REFERENCES

Abraham,J.A., Whang,J.L., Tumolo A., Mergia,A., and Fiddes,J.C.,1987, Human basic fibroblast growth factor: nucleotide sequence genomic organization and expression in mammalian cells, in: Molecular Biology of Homo Sapiens, Cold Spring Harbor, New York, Vol.51, pp. 657-668.

Abraham,J.A., Whang,J.L., Tumolo,A., Mergia,A., Friedman,J., Gospodarowicz,D., and Fiddes,J.C., 1986a, Human basic fibroblast growth factor: nucleotide sequence and genomic organization, EMBO (Europ. Molec. Biol. Org.) J., 5:2523-2528.

Abraham,J.A., Mergia,A. Whang,J.L., Tuomolo,A., Friedman,J., Hjerrild, K.A., Gospodarowicz,D., and Fiddes,J.C.,1986b, Nucleotide sequence of a bovine clone encoding the angiogenic protein, basic fibroblast growth factor, Science, Washington D.C.,233:545-548.

Baird,A., Durkin,T.,1986, Inhibition of endothelial cell proliferation by type B transforming growth factor: interactions with acidic and basic fibroblast factors, Biochem. Biophys. Res. Commun., 138:476-482.

Baird,A., Esch,F., Mormede,P., Ueno,N., Ling,N., Böhlen,P., Ying,S.Y., Wehrenberg,W.B., and Guillemin,R.,1986, Molecular characterization of fibroblast growth factor: distribution and biological activities in various tissues, Recent Prog. in Hormone Research, 42:143-205.

Dickson,C., and Peters,G.,1987, Potential oncogene product related to growth factors, Nature, 326:833.

Dickson,C., Smith,R., Brookes,S., and Peters,G.,1984, Tumorogenesis by mouse mammary tumor virus: Proviral activation of a cellular gene in the common integration region Int-2, Cell, 37:529-536.

Esch,F., Ueno,N., Baird,A., Hill,F., Denoroy,L., Ling,N., Gospodarowicz,D., and Guillemin,R.,1986, Primary structure of bovine brain acidic fibroblast growth factor (FGF), Biochem. Biophys. Res. Commun., 133:554-562.

Ferrara,N., Schweigerer,L., Neufeld,G., and Gospodarowicz,D. A new function for pituitary follicular cells: The production of basic Fibroblast groth factor, Proc. Natl. Acad. Sci., USA.In press.

Frater-Schroder,M., Muller,G., Burchmeier,W., and Böhlen,P.,1986, Transforming growth factor B inhibits endothelial cell proliferation, Biochem. Biophys. Res. Commun., 137:295-302.

Gimenez-Gallego,G., Conn,G., Hatcher,V.B., and Thomas,K.A.,1986, Human brain-derived acidic and basic fibroblast growth factors: amino terminal sequences and specific mitogenic activities, Biochem. Biophys. Res. Commun., 135:561-566.

Gospodarowicz,D.,1974,Localization of a fibroblast growth factor and its effect alone and with hydrocortisone on 3T3 cell growth, Nature, London 249:123-127.

Gospodarowicz, D.,1979, Fibroblast and epidermal growth factors: their uses in vivo and in vitro in studies on cell functions and cell

transplantation, Mol. Cell. Biochem., 25:79-110.
Gospodarowicz,D.,1983, The control of mammalian cell proliferation by growth factors, extracellular matrix and lipoproteins, J. Inv. Derm., 81: 41-50.
Gospodarowicz,D.,1985, Biological activity in vivo and in vitro of pituitary and brain fibroblast growth factor, in: "Mediators in Cell Growth and Differentiation," R.J. Ford, A.L. Maizel,eds, Raven Press,New York,pp.109-134.
Gospodarowicz,D.,1987, Purification of brain and pituitary FGF, in: "Methods in Enzymology: Peptide growth factors," D.Barnes and D. Sirbasku, eds., Academic Press, Orlando, FL,147A;106-119.
Gospodarowicz,D., and Cheng,J.,1986, Heparin protects basic and acidic FGF from inactivation, J. Cell. Physiol.,128:475-484.
Gospodarowicz,D., and Greenburg,G.,1981, Growth control of mammalian cells. Growth factors and extracellular matrix, in:"The Biology of Normal Human Growth," M.Ritzen, A.Aperia, K.Hall, A.Larsson, A.Zetterberg, R.Zetterstrom, eds., Raven Press, New York, pp.1-21.
Gospodarowicz,D., and Mescher,A.L.,1981, Fibroblast growth factor and vertebrate regeneration, in: "Advances in Neurology: Neurofibromatosis," V.M.Riccardi, and J.J.Mulvihill, eds., Raven Press,New York,Vol.29,pp.149-171.
Gospodarowicz,D.,and Moran,J.,1974, Effect of a fibroblast growth factor, insulin, dexamethasone, and serum on the morphology of BALB/c 3T3 cells, Proc.Natl.Acad.Sci. USA, 71:4648-4652.
Gospodarowicz,D., and Tauber,J.-P.,1980, Growth factors and extracellular matrix, Endocrine Review, 1:201-227.
Gospodarowicz,D., Cohen,D.C., and Fujii,D.K.,1982, Regulation of cell growth by the basal lamina and plasma factors: relevance to embryonic control of cell proliferation, in:"Cold Spring Harbor Conferences on Cell Proliferation Vol.9: Growth of cells in hormonally deficient media," G.Sato, A.Pardee, and D.Sirbasku, eds., Cold Spring Harbor, New York, pp.95-124.
Gospodarowicz,D., Greenburg,G., and Bialecki,H.,1978, Factors involved in the modulation of cell proliferation in vivo and in vitro: the role of fibroblast and epidermal growth factors in the proliferative response of mammalian cells, In Vitro,14:85-118.
Gospodarowicz,D., Neufeld,G., and Schweigerer,L. ,1986a, Fibroblast growth factor, Mol.Cell.Endocrin., 46:187-206.
Gospodarowicz,D., Neufeld,G., and Schweigerer,L.,1986b, Molecular and biological characterization of fibroblast growth factor: an angiogenic factor which also controls the proliferation and differentiation of mesoderm and neuroectoderm-derived cells, Cell. Differ., 19:1-17.
Gospodarowicz,D., Weseman,J., and Moran,J.,1975, Presence in the brain of a mitogenic agent distinct from fibroblast growth factor that promotes the proliferation of myoblasts in low density culture, Nature, London, 256:216-220.
Gospodarowicz,D., Ferrara,N., Schweigerer, L., and Neufeld,G.,1987, Structural characterization and biological functions of fibroblast growth factor, Endocrine Review, 8:1-20.
Gospodarowicz,D., Weseman,J., Moran,J., and Lindstrom,J.,1976, Effect of fibroblast growth factor on the division and fusion of bovine myoblasts,J.Cell Biol., 70:395-405.
Gospodarowicz,D., Cheng,J., Lui,G.M., Baird,A., Esch,F., and Böhlen,P., 1985, Corpus luteum angiogenic factor is related to fibroblast growth factor, Endocrinology,117:2283-2291.
Gospodarowicz,D., Vlodavsky,I., Greenberg,G., Alvarado.J., Johnson,L.K., and Moran,J.,1979, Cellular shape is determined by the extracellular matrix and is responsible for the control of cellular growth and function, in: "Cold Spring Harbor Conferences on Cell Proliferation, Vol 6: Hormones and Cell Culture,"

R.Ross,G.Saro,eds., Cold Spring Harbor, New York, pp.561-592.
Hauschka,P.V., Mavrakas,A.E., Iafrati,M.D., Doleman,S.E.,and Klagsbrun,N.,1986, Growth factors in bone matrix: isolation of multiple types by affinity chromatography on heparin-Sepharose, J.Biol.Chem., 261:12665-12674.
Jaye,M., Howk,R., Burgess, W., Ricca,G.A., Chiu,I.M., Ravera,M.W., O'Brien,S.J., Modi,W.S., Maciag,T., and Drohan,W.N.,1986, Human endothelial cell growth factor: cloning, nucleotide sequence, and chromosome localization, Science, 233:541-544.
Jakobovits,A., Shackleford,G.M., Varmus,H.E., and Martin,G.R.,1986, Two proto-oncogenes implicated in mammary carcinogenesis, Int-1 and Int-2, are independently regulated during mouse development, Proc.Natl.Acad.Sci.,USA,83:7806-7810.
Klagsbrun,M., Sasse,J., Sullivan,R., and Smith,J.A.,1986, Human tumor cells synthesize an endothelial cell growth factor that is structurally related to basic fibroblast growth factor, Proc.Natl.Acad.Sci., USA,83:2448-2452.
Lathrop,B., Olson,E., Glaser,L.,1985, Control of myogenic differentiation by fibroblast growth factor is mediated by position in the Gl phase of the cell cycle, J. Cell Biol., 101:2194-2208.
Lemmon,S.K., Rielly,M.C., Thomas,K.A., Hoover,G.A., Maciag,T.,and Bradshaw,R.,1982, Bovine fibroblast growth factor: comparison of brain and pituitary preparations, J. Cell. Biol., 95:162-169.
Maciag,R., Cerundolo,J., Isley,S., Kelley,P.R., and Forand,R.,1979, An endothelial cell growth factor from bovine hypothalamus: identification and partial characterization, Proc.Natl.Acad.Sci.,USA, 76:5674-5678.
Mergia,A., Tumolo,A., Haaparanta,T., Whang,J.L., Gospodarowicz,D., Abraham,J.A., and Fiddes,J.C.,1987, Isolation and characterization of the human gene for acidic FGF, In preparation.
Montesano,R., Vassali,J.D., Baird,A., Guillemin,R., and Orci,L.,1986, Basic fibroblast growth factor induces angiogenesis in vitro, Proc. Natl.Acad.Sci.,USA, 83:7297-7301.
Neufeld,G., and Gospodarowicz,D.,1986, Basic and acidic fibroblast growth factor interact with the same cell surface receptor, J.Biol.Chem., 261:5631-5637.
Neufeld, G., and Gospodarowicz, D., Protamine Sulfate Inhibits the mitogenic activities of the extracellular matrix and FGF, but potentiates that of epidermal growth factor, J. Cell Physiol. In press.
Nieuwkoop,P.,1969,The formation of mesoderm in Urodelean amphibians. I.Induction by the endoderm, Wilhelm Roux' Arch. Entw. Mech. Org., 162:341-373.
Nilsen-Hamilton,M., and Hamilton,R.T. ,1987, Detection of proteins induced by growth regulators, Methods in Enzymology, 147:427-444, Academic Press, Orlando, Florida.
Nilsen-Hamilton,M., Hamilton,R.T., and Alvarez-Azaustre,E.,1986, A relationship between mitogen regulated protein (MRP) and proliferin, a member of the prolactin/growth hormone family, Gene, 51:163-170.
Pettman,B., Labourdette,G., Weibel,M., and Sensenbrenner,M.,1986, The brain fibroblast growth factor (FGF) is localized in neurons, Neurosci. Letters, 68:175-179.
Risau,W.,1986, Developing brain produces an angiogenesis factor, Proc.Natl.Acad.Sci.,USA,83:3855-3859.
Schweigerer,L., Neufeld,G., Friedman,J., Abraham,J.A., Fiddes,J.C., and Gospodarowicz,D.,1987a, Capillary endothelial cells express basic fibroblast growth factor, a mitogen that stimulates their own growth, Nature, London,325:257-259.
Schweigerer,L., Neufeld,G., Mergia,A., Abraham,J.A., Fiddes,J.C., and Gospodarowicz,D.,1987b, Basic fibroblast growth factor in human

rhabdomyosarcoma cells: implications for the proliferation and neovascularization of myoblast-derived tumors, Proc.Natl.Acad.Sci., USA, 84:842-846

Slack,J.M. ,1983, "From egg to embryo: determinative events in early development," Cambridge University Press, Cambridge and London.

Slack,J.M., Darlington,B., Heath,H., and Godsave,S.,1987, Heparin binding growth factors as agents of mesoderm induction in early Xenopus embryo, Nature, 326:197-200.

Sporn,M.B., and Roberts,A.B.,1985, Autocrine growth factor and cancer, Nature, 313:745-747.

Ueno,K., Baird,A., Esch,F., Ling,N., and Guillemin,R.,1986, Isolation of an amino acid terminal extended form of basic fibroblast growth factor, Biochem. Biophys. Res. Commun., 138:580-588.

Vlodavsky,I., Folkman,J. Sullivan,R., Frieman,R., Ishai,R., Michaeli, Sasse,J., and Klagsbrun,M.,1987, Endothelial cells derive basic fibroblast growth factor: Synthesis and deposition into subendothelial extracellular matrix, Proc. Natl. Acad. Sci., USA, 84:2282-2296.

CHEMICAL AND BIOCHEMICAL PROPERTIES OF HUMAN ANGIOGENIN

Bert L. Vallee and James F. Riordan

Center for Biochemical and Biophysical
Sciences and Medicine
Harvard Medical School
Boston, Massachusetts

INTRODUCTION

It is now widely appreciated that neovascularization is essential for the growth and progression of solid tumors. Indeed, this requirement has led to the concept of tumor-induced angiogenesis whereby tumor cells secrete a substance or substances that promote the proliferation of neighboring blood vessels. Such a substance was, in fact, thought to be not only obligatory but also specific and peculiar to malignant cells. As a consequence, a great deal of attention has been directed toward the isolation and characterization of such a material. Containment of neoplastic growth through specific control of the tumor-induced angiogenic process has been suggested to be a potential therapeutic approach and this has added additional impetus to the effort.

The starting materials in which angiogenic activity has been detected and, in some instances, reported to be isolated have varied from homogenates of solid tumors and tumor cell suspensions maintained as ascites or in tissue culture to tumor conditioned medium. Early experiments demonstrated that tumor cells separated from the vasculature of the host by a Millipore filter could still induce the growth of new blood vessels (Greenblatt and Shubik, 1968; Ehrmann and Knoth, 1968). This indicated that the factor responsible for vessel proliferation was extracellular and diffusible, characteristics that were consistent with the concept of tumor-induced angiogenesis. It was thought that identical factors were not present in normal tissue, a view no longer tenable based on experimental facts. Most important, this realization has re-established interest in the much broader issue of organ development in general, because it points to the possible existence of substances that could rightly be called organizers.

During the early part of this century experimental embryologists were intrigued by the problem of organ development. In 1924 Spemann proposed the organizer hypothesis (Spemann and Mangold, 1924) to account for his observations with salamanders.

ANGIOGENESIS

Some fifteen years ago we became interested in this biochemical problem and again asked the question whether or not organogenic molecules exist. If so, how might they be identified, isolated, and characterized? We took for granted that such organogenic molecules would occur in vanishingly small quantities which would have escaped detection by earlier approaches and methodology. Moreover, stating the question in this manner broadened the approach to angiogenesis from the more restrictive neoplasic related ideas to general embryology and biology.

Baltzer (1928) had postulated that Spemann's "organizers" exist as chemically distinct and specific molecules whose primary function is to induce organ development. This hypothesis, while welcomed enthusiastically at first, subsequently lost favor for a long time. We chose to reexamine it by focusing on potential organizer molecules that might induce the development of the cardiovascular system. The vasculature component of this system is an organ that is readily accessible, can regenerate to a considerable extent, and under certain conditions undergoes renewal and repair, even in higher organisms. Importantly, the methodology and technology to recognize and trace both appearance and suppression of angiogenesis (blood vessel formation) were more highly developed then than those for different organ systems that might have been chosen as alternative objectives for such experiments.

Blood vessel proliferation occurs during embryonic growth and subsequent maturation, during wound healing, organ and limb regeneration, in the cyclical development of the endometrium, and in a number of pathological states, particularly the growth of solid tumors.

A number of substances have been reported to be capable of inducing blood vessel formation (Table 1). Among these are angiogenin (Fett et al., 1985) and other proteins such as acidic and basic fibroblast growth factors, and the α and β transforming growth factors (Lobb et al., 1986; Folkman and Klagsbrun, 1987). Still others are of relatively low molecular weight and include prostaglandins of the E type (Ziche et al., 1982) and the recently discovered nicotinamide (Kull et al., 1987). Some have only been characterized partially (Vallee et al., 1985; Wissler et al., 1987). In almost all cases the amount of material required to induce blood vessel formation is exceedingly small and in no case has the mechanism of induction been established. Significantly, these substances have been found both in normal and malignant tissues.

ISOLATION OF ANGIOGENIN

Our efforts of isolate an angiogenic substance were initially directed toward the male mouse submaxillary gland since this tissue was known to be a source of other growth factors. However, the presence of inflammatory materials made for unexpected complications and we therefore turned to neoplastic tissue largely because tumors were thought to be potentially the richest source. Preliminary studies were carried out with the Walker 256 carcinoma of the rat and these provided much helpful information (Vallee et al., 1985). Later our attention turned to a human cell line, the HT-29 colon adenocarcinoma.

One of the most critical factors leading to the eventual isolation and characterization of the first pure angiogenic molecule from a human tumor was the definition of conditions for cell growth in the absence of exogenous proteins (Alderman et al., 1985). We were looking for a

Table 1: Some Substances Capable of Inducing Blood Vessel Formation

Angiogenin	Fett et al., 1985
Acidic fibroblast growth factor	Lobb et al., 1985
Basic fibroblast growth factor	Gospodarowicz et al., 1979
Endothelial cell growth factor	Maciag, et al., 1979
Transforming growth factor α	Schreiber et al., 1986
Transforming growth factor β	Roberts et al., 1986
Prostaglandin E_1 and E_2	Ziche et al., 1982
Copper-ribonucleopeptide	Wissler et al., 1987
Nicotinamide	Kull et al., 1987

molecule that would be secreted by tumor cells, a messenger molecule that would ordinarily induce the directed growth of neighboring capillaries. Since it seemed likely that this molecule would be active at very low concentrations, the presence of large amounts of the protein in the growth medium would only cause problems. We therefore used a procedure in which the HT-29 cells were cultured and maintained in a serum-free and exogenous protein-free, defined growth medium. While the cells do not proliferate under these conditions they actively secrete an angiogenic molecule which we have called angiogenin.

All procedures for isolation of angiogenic substances require an appropriate bioassay. One of the most useful involves the implantation of the test substrate in the avascular corneal stroma of the rabbit eye (Langer and Folkman, 1976). If the substance is angiogenic it will induce new blood vessels to grow out from and perpendicular to the limbus. Quantitation is based on the rate of vessel growth or number of new vessels.

A more convenient assay for screening large numbers of samples was developed by Knighton et al. (1977). The sample is applied to the chorioallantoic membrane of the chick embryo through a window cut in the shell. Several variations of the method have been employed which differ primarily in the mode of sample application. The version adopted for use in our laboratory involves 13-mm transparent thermanox discs on which a 5 μl aliquot of an aqueous solution of the test substance has been evaporated to dryness. The disc is inverted, placed on the 9 day membrane and the result assessed at 24 hour intervals. A sample is designated as positive if at least 60% of the test eggs exhibit neovascularization on the third day with the new blood vessels growing toward the site of application. There are numerous pitfalls associated with all assays of angiogenesis and the need for cautious interpretation has been emphasized (Vallee et al., 1985).

Conventional cell culture techniques provide relatively small volumes of conditioned growth medium. Therefore, in order to obtain adequate amounts of material for further processing large scale cell factories were employed. These could yield 4-5 liters per week. A typical purification from 10 liters of pooled, serum-free conditioned medium is summarized in Table 2. A total of 5 μg of pure product is obtained by a combination of ion-exchange chromatography and reversed-phase high performance liquid chromatography.

Statistical evaluation demonstrates that angiogenin displays activity on the chorioallantoic membrane with as little as 35 fmol per

Table 2: Purification of Angiogenin from HT-29 cell Conditioned Medium

Step	Recovery, mg	Purification
Conditioned medium, 10 liters	16	1
Acidification, freeze-thaw	6.3	2.5
CM-Cellulose, unbound	3.2	-
CM-Cellulose, bound	2.3	7.0
RP-HPLC	0.005	32000

egg. It requires 3.5 pmol to induce extensive blood vessel growth in the rabbit cornea (Fett et al., 1985). Angiogenin is neither a growth factor nor a mitogen in the sense of the present usage of these words.

The isolation scheme for angiogenin was worked out, first in broad outline with the rat Walker 256 carcinoma cell line and in eventual detail with HT-29 cells, over a period of eight years. In the course of the subsequent two years we ultimately produced about 1 mg of physicochemically homogeneous protein. Since we could only produce it in limited quantities, we had to be quite prudent in defining and judging our most important experimental goals in terms of their intrinsic significance relative to the amount of material required. We considered the establishment of its biochemical properties and chemical structure to be of paramount importance. Once known, these would likely yield more knowledge of predictive value for future experimentation and understanding than anything else we could imagine. Hence, we elected to establish the primary structure of angiogenin as the most important and immediate scientific objective.

The amino acid sequence (Figure 1) and disulfide bond pairing of angiogenin was determined by conventional sequencing techniques adapted and applied to nanomole and subnanomole levels of material. Angiogenin, obtained from HT-29 cell conditioned media, is a single-chain protein consisting of 123 amino acids (pI > 9.5) with a molecular weight of about 14,400. It contains one tryptophan, one methionine, six cysteines, seven lysines and thirteen arginines. Basic residues outnumber acidics by two to one. There are also six histidines and the ultraviolet absorption spectrum reflects the single tryptophan, four tyrosine and five phenylalanine residues. The determination of the

<Glu-Asp-Asn-Ser-Arg-Tyr-Thr-His-Phe-Leu-Thr-Gln-His-Tyr-Asp
-Ala-Lys-Pro-Gln-Gly-Arg-Asp-Asp-Arg-Tyr-Cys-Glu-Ser-Ile-Met
-Arg-Arg-Arg-Gly-Leu-Thr-Ser-Pro-Cys-Lys-Asp-Ile-Asn-Thr-Phe
-Ile-His-Gly-Asn-Lys-Arg-Ser-Ile-Lys-Ala-Ile-Cys-Glu-Asn-Lys
-Asn-Gly-Asn-Pro-His-Arg-Glu-Asn-Leu-Arg-Ile-Ser-Lys-Ser-Ser
-Phe-Gln-Val-Thr-Thr-Cys-Lys-Leu-His-Gly-Gly-Ser-Pro-Trp-Pro
-Pro-Cys-Gln-Tyr-Arg-Ala-Thr-Ala-Gly-Phe-Arg-Asn-Val-Val-Val
-Ala-Cys-Glu-Asn-Gly-Leu-Pro-Val-His-Leu-Asp-Gln-Ser-Ile-Phe
-Arg-Arg-Pro -OH

Figure 1. Amino acid sequence of human angiogenin (Strydom et al., 1985).

Table 3: Comparison of Sequence Homologies of Human Angiogenin and Pancreatic Ribonuclease

	Residues
Identical to human RNase	43
Identical to other RNases	25
Conservative replacements	16
Homology: 84/123	68%

amino acid sequence was aided greatly by the large number of trypsin susceptible lysyl and arginyl bonds and the single methionine, which allowed chemical cleavage (Strydom et al., 1985). The amino terminus is blocked by a pyroglutamyl residue and the carboxyl-terminal residue is proline. Three disulfide bonds link the half-cysteinyl residues 26-81, 39-92, and 57-107.

The cDNA coding for angiogenin was isolated from a cDNA library prepared from normal fetal human liver poly(A) mRNA employing a synthetic oligonucleotide as a hybridization probe. Of the cDNAs detected, the largest insert (697 base pairs) contained a short 5′-noncoding sequence followed by a sequence coding for a signal peptide of 24 (or 22) amino acids, 369 nucleotides coding for the mature protein of 123 amino acids, a stop codon, a 3′-noncoding sequence of 175 nucleotides, and a poly(A) tail. The gene coding for human angiogenin was then isolated from a genomic λ Charon 4A bacteriophage library employing the cDNA as a probe. The nucleotide sequence of the gene and the adjacent 5′- and 3′-flanking regions (4688 base pairs) was then determined. The coding and 3′-noncoding regions of the gene for human angiogenin were found to be free of introns, and the DNA sequence for the gene was in agreement with that of the cDNA. The amino acid sequence of human angiogenin predicted from the gene sequence coincided with that determined by amino acid sequence analysis (Kurachi et al., 1985). These circumstances established the presence of the angiogenin gene in normal fetal liver.

RIBONUCLEASE HOMOLOGY

The primary sequence of angiogenin is remarkably homologous to that of the family of pancreatic ribonucleases, enzymes that cleave ribonucleic acids at their pyrimidine bases. Of the 123 amino acids in human angiogenin 43 (35%) are identical with those at the corresponding positions in human pancreatic RNase (Beintema, 1984) (Table 3) and another 41 either are identical with residues in other pancreatic RNases (Blackburn and Moore, 1982; Beintema, 1985) or are conservative replacements, constituting an overall homology of 68%.

The primary sequence of pancreatic ribonucleases from at least 41 species are known (Blackburn and Moore, 1982; Beintema, 1985), and the homology of angiogenin holds throughout. Detailed comparisons and analogies can be made to compare their structure, function and the interdependence of all of these, and to examine the evolutionary relationships of these proteins. In particular, it will be of great interest to determine when during the course of evolution angiogenin and the pancreatic RNases diverged. Such studies would be furthered greatly by the identification and characterization of angiogenin from other species such as the pig, cow, and horse now underway in our laboratory.

A preliminary three-dimensional structure of angiogenin has been computed, based on its homology to bovine pancreatic ribonuclease A. A standard-geometry structure of ribonuclease was first obtained from its x-ray coordinates. The fit of the backbone of angiogenin to that of ribonuclease was then optimized by taking account of amino acid deletions and by minimizing its conformational energy-plus-a-penalty distance function constraining its backbone to that of ribonuclease. Side-chain and backbone dihedral angles were allowed to vary throughout the cycles of energy minimization. In the last stages of minimization, the penalty distance function was removed and a low-energy structure resembling ribonuclease was obtained (Figure 2) (Palmer et al., 1986).

RIBONUCLEOLYTIC ACTIVITY

It is indeed remarkable that, in angiogenin, His 12, His 119, and Lys 41 of pancreatic ribonuclease are all conserved. Based on chemical modifications (Stein and Barnard, 1959; Crestfield et al., 1963; Hirs et al., 1965) and x-ray studies of pancreatic ribonuclease (Richards and Wyckoff, 1973), all three are known to be at its active site. One might expect, therefore, that angiogenin might exhibit ribonucleolytic activity.

Angiogenin does not hydrolyze any of the conventional substrates of pancreatic ribonuclease (Table 4). In particular, it does not produce detectable amounts of acid-soluble fragments from high molecular weight wheat germ RNA, poly(C), or poly(U), nor does it hydrolyze cytidine or uridine cyclic 2′,3′-phosphate (Shapiro et al., 1986). Nevertheless, the structural homology of the two types of proteins implies that both might utilize RNA as a substrate, though perhaps the sites and/or nature of cleavage might differ. This inference has proven to be correct. Angiogenin, indeed, exhibits endonucleolytic activity toward 28S and 18S rRNA as determined by agarose gel electrophoresis. However, unlike pancreatic RNase, angiogenin generates relatively large fragments, containing from 100 to 500 nucleotides, indicating only limited cleavage within each RNA species (Figure 3). Moreover, the amount of angiogenin required to generate these fragments in a given period of time is

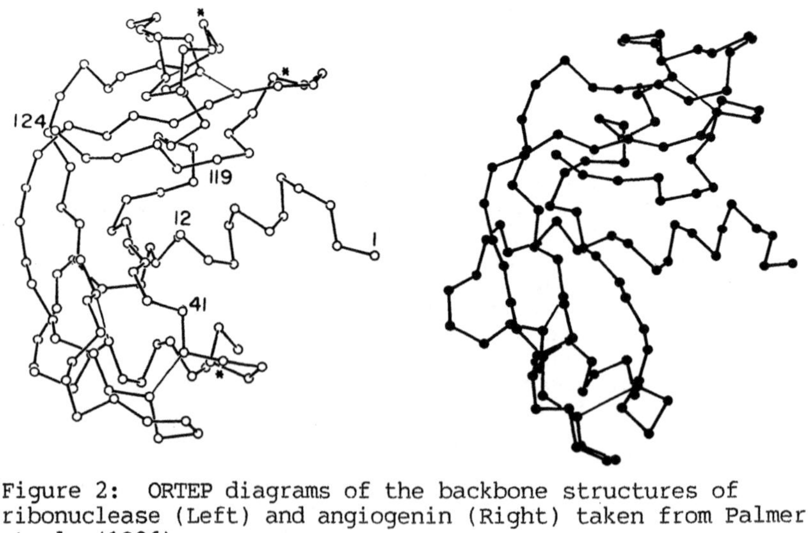

Figure 2: ORTEP diagrams of the backbone structures of ribonuclease (Left) and angiogenin (Right) taken from Palmer et al. (1986).

Table 4: Conventional Ribonuclease Substrates and Other Nucleotides that are not Hydrolyzed by Angiogenin

RNase Substrates	Non-RNase Substrates
Polynucleotides	Polynucleotides
polyC, polyU	polyA, G, IC, AU, AdT CdI, GdC, UdA
Dinucleotides	Dinucleotides
CU, UU, CA, UA CG, UG, CC, UC	AU, GU, AA, GA, AG, GG, AC, GC, 2'5'-CA, 2'5'-CC, 2'5'-UA
Cyclic nucleotides	Nucleotides
C 2',3' cyclic P U 2',3' cyclic P	ADP, ATP, cAMP, 5'AMP 3'-AMP, 3'-CMP, 3'-UMP cGMP, GDP, GTP

several orders of magnitude greater than the amount of pancreatic RNase needed to degrade the RNA completely. The failure of angiogenin to degrade these rRNA fragments further suggests either that they have a secondary structure that protects them from angiogenin or that the cleavage sites are an expression of an extraordinary specificity. As a third alternative, product inhibition could prevent complete degradation when there is secondary structure. In any event, the precise features of the sites where cleavage occurs, optimum conditions of assay, and kinetic parameters remain to be established. Similarly, the potential functional implications of this enzymatic activity must also be defined; it could be that the nucleolytic activity of angiogenin may be even more specific and selective toward yet untested RNA substrates.

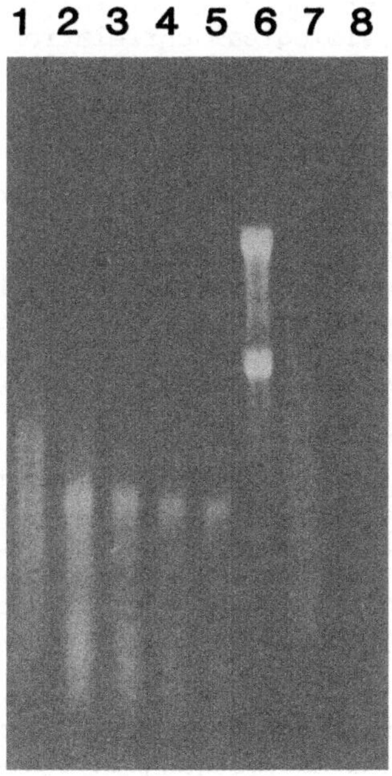

Figure 3. Time course for degradation of HT-29 cell total RNA by angiogenin and bovine pancreatic ribonuclease. (Lanes 1-5) Samples containing 1.9 μM angiogenin incubated for 3, 10, 25, 60, and 255 min. respectively; (Lane 6) Control RNA; (Lanes 7 and 8) Samples containing 0.13 nM ribonuclease for 3 and 30 min. Samples were electrophoresed On 1.1% agarose gel under denaturing conditions.

One indication that this might indeed be the case has been obtained by studying the effect of angiogenin on a cell-free protein synthesizing system that utilizes a rabbit reticulocyte lysate and globin messenger RNA. As shown in Table 5 angiogenin is a potent inhibitor of protein synthesis. In fact, it is as good if not better than RNase at abolishing the incorporation of [^{35}S]-methionine into products precipitable with trichloroacetic acid. Preliminary evidence (St. Clair, 1987) suggests that this loss of activity is due to degradation of ribosomal RNA and that the remarkable specificity of this action only occurs when the rRNA is present in the intact ribosome and is not observed with naked rRNA.

It is not known whether or not this activity of angiogenin is the basis for its ability to induce blood vessel formation. Nevertheless, it is important to note that the high degree of sequence homology between angiogenin and the pancreatic ribonucleases, in general, and of the three catalytic residues, His-12, Lys-41, and His-119, in particular, has led to the recognition of the chemical nature of potential angiogenin substrates--insights that may well bear importantly on the physiological function of angiogenin.

A possible connection between the angiogenic and ribonucleolytic activities of angiogenin has been obtained by means of chemical modification. Patterned on the extensive work carried out to identify the active site residues of pancreatic ribonuclease (Barnard and Stein, 1964; Richards and Wyckoff, 1971), angiogenin is inactivated by reagents that modify histidine, lysine, or arginine residues, but not by those that modify tyrosine, aspartate, glutamate, or methionine. Thus, for example, bromoacetate inactivates by carboxymethylation of His-13 and His-114 at pH 5.5, whereas inactivation by fluorodinitrobenzine (FDNB), diethylpyrocarbonate and formaldehyde/cyanoborohydride involves lysyl residue(s), though the specific one(s) have not yet been determined (Table 6). Importantly, both the angiogenic and the ribonucleolytic activities of angiogenin appear to be abolished by such chemical modifications. Much detailed work remains to be carried out to establish the chemical basis for loss of these activities and to define the extent to which they connote a physiological correlation.

IMMUNOLOGY AND DIAGNOSTIC POTENTIAL

The availability of both pure angiogenin and information regarding its amino acid sequence has made possible the development of immunoassay methods for its detection and quantitation. Antibodies raised by immunizing rabbits with the protein covalently linked to keyhole limpet hemocyanin have been used successfully to monitor the production and isolation of angiogenin. Both a radioimmunoassay and an inhibition ELISA method allow detection of angiogenin in the ng/ml range (Olson, 1987).

One of the objectives of developing an immunoassay was to be able to examine the diagnostic potential of blood serum angiogenin concentration measurements. As a preliminary to screening sera from, e.g., cancer patients, normal human serum samples were tested for the presence or absence of angiogenin. Surprisingly, the serum samples were all positive indicating that serum might be a potential source of the protein.

We have now isolated angiogenin from normal human plasma and purified it to homogeneity. This plasma-derived angiogenin exhibits the same angiogenic and ribonucleolytic activities, amino acid composition,

Table 5: Inhibition of Protein Synthesis by Angiogenin

[Ang], nM	[^{35}S]-Met, cpm	% Inhibition
0	34,500	0
43	13,500	61
64	0	100

Conditions: Rabbit reticulocyte lysate was preincubated at 30° with angiogenin at the concentrations shown for 15 min. An equal concentration of placental ribonuclease inhibitor was added and protein synthesis was initiated by addition of [^{35}S]-Met. Protein synthesis was determined from the amount of radioactivity in products precipitable by 10% trichloroacetic acid.

Table 6: Effects of Chemical Modifications on the Ribonucleolytic Activity of Angiogenin

Residue Modified	Reagent	Ribonucleolytic Activity (v/vc x 100)
His	Bromoacetate	2
Lys	FDNB	9
Arg	Phenylglyoxal	3
Tyr	N-Acetylimidazole	85
Glu/Asp	Carbodiimide	90
Met	DMSO/HCl	87

molecular weight, immunoreactivity, and chromatographic behavior as the tumor cell-derived protein. Peptide mapping and sequencing studies show them to be chemically identical (Shapiro et al., 1987). The yield of angiogenin from either plasma or serum is 60-150 μg/l. Thus angiogenin is not a tumor-specific product in accord with the isolation of the gene from normal liver. This provides further opportunities for the investigation of the role and mechanism of action of angiogenin and its potential diagnostic or prognostic utility.

PLACENTAL RIBONUCLEASE INHIBITOR

Homology considerations have also provided the opportunity to examine the capacity of known inhibitors of pancreatic RNase to inactivate angiogenin. Protein inhibitors of pancreatic and other RNases have been identified in the cytoplasm of a wide variety of mammalian tissues (Blackburn and Moore, 1977). In particular, the human placental inhibitor, PRI, has been purified and isolated: it very effectively inhibits pancreatic RNase (Table 7). PRI is also a potent antagonist of both the angiogenic and ribonucleolytic activities of

Table 7: Some Properties of Human Placental Ribonuclease Inhibitor

Molecular weight	51,000
Polypeptide chains	1
Cysteine residues	30
Leucine residues	95
K_i (RNase A), fM	41.0
K_i (angiogenin), fM	0.7

angiogenin. In fact, it is almost 60 times more effective against angiogenin and the stoichiometry of the interaction is 1:1. Analogous to bovine pancreatic RNase, this inhibition is reversible, and p-chloromercuribenzoate (PHMB) dissociates the complex readily to yield active angiogenin, chromatographically identical to the native protein. A slight molar excess of PRI suffices to inhibit enzymatic activity completely (Shapiro and Vallee, 1987). Further, carboxymethylation of Lys-41 of pancreatic RNase substantially reduces binding strength, indicating that this residue is essential to the interaction between PRI and RNase A (Lee, 1987).

These findings represent a hitherto unsuspected function for PRI: regulation of angiogenesis. This could lead to its therapeutic use in the management of pathological conditions characterized by or dependent on abnormal neovascularization. Inhibition of both the biological and enzymatic activities of angiogenin by human PRI may therefore have important mechanistic, physiological and pharmacologic implications.

ANGIOGENIN/RIBONUCLEASE HYBRIDS

The sequence homology of pancreatic ribonuclease with angiogenin provides the basis for a variety of other experimental approaches which should prove useful in the elucidation of the chemical mechanism of angiogenin's action. Angiogenin/RNase hybrid enzymes have been prepared (Harper et al., 1987) based on the cleavage of pancreatic RNase by subtilisin into S-peptide and S-protein and the restoration of activity on mixing the products (Richards and Wyckoff, 1971), an approach to explore the contributions of various segments of the primary structure of angiogenin to its function. Two regions of angiogenin which are expected to play prominent roles in activity are located at the N- and C-termini, and contain His-13 and His-114, respectively (angiogenin numbering system). Readily accessible starting points for the preparation of hybrid enzymes which examine the functional potential of the N- and C-terminal regions of angiogenin are provided by two proteolytic fragments of pancreatic RNase-- the S-protein and RNase lacking the last 6 residues (RNase 1-118). Inactive S-protein, derived from RNase S, can be reactivated by addition of the S-peptide (Richards and Vithayathil, 1959; Richards and Wyckoff, 1971). Similarly, RNase lacking the last six residues can be reactivated by addition of a synthetic C-terminal RNase peptide from residue number 111 to 124) (Lin et al., 1970; Levit and Berger, 1976).

Hence, either the N-terminal S-peptide or the C-terminal hexapeptide of pancreatic RNase has been replaced by adding the corresponding one of angiogenin prepared by Merrifield synthesis (Harper et al., 1987). In both instances noncovalent interactions restore significant pancreatic RNase but not angiogenic activities. Thus, it is imperative to examine the "core protein", the angiogenin equivalent of pancreatic RNase residues 21 to 118, regarding its effects on angiogenin. This can be accomplished by either recombinant DNA techniques, chemical synthesis, or both.

Overall such studies might be expected to result in agonists and/or antagonists of the action of angiogenin tailored for specific therapeutic objectives. Moreover, it will permit the examination of the relationship, if any, of angiogenin to the role of other angiogenic substances in the processes of neovascularization.

In summary, the complete chemical characterization of a unique human organogenic messenger molecule, i.e., one that induces organ formation, accomplishes a first major objective in a long-term investigation of organogenesis in general and angiogenesis in particular. The unexpected homology to ribonuclease A allows the use of novel approaches to the investigation of the biological process of angiogenesis and may be relevant to the evolution of organogenic molecules. Importantly, it provides a chemical basis for the prediction of angiogenic function. These provocative findings are thought to have important physiological implications.

REFERENCES

Baltzer, F., 1928, Über metagame Geschlechtsbestimmung and ihre Beziehung zu einigen Problemen der Entwicklungsmechanik und Vererbung (auf Grund von Versuchen an Bonellia). Verh. dt. zool. Ges., 32:273.

Barnard, E.A., and Stein, W.D., 1959, The histidine residue in the active centre of ribonuclease I. A specific reaction with bromoacetic acid, J. Mol. Biol., 1:339.

Beintema, J.J., Wietzes, P., Weickmann, J.L., and Glitz, D.G., 1984, The amino acid sequence of human pancreatic ribonuclease, Anal. Biochem., 136:48.

Beintema, J.J., Broos, J., Menlenberg, J., and Schuller, C., 1985, The amino acid sequence of snapping turtle (Chelydra serpentina) ribonuclease, Eur. J. Biochem., 153:305.

Blackburn, P., and Moore, S., 1982, Pancreatic Ribonuclease, in: The Enzymes (3rd Ed.) Vol 15. P.D. Boyer, ed. Academic Press, New York.

Crestfield, A.M., Stein, W.H., and Moore, S., 1963, Alkylation and identification of the histidine residues at the active site of ribonuclease, , J. Biol. Chem., 238:2413.

Ehrmann, R.L., and Knoth, M., 1968, Choriocarcinoma: transfilter stimulation of vasoproliferation in the hamster check pouch studied by light and electron microscopy, J. Natl. Cancer Inst., 41:1329.

Fett, J.W., Strydom, D.J., Lobb, R.R., Alderman, E.M., Bethune, J.L., Riordan, J.F., and Vallee, B.L., 1985, Isolation and characterization of angiogenin, an angiogenic protein from human carcinoma cells, Biochemistry, 24:5480.

Folkman, J., and Klagsbrun, M., 1987, Angiogenic factors, Science, 235:442.

Gospodarowicz, D., Bialecki, H., and Thakral, T.K., 1979, The angiogenic activity of the fibroblast and epidermal growth factor, Exp. Eye Res. 28:501.

Greenblatt, M., and Shubik, P., 1968, Tumor angiogenesis: transfilter diffusion studies in the hamster by the transparent chamber technique, J. Natl. Cancer Inst., 41:111.

Harper, J.W., Auld, D.S., and Vallee, B.L., 1987, in preparation.

Hirs, C.H.W., Halmann, M., and Kycia, J.H., 1965, Dinitrophenylation and

inactivation of bovine pancreatic ribonuclease A, Arch. Biochem. Biophys., 111:209.
Knighton, D., Ausprunk, D., Tapper, D., and Folkman, J., 1977, Avascular and vascular phases of tumor growth in the chick embryo, Br. J. Cancer, 35:347.
Kull, F.C., Jr., Brent, D.A., Parikh, I., and Cuatrecasas, P., 1987, Chemical identification of a tumor-derived angiogenic factor, Science, 236:843.
Kurachi, K., Davie, E.W., Strydom, D.J., Riordan, J.F., and Vallee, B.L., 1985, Sequence of the cDNA and gene for angiogenin, a human angiogenesis factor, Biochemistry, 24:5494.
Langer, R., and Folkman, J., 1976, Polymers for the sustained release of proteins and other macromolecules, Nature, 263:797.
Lee, R., 1987, unpublished experiments.
Levit, S., and Berger, A., 1976, Ribonuclease S-peptide, a model for molecular recognition, J. Biol. Chem., 247:4768.
Lin, M.C., Gutte, B., Caldi, D.G., Moore, S., and Merrifield, R.B., 1972, Reactivation of des(119-124) ribonuclease A by mixture with synthetic COOH-terminal peptides: the role of phenylalanine 120, J. Biol. Chem., 247:4768.
Lobb, R.R., Alderman, E.M., and Fett, J.W., 1985, Induction of angiogenesis by bovine brain derived class 1 heparin binding growth factor, Biochemistry, 24:4969.
Lobb, R.R., Sasse, J., Sullivan, R., Shing, Y., D'Amore, P., Jacobs, J., and Klagsbrun, M., 1986, Purification and characterization of heparin-binding endothelial cell growth factors, J. Biol. Chem., 261:1924.
Maciag, T., Cerundolo, I.S., Kelley, P.R., and Forand, R., 1979, An endothelial cell growth factor from bovine hypothalamus: Identification and partial characterization, Proc. Natl. Acad. Sci. USA, 76:5674.
Olson, K.A., 1987, unpublished experiments.
Palmer, K.A., Scheraga, H.A., Riordan, J.F., and Vallee, B.L., 1986, A preliminary three-dimensional structure of angiogenin, Proc. Natl. Acad. Sci. USA, 83:1965.
Richards, F.W., and Vithayathil, P., 1959, The preparation of subtilisin-modified ribonuclease and the separation of the peptide and protein components, J. Biol. Chem., 234:1459.
Richards, F.M., and Wyckoff, H.W., 1971, Bovine pancreatic ribonuclease, in The Enzymes (3rd ed.) Vol 4., P.D. Boyer, Ed. Academic Press, New York.
Richards, F.W., and Wyckoff, H.W., 1973, Ribonuclease S, in Atlas of Molecular Structures in Biology, D.C. Phillips and F.M. Richards, eds. Oxford Univ. Press, London.
Roberts, A.B., Sporn, M.B., Assoian, R.K., Smith, J.M., Roche, N.S., Wakefield, L.M., Heine, U.I., Liotta, L.A., Falanga, V., Kehrl, J.H., and Fauci, A.S., 1986, Transforming growth factor β: Rapid induction of fibrosis and angiogenesis in vivo and stimulation of collagen formation in vitro, Proc. Natl. Acad. Sci. U.S.A., 83:4167.
St. Clair, D.K., 1987, unpublished experiments.
Schreiber, A.B., Winkler, M.E., and Derynck, R., 1986, Transforming growth factor α: A more potent angiogenic mediator than epidermal growth factor, Science 232:1250.
Shapiro, R., Riordan, J.F., and Vallee, B.L., 1986, Characteristic ribonucleolytic activity of human angiogenin, Biochemistry, 25:3527.
Shapiro, R., and Vallee, B.L., 1987, Human placental ribonuclease inhibitor abolishes both angiogenin and ribonucleolytic activities of angiogenin, Proc. Natl. Acad. Sci. USA, 84:2238.
Shapiro, R., Strydom, D.J., Olson, K.A., and Vallee, B.L., 1987, Isolation of angiogenin from normal human plasma, Biochemistry, in press.

Spemann, H., 1938, "Embryonic Development and Induction", Yale University Press, New Haven.
Spemann, H., and Mangold, H., 1924, Über induktion von embryonalanlagen durch implantation artfremder organisatoren, Arch. Mikrosk. Anat. Entwicklungsmech., 100:599.
Stein, W.H., and Barnard, E.A., 1959, The histidine residue in the active centre of ribonuclease II. The position of this residue in the primary protein chain, J. Mol. Biol., 1:350.
Strydom, D.J., Fett, J.W., Lobb, R.R., Alderman, E.M., Bethune, J.L., Riordan, J.F., and Vallee, B.L., 1985, Amino acid sequence of human tumor derived angiogenin, Biochemistry, 24:5486.
Vallee, B.L., Riordan, J.F., Lobb, R.R., Higachi, N., Fett, J.W., Crossley, G., Bühler, R., Budzik, G., Breddam, K., Bethune, J.L., and Alderman, E.M., 1985, Tumor-derived angiogenesis factors from rat Walker 256 carcinoma: an experimental investigation and review, Experientia, 41:1.
Wissler, J.H., Logemann, E., Meyer, H.E., Krützfeldt, B., Höckel, M., and Heilmeyer, L.M.G. jr., 1987, Bioactive copper-ribonucleo-polypeptide complexes: angio-morphogens of porcine monocytes, Federation Proc., 46:2192.
Ziche, M., Jones, J., and Gullino, P.M., 1982, Role of prostaglandin E, and copper in angiogenesis, J. Natl. Cancer Inst., 69:475.

STRUCTURE-FUNCTION RELATIONSHIPS IN CELLULAR AND VIRAL fps/fes CYTOPLASMIC PROTEIN-TYROSINE KINASES

Tony Pawson, Peter Greer, Michael Moran, Kelly Meckling-Hansen, Angela Brooks-Wilson and Ivan Sadowski

Division of Molecular and Developmental Biology
Mt. Sinai Hospital Research Institute
600 University Ave.
Toronto, Ont. M5G 1X5
Canada

CYTOPLASMIC AND TRANSMEMBRANE PROTEIN-TYROSINE KINASES

A majority of retroviral oncogenes encode protein-tyrosine kinases (PTKs). Both oncogenic PTKs and their normal counterparts can be divided into two classes : receptor-like polypeptides that possess an extracellular region linked by a transmembrane segment to an internal catalytic domain, and cytoplasmic PTKs that are entirely confined to the inside of the cell (see fig. 1). Cytoplasmic PTKs are usually found in association with the plasma membrane and cytoskeleton (Rohrschneider and Gentry, 1984). The cytoplasmic PTKs can be further subdivided into 3 families corresponding to a group of proteins of approximately 60 kDa for which $p60^{src}$ is the prototype, fps/fes gene products, and abl proteins respectively (Hunter and Cooper, 1985).

The normal functions of cytoplasmic PTKs are less clear than those of growth factor receptors for which specific activating ligands can be identified. Members of the src and fps/fes families are generally expressed in a restricted set of cell types, suggesting that they function in lineage-specific events (Brugge et al., 1985; Golden et al., 1986; Feldman et al., 1985; MacDonald et al., 1985). If cytoplasmic PTKs are involved in the transfer of developmental or mitogenic signals initiated by hormonal stimuli they must function as intermediaries, since they themselves have no extracellular ligand-binding domain. One possibility is that cytoplasmic PTKs function as catalytic subunits of receptors with no intrinsic cytoplasmic effector domain (see fig. 1). Alternatively their activity might be directly or indirectly regulated by transmembrane PTKs.

Cytoplasmic PTKs are complex multi-domain polypeptides. To investigate structure-function relationships in an oncogenic PTK we have employed the 130 kDa oncoprotein ($P130^{gag-fps}$) of Fujinami avian sarcoma virus (FSV).

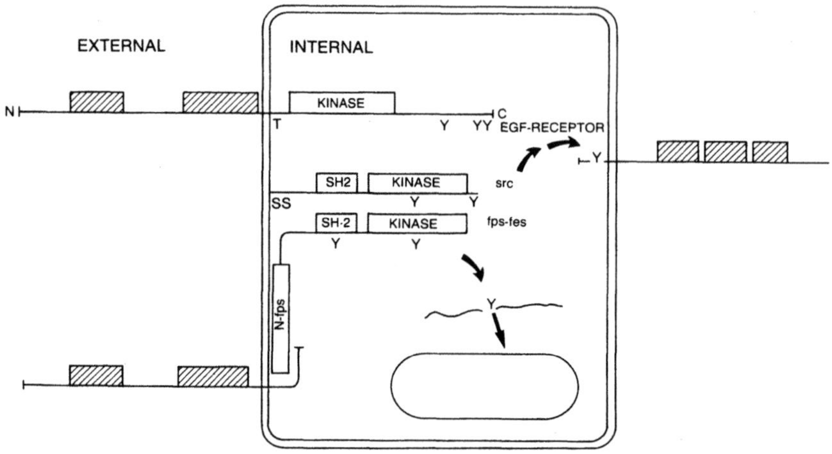

Fig. 1 Transmembrane and cytoplasmic protein-tyrosine kinases. The EGF-receptor, and the products of the c-src and c-fps/fes genes are illustrated. A hypothetical transmembrane receptor is depicted interacting with the c-fps/fes protein. Some sites of serine (S), threonine (T) and tyrosine (Y) phosphorylation are shown. Domains (N-fps, SH2, kinase) are described in the text. Possible PTK targets are indicated by arrows.

CELLULAR AND VIRAL fps/fes PROTEINS

The fps/fes gene was originally identified as the transforming element of a number of avian and feline sarcoma viruses, and is the most frequently isolated retroviral oncogene. The transforming PTKs encoded by these viruses are hybrids composed of N-terminal retroviral gag coat protein sequences linked to C-terminal fps-fes domains (fig. 2). Although the avian form of this gene was termed fps while the feline oncogene was termed fes, their normal counterparts are cognate gene (Groffen et al., 1983). The avian cellular fps/fes gene encodes a 98 kDa PTK ($p98^{c-fps}$) (Mathey-Prevot et al., 1982), while the human c-fps/fes locus on chromosome 15 specifies a closely related 92 kDa protein ($p92^{c-fes}$) (Feldman et al., 1985; MacDonald et al., 1985; Greer et al., 1987). The structural relationshps between normal and oncogenic fps/fes proteins are indicated in Fig. 2. Fps/fes expression is normally limited to hematopoietic cells, particularly immature and differentiated myeloid cells, suggesting a role for c-fps/fes PTKs in myelopoiesis (Feldman et al., 1985; MacDonald et al., 1985; Samarut et al., 1985). Ectopic expression of $p98^{c-fps}$ in chicken embryo fibroblasts and of human $p92^{c-fes}$ in rat fibroblasts to levels much greater than those normally observed in myeloid cells does not induce neoplastic transformation, apparently because kinase activity of these normal fps/fes proteins is restrained in vivo by cellular regulatory elements (Foster et al., 1985; Greer et al., 1987). Thus structural alterations are required for oncogenic activation of fps/fes polypeptides. The function of $p92^{c-fes}$ in myeloid cells is unknown; it seems likely that its kinase activity is transiently activated to perform a specific function in myeloid differentiation.

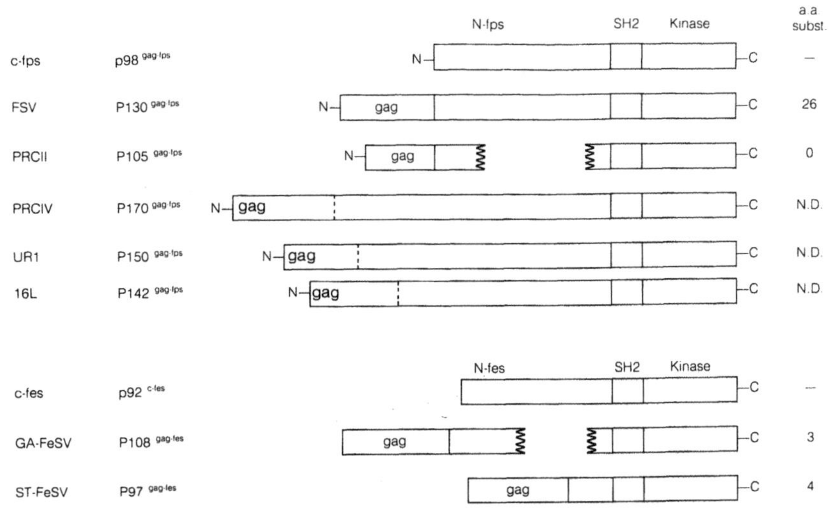

Fig. 2 Structural relationships between viral and cellular fps/fes proteins. The product of the avian c-fps locus is compared with avian v-fps oncoproteins; the feline c-fes protein is compared with the transforming proteins of Gardner-Arnstein (GA-) and Snyder-Theilen (ST-) feline sarcoma viruses (FeSV).

In contrast the oncogenic fps/fes kinases are enzymatically very active in fibroblasts (Pawson et al., 1980). Three striking differences have been noted between viral and cellular fps/fes proteins that may account for the constitutive in vivo activity and oncogenic properties of the viral forms. The transforming proteins autophosphorylate at a tyrosine within the kinase domain that is conserved among all PTKs (tyr-1073 of P130$^{gag-fps}$, tyr-416 of p60^{v-src}), while p98^{c-fps} and p92^{c-fes} are generally not phosphorylated at this site (Patschinsky et al., 1982; Weinmaster et al., 1983; Greer et al., 1987). Two lines of evidence indicate that tyr-1073 of P130$^{gag-fps}$ is a regulatory site whose occupancy by phosphate stimulates kinase activity. First, replacement of tyr-1073 by phe, ser, thr, gly, his or glu (none of which are substrates for the autophosphorylation reaction) using site-directed mutagenesis results in decreased kinase activity and oncogenic potential (Weinmaster et al., 1984; Weinmaster and Pawson, 1986; M.M. and T.P., unpublished observation); second, manipulation of the extent of tyr-1073 phosphorylation using phosphatases and their inhibitors confirms that modification of this residue enhances v-fps kinase activity (Meckling-Hansen et al., 1987). In addition v-fps/fes oncoproteins have a stronger association with the plasma membrane and/or cytoskeleton than do their normal counterparts (Young and Martin, 1984). Finally v-fps/fes proteins are expressed to unphysiologically high levels in virus-transformed cells. Therefore changes in subcellular localization and expression levels are accompanied in v-fps/fes-transformed cells by autophosphorylation and consequent stimulation of fps/fes kinase activity for exogenous substrates.

DOMAIN STRUCTURE OF P130$^{gag-fps}$

Experimental Approaches

We have used several distinct approaches to determine the organization of the P130$^{gag-fps}$ oncoprotein. As a preliminary, the predicted amino acid sequence of P130$^{gag-fps}$ was compared with those of other protein kinases and nucleotide-binding proteins. Using the cloned FSV viral genome (Shibuya and Hanafusa, 1982), in-frame linker-insertion mutagenesis was developed as a technique for the genetic dissection of this complex polypeptide (Stone et al., 1984). A dipeptide insertion within a specific domain has the general property of disrupting the conformation of that domain without affecting the structure or activity of adjacent but separately folding domains in the same polypeptide (Stone and Pawson, 1985). By analysis of a library of FSV linker-insertion mutants it was possible to establish in a crude way the domain structure of P130$^{gag-fps}$ (Stone et al., 1984). Classically partial proteolysis has been used to define protease-resistant globular domains within large proteins. A conceptual extension of this procedure is to express various portions of a protein using, in our case, bacterial expression vectors as a means to precisely define the boundaries of a specific functional domain. Both these techniques have been applied to P130$^{gag-fps}$ (Weinmaster et al., 1983, Sadowski and Pawson, 1987). Furthermore by creating chimeric coding sequences, a domain of known function from one kinase can be tested for its ability to functionally complement a domain of unknown activity in a distinct kinase. Once the overall domain structure of a protein such as P130$^{gag-fps}$ is appreciated, the role of smaller structural motifs and specific amino acids within a domain can be investigated by biochemical and biophysical approaches coupled with oligonucleotide-directed mutagenesis.

The domain structure of P130$^{gag-fps}$ revealed by these experiments is illustrated in fig. 3.

The Catalytic Domain

P130$^{gag-fps}$ has approximately 260 amino acids located at its extreme C-terminus that are homologous with sequences present in all known protein kinases (Weinmaster et al., 1983). That these residues form the v-fps tyrosine kinase catalytic domain has been shown in two ways. Partial trypsinolysis of P130$^{gag-fps}$ ultimately yields a 29 kDa C-terminal polypeptide that retains kinase activity (Weinmaster et al., 1983). More conclusively, expression of partial v-fps polypeptides in Escherichia coli has shown that residues 919-1182 of P130$^{gag-fps}$ are able to autophosphorylate and phosphorylate exogenous substrates such as enolase and poly(glu,tyr), but that removal of an additional 6 N-terminal amino acids to residue 925 results in loss of catalytic activity (Sadowski and Pawson, 1987). The proposed ATP-binding fold of P130$^{gag-fps}$ starts at residue 922 (Weinmaster et al., 1986), and it therefore seems likely that the minimal v-fps catalytic domain encompasses 260 residues starting with the ATP-binding fold. The extent of sequences within this domain actually required for ATP-binding is not known. However residues 922-950 show considerable sequence homology with the adenine nucleotide-binding sites of a number of dehydrogenases whose crystal structures have been solved (Weinmaster et al., 1986). In addition residue 950 corresponds to an invariant lysine that in other protein kinases is covalently modified by the ATP analogue fluorosulfonylbenzoyl 5'-adenosine (Zoller and Taylor, 1979; Kamps et al., 1984). Indeed conversion of the codon for P130$^{gag-fps}$ lys-950 to arginine or glycine codons results in complete loss of P130$^{gag-fps}$ kinase activity and oncogenicity, suggesting that the ε-amino group of lys-950 may be located in the active site and participate in the phosphotransfer reaction (Weinmaster et al., 1986).

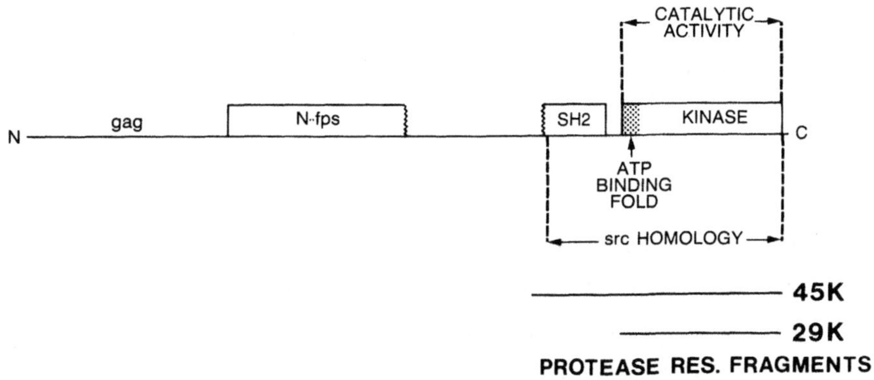

Fig. 3 $P130^{gag-fps}$ domain structure.

Within the v-fps catalytic domain two motifs can be identified that are homologous not only with tyrosine-specific but also with serine/threonine-specific protein kinases. These are the ATP-binding site, and a second region surrounding the main autophosphorylation site (tyr-1073) (indicated by diagonal shading in fig. 4). These regions are separated by sequences that are more variable between different kinases. Presumably the highly conserved regions are crucial for the proper folding and biochemical activity of the protein kinase catalytic domain.

The SH2 Domain

Dipeptide insertions within the v-fps catalytic domain tend, not surprisingly, to destroy activity. However two v-fps linker-insertion mutants (RX15m and AX9m) with dipeptide insertions about 80 amino acids N-terminal to the ATP-binding site (Fig. 4) were also unable to transform Rat-2 cells (Sadowski et al., 1986). Mutant $P130^{gag-fps}$ proteins expressed in Rat-2 cells had decreased kinase activity. The region of $P130^{gag-fps}$ immediately N-terminal to the catalytic domain is clearly not required for enzymatic activity (see above), but the phenotype of the RX15m and AX9m FSV mutants suggests that it can interact with the kinase domain. Examination of the relevant amino acid sequences shows that this non-catalytic domain, spanning approximately 100 amino acids N-terminal to the minimal kinase domain, is conserved among all cytoplasmic PTKs. The homology is highest within 40 distal residues displayed in fig. 4. This non-catalytic domain has been termed SH2 for src homology 2. When the predicted amino acid sequences of the Drosophila and human abl proteins are compared, the SH2 domain is found to be more highly conserved (77% identical residues) than the catalytic domain (76% identity) (M. Hoffman, personal communication). Thus the SH2 domain is not only homologous between different cytoplasmid PTKs, but is also highly conserved during the evolution of a specific kinase.

Direct evidence that the SH2 domain folds against the catalytic domain comes from partial proteolysis of $P130^{gag-fps}$. The principal protease-resistant fragment generated by partial digestion of $P130^{gag-fps}$ is a 45 kDa C-terminal polypeptide that contains both SH2 and catalytic domains (Weinmaster et al., 1983). In addition the SH2 domain contains an autophosphorylation site, consistent with the view that SH2 makes direct contact with the catalytic domain (Sadowski and Pawson, 1987).

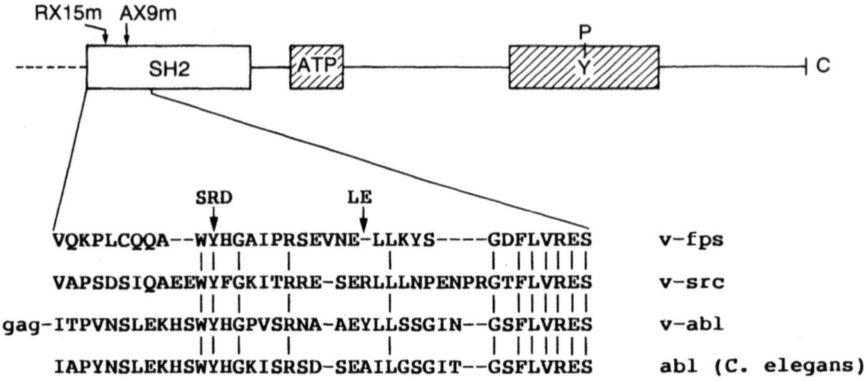

Fig. 4 Linker insertions in the SH2 domain of $P130^{gag-fps}$. The SH2 domain is shown relative to the ATP-binding fold and a second conserved region within the catalytic domain (striped boxes). The positions of the RX15m and AX9m mutations are shown with the corresponding amino acid substitutions (below). The sequences of 3 distinct cytoplasmic PTKs within the most highly conserved segment of SH2 are compared. Identical residues are connected.

The function of SH2 is unclear, but preliminary evidence suggests that it forms a binding site for cellular proteins that regulate, or possibly serve as targets for, the kinase domain. This is inferred from the observation that the effects of SH2 mutations depend on the cellular environment in which the mutant proteins are expressed. The RX15m and AX9m mutations have little effect on v-fps kinase activity in E. coli (in contrast to Rat-2 cells) (Sadowski et al., 1986). In addition the AX9m FSV mutant is able to transform chicken embryo fibroblasts in a temperature-dependent fashion, although it is entirely defective in Rat-2 cells (DeClue et al., 1987). Thus AX9m-FSV is a host-range mutant; it is likely that the host-dependence of AX9m can be used to identify cellular proteins that interact with SH2 and are critical to v-fps transforming activity.

N-fps/fes

The N-terminal gag sequences of $P130^{gag-fps}$ are dispensible for transforming activity (Foster and Hanfusa, 1984). However the 250 N-terminal v-fps amino acids form a structural and functional domain (termed N-fps) which is important for $P130^{gag-fps}$ oncogenicity. Insertions, deletions or amino acid substitutions with N-fps have no detectable effect on kinase activity, but impair or abolish FSV transforming activity (Stone et al., 1984; Stone and Pawson, 1985; Ariizumi and Shibuya, 1985; Chen et al., 1986). A likely function of N-fps is in subcellular localization of $P130^{gag-fps}$ (Woolford and Beemon, 1984). The transforming proteins of GA- and ST-FeSV have deletions within this domain (Fig. 2); however the N-terminal glycine of mammalian gag is myristylated, a modification which assists membrane attachment and which is absent from avian gag (Henderson et al., 1983). Thus feline gag

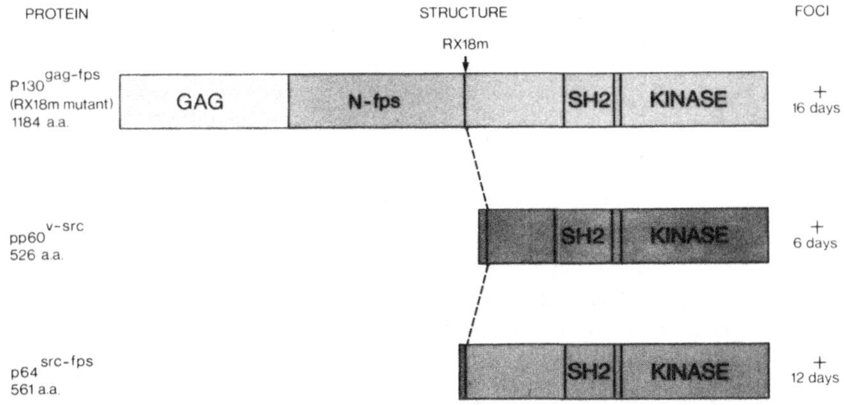

Fig. 5 Transforming ability of a src-fps hybrid in Rat-2 cells. The replacement of the N-fps domain of $P130^{gag-fps}$ with the membrane-localizing signal of $p60^{v-src}$ gives a src-fps hybrid polypeptide which induces focus formation on Rat-2 cells more rapidly than $P130^{gag-fps}$.

elements may override the necessity for an intact N-fps/fes domain in the GA- and ST-FeSV oncoproteins. To test this hypothesis we replaced the gag-N-fps sequences of $P130^{gag-fps}$ with the N-terminal 14 amino acids of $p60^{v-src}$ (fig. 5). These 14 v-src residues are sufficient to direct incorporation of heterologous proteins into the plasma membrane (Pellman et al., 1985). The hybrid v-src-fps protein transforms Rat-2 cells with greater efficiency than wild type $P130^{gag-fps}$, even though it lacks the (otherwise essential) N-fps domain. We tentatively conclude that the N-fps domain is required for subcellular localization, thereby allowing $P130^{gag-fps}$ access to critical targets, and that it can be replaced by the membrane localizing signal of $p60^{v-src}$.

CONCLUSIONS

The domain structure of the $P130^{gag-fps}$ oncogenic cytoplasmic PTK has been elucidated using a variety of techniques. Three distinct domains appear crucial to the transforming ability of this protein:

1) An N-terminal domain (N-fps) probably required for localization.

2) A conserved non-catalytic domain (SH2) that apparently folds against the catalytic domain. This domain is suggested to function as a binding site for cellular proteins that interact with the catalytic domain either as regulators or substrates.

3) A C-terminal catalytic domain. The activity of this domain is regulated by autophosphorylation at tyrosine and is required for transformation.

Work in the author's laboratory is supported by grants from the National Cancer Institute of Canada, the Medical Research council of Canada, the Natural Science and Engineering Research Council and the Leukemia Research Fund (Toronto). T.P. is a Senior Scientist, M.M. is a postdoctoral fellow, and K.M.-H. and I.S. are student fellows of the NCIC.

REFERENCES

Ariizumi, K., and Shibuya, M., 1985, Construction and biological analysis of deletion mutants of Fujinami sarcoma virus: 5'-fps sequence has a role in the transforming activity, J. Virol., 55:660.

Brugge, J.S., Cotton, P.C., Queral, A.E., Barrett, J.N., Nonner, D., and Keane, R.W., 1985, Neurones express high levels of a structurally modified, activated form of $p60^{c-src}$, Nature, 316:554.

Chen, L.-H., Hatada, E., Wheatley, W., and Lee, W.-H., 1986, Single amino acid substitution from glu^{1025} to asp, of the fps oncogenic protein causes temperature sensitivity in transformation and kinase activity, Virology, 155:106.

DeClue, J., Sadowski, I., Martin, G.S., and Pawson, T., 1987, A conserved domain regulates interactions of the v-fps protein-tyrosine with the host cell, Proc. Natl. Acad. Sci. USA, In Press.

Feldman, R.A., Gabrilove, J.L., Tam, J.P., Moore, M.A.A., and Hanafusa, H., 1985, Specific expression of the normal celluar fps/fes-encoded protein NCP92 in normal and leukemic myeloid cells, Proc. Natl. Acad. Sci, USA, 82:2379.

Foster, D.A., and Hanafusa, H., 1984, A fps gene without gag gene sequences transforms cells in culture and induces tumors in chickens, J. Virol., 48:744.

Foster, D.A., Shibuya, M., and Hanafusa, H., 1985, Activation of the transformation potential of the cellular fps gene, Cell, 42:105.

Golden, A., Nemeth, S.P., and Brugge, J.S., 1986, Blood platelets express high levels of the $p60^{c-src}$-specific tyrosine kinase activity, Proc. Natl. Acad. Sci. USA, 83:852.

Greer, P., Meckling-Hansen, K., and Pawson, T., 1988, Human c-fps/fes gene product expressed ectopically in rat fibroblasts is non-transforming and has restrained protein-tyrosine kinase activity, Mol. Cell. Biol., In Press.

Groffen, J., Heisterkamp, N., Shibuya, M., Hanafusa, H., and Stephenson, J.F., 1983, Transforming genes of avian (v-fps) and mammalian (v-fes) retroviruses correspond to a common cellular locus, Virology, 125:480.

Henderson, L.E., Krutzsch, H.C., and Oroszlan, S., 1983, Myristyl amino-terminal acylation of murine retrovirus proteins: An unusual post-translational protein modification, Proc. Natl. Acad. Sci. USA, 80:339.

Hunter, T., and Cooper, J.A., 1985, Protein-tyrosine kinases, Annu. Rev. Biochem., 54:897.

Kamps, M.P., Taylor, S.S., and Sefton, B.M., 1984, Direct evidence that oncogenic tyrosine kinases and cyclic AMP-dependent protein kinases have homologous ATP-binding sites, Nature, 310:589.

MacDonald, I., Levy, J., and Pawson, T., 1985, Expression of the mammalian c-fes protein in hematopoietic cells and identification of a distinct fes-related protein, Mol. Cell. Biol., 5:2543.

Mathey-Prerot, B., Hanafusa, H., and Kawai, S., 1982, A cellular protein is immunologically cross-reactive with and functionally homologous to the Fujinami sarcoma virus transforming protein, Cell, 28:897.

Meckling-Hansen, K., Nelson, R., Branton, P., and Pawson, T., 1987, Enzymatic activation of Fujinami sarcoma virus gag-fps transforming proteins by autophosphorylation at tyrosine, EMBO J., 6:659.

Patschinsky, T., Hunter, T., Esch, F., Cooper, J., and Sefton, B., 1982, Analysis of the sequence of amino acids surrounding sites of tyrosine phosphorylation, Proc. Natl. Acad. Sci. USA, 79:973.

Pawson, T., Guyden, J., Kung, T.-H., Radke, K., Gilmore, T., and Martin, G.S., 1980, A strain of Fujinami sarcoma virus which is temperature-sensitive in protein phosphorylation and cellular transformation, Cell, 22:767.

Pellman, D., Garber, E.A., Cross, F.R., and Hanafusa, H., 1985, An N-terminal peptide from $p60^{src}$ can direct myristylation and plasma membrane localization when fused to heterologous proteins, Nature, 314:374.

Rohrschneider, L.R., and Gentry, L.E., 1984, Subcellular location of retroviral transforming proteins define multiple mechanisms of transformation, Adv. Virol. Oncol., 4:269.

Sadowski, I., and Pawson, T., 1987, Catalytic and non-catalytic domains of the Fujinami sarcoma virus $P138^{gag-fps}$ protein-tyrosine kinase distinguished by the expression of v-fps polypeptides in Escherichia coli, Oncogene, 1:181.

Sadowski, I., Stone, J.C., and Pawson, T., 1986, A noncatalytic domain conserved among cytoplasmic protein-tyrosine kinases modifies the kinase function and transforming activity of Fujinami sarcoma virus $P130^{gag-fps}$, Mol. Cell. Biol., 6:4396.

Samarut, J., Mathey-Prevot, B., and Hanafusa, H., 1985, Preferential expression of the c-fps protein in chicken macrophages and granulocytic cells, Mol. Cell. Biol., 5:1067.

Shibuya, M., and Hanafusa, H., 1982, Nucleotide sequence of Fujinam sarcoma virus: evolutionary relationship of its transforming gene with transforming genes of other sarcoma viruses, Cell, 30:787.

Stone, J.C., Atkinson, T., Smith, M., and Pawson, T., Identification of functional regions in the transforming protein of Fujinami sarcoma virus by in-phase insertion mutagenesis, 1984, Cell, 37:549.

Stone, J.C., and Pawson, T., 1985, Correspondence between immunological and functional domains in the transforming protein of Fujinami sarcoma virus, J. Virol., 55:721.

Weinmaster, G., Hinze, E., and Pawson, T., 1983, Mapping of multiple phosphorylation sites within structural and catalytic domains of the Fujinami avian sarcoma virus transforming protein, J. Virol., 46:29.

Weinmaster, G., and Pawson, T., Protein kinase activity of FSV $P130^{gag-fps}$ shows a strict specificity for tyrosine residues, J. Biol. Chem., 261:328.

Weinmaster, G., Zoller, M.J., Smith, M., Hinze, E., and Pawson, T., 1984, Mutagenesis of Fujinami sarcoma virus: Evidence that tyrosine phosphorylation of $P130^{gag-fps}$ modulates its biological activity, Cell, 37:559.

Weinmaster, G., Zoller, M.J., and Pawson, T., 1986, A lysine in the ATP-binding site of $P130^{gag-fps}$ is essential for protein-tyrosine kinase activity, EMBO, 5:69.

Woolford, J., and Beemon, K., 1984, Transforming proteins of Fujinami and PRCII avian sarcoma viruses have different subcellular locations, Virology, 135:168.

Young, J.C., and Martin, G.S., 1984, Cellular localization of c-fps gene product NCP98, J. Virol., 52:913.

Zoller, M.J., and Taylor, S.S., 1979, Affinity labeling of the nucleotide binding site of the catalytic subunit of cAMP-dependent protein kinase using p-fluorosulfonyl-[^{14}C] benzoyl-5'-adenosine, J. Biol. Chem., 17:8368.

REGULATION OF CELL GROWTH AND TRANSFORMATION BY THE EPIDERMAL GROWTH FACTOR RECEPTOR

Joseph Schlessinger

Rorer Biotechnology, Inc.
4 Research Court, Rockville, Maryland 20850 USA and
Department of Chemical Immunology, The Weizmann
Institute of Science, Rehovot 76100, Israel

Epidermal growth factor (EGF) is a small peptide growth factor which mediates its response by a membrane glcoprotein denoted as the EGF-receptor,(reviewed in 1-2). Following the purification of the EGF-receptor,[3] and its partial sequencing,[4] the complete primary structure of the EGF-receptor was deduced from nucleotide sequence of cDNA clones.[5] The mature receptor is composed of 1,186 amino acid residues that are preceded at the NH_2-terminal end by a signal peptide of 24 hydrophobic amino acids. The signal peptide is cleaved after insertion of the nascent receptor into the membrane of the endoplasmic reticulum. Cotranslationally, the receptor is glycosylated and transported through the Golgi apparatus to the plasma membrane. The mature receptor is composed of three major structural elements. An extracellular EGF-binding domain composed of 621 amino acid residues, anchored to the plasma membrane by a single transmembrane region of 23 hydrophobic amino acids. The transmembrane region is followed by a sequence of mostly basic residues; a feature common to many membrane proteins. The cytoplasmic domain of EGF-receptor is composed of 542 amino acids: it contains a region of ~300 amino acid residues that is homologous to the catalytic domain of the protein tyrosine kinase encoded by the src gene family of oncogenes.[6] Like the other protein tyrosine kinases, the catalytic domain of EGF-receptor kinase contains a lysine (Lys 721) residue that is located 15 residues to the carboxy-terminal side of a consensus sequence, Gly-X-Gly-X-phe-Gly-X-Val. The lysine residue, together with the consensus sequence, probably functions as part of the ATP binding site.[6]

The binding of EGF to the receptor induces the activation of the protein tyrosine kinase, which phosphorylates various cellular proteins as well as the EGF-receptor itself. In intact cells, autophosphorylation occurs mainly on tyr1173; however, at least two additional tyrosine residues located at the carboxy-terminal end of EGF-receptor are phosphorylated when EGF is added to solubilized membranes or to the pure receptor.[6] It was suggested that the autophosphorylation of EGF-receptor regulates its capacity to phosphorylate exogenous substrates.[7]

A hallmark of the extracellular domain of EGF-receptor is a high proportion of cysteine residues clustered in two regions, each 160 residues long. The two cysteine-rich domains were also found in the insulin receptor,[8,9] and in the HER2/neu protein, which probably functions as a membrane receptor for an as yet unknown growth factor. Hence, the cysteine-rich domain of growth factor receptors probably evolved from a common ancestral gene.[10]

We have shown that the v-erbB oncogene of avian erythroblastosis virus encodes a truncated EGF-receptor and proposed that the v-erbB protein induces transformation by functioning as an activated growth factor receptor.[4] The v-erbB protein[11] is devoid of most of the extracellular domain of EGF-receptor and also of 32 amino acid residues at the carboxy-terminal end of the receptor, thus losing the major autophosphorylation site of the native receptor. Nevertheless, the v-erbB protein, $gp74^{v\text{-}erbB}$, possesses intrinsic protein tyrosine kinase activity towards exogenous substrates and also undergoes autophosphorylation.[12] Chicken erythroblastosis is also caused by avian leukosis virus, which transforms these cells by activating the c-erbB/EGF-receptor gene through a promoter insertion mechanism. Proviral integration leads to the expression of a truncated EGF-receptor with intrinsic protein tyrosine kinase activity.[13] Interestingly, the v-erbB-like proteins expressed in these leukemias contain the coding information for 34 amino acids at the COOH terminus that are homologous to the respective sequence of the human EGF-receptor, but which are deleted from v-erbB protein,[14] thus suggesting that the COOH-terminal deletion is not required for erythroblastosis. Numerous studies suggest that the EGF-receptor may play a role in oncogenesis through an autocrine mechanism. Various animal and human tumor cells produce a growth factor called transforming growth factor-alpha (alpha-TGF).[15] This growth factor is a member of the EGF gene family. It binds to cells bearing EGF-receptor with an affinity similar to that of EGF and stimulates their proliferation. Todaro et al.[15] suggested that alpha-TGF plays a role in oncogenesis by stimulating the growth of cells bearing EGF-receptors through an autocrine mechanism.[15] Finally, it was shown that the EGF-receptor gene is amplified and rearranged in many human brain tumors of glial origin.[16] The resultant overexpression of the EGF-receptor may play a role in the development or progression of these tumors.

Binding experiments of ^{125}I-EGF to EGF-receptor on living cells reveal two distinct affinity states of EGF-receptor.[17] High affinity sites usually represent 5-10%

of the total receptor population, depending on cell type. This correlates well with the optimal receptor occupancy required for the initiation of DNA synthesis, and, therefore, it was suggested that the high affinity receptor plays a role in the mitogenic signalling process. The treatment of fibroblasts and A-431 cells with the tumor promoter 12-O-tetradecanoyl phorbol-13-acetate (TPA) abolished the high affinity state of the EGF-receptor[18,19] and reduced its protein tyrosine kinase activity.[20,21] Moreover, the addition of TPA to cells also blocks the stimulation of Na+/H+ exchange induced by EGF and by other growth factors. TPA activates the Ca^{+2}-sensitive protein kinase C, which in turn phosphorylates the EGF-receptor on several sites. One of these sites in Thr654 of the EGF-receptor;[22,23] a residue located 10 amino acid residues from the membrane in the cytoplasmic domain of EGF-receptor. Based on these results, it was suggested that the phosphorylation of EGF-receptor on Thr654 regulates the affinity of the extracellular domain towards the ligand and the enzymatic activity of the protein tyrosine kinase domain.

An interesting conclusion drawn from the primary structure of EGF-receptor[5] is that the extracellular ligand binding domain is connected to the cytoplasmic kinase region by a single transmembrane region. This is in contrast to other transmembrane proteins that are also involved in signal transduction across the plasma membrane and that usually traverse the lipid bilayer more than once. Two distinct mechanisms can be reasoned for the transmembrane signalling of membrane proteins with a single transmembrane region such as the EGF-receptor.[24,25] In an intramolecular model a vertical dislocation of the membrane hydrophobic stretch of EGF-receptor is required for the transfer of a conformational change from the ligand binding domain to the cytoplasmic kinase region. However, the juxtamembranal sequence of EGF-receptor contains charged amino acids at both faces of the plasma membrane, which may impose a high energy barrier for a putative vertical conformational change from the extracellular domain to the cytoplasmic region.

An alternative mechanism involves an intermolecular allosteric process in which ligand-induced receptor oligomerization leads to the activation of the protein tyrosine kinase domain by subunit interaction between neighboring cytoplasmic domains, thus bypassing the requirement for a conformational change to propagate through the transmembrane region. Recent results indicate that EGF-induced activation of its receptor kinase may involve an intermolecular step. However, once the receptor is activated, autophosphorylation seems to proceed by an intramolecular process. It was shown that EGF-induced receptor self-phosphorylation has a parabolic dependence on the concentration of EGF-receptor and that cross-linking of EGF-receptor by antibodies or lectins stimulates receptor self-phosphorylation. Moreover, immobilization of EGF-receptor on various solid matrices prevents EGF from activating the kinase function. It was also shown that in native gels, EGF-receptor migrates in two forms: a fast-migrating form and an EGF-induced slow-migrating form. Based on various control and calibration experiments it was concluded that the low form represents the monomeric 170-kD EGF-receptor and the high form represents an EGF-receptor dimer. The binding of EGF causes a rapid, temperature-sensitive dimerization of EGF-receptors. Receptor dimerization is fully reversible and involves saturable, noncovalent interactins that are stable at neutral pH and in nonionic detergents.

Like many other serum proteins, EGF is internalized by receptor-mediated endocytosis. Unlike the low density lipoprotein (LDL) receptor, which is preclustered in coated pits,[26] EGF-receptors are randomly dispersed on the cell surface (reviewed in 2). The occupied EGF-receptors cluster in coated pits, and, after internalization, both EGF and its receptor are degraded by lysosomal enzymes. Hence, in contrast to the recycling of LDL receptor that is ligand-independent, EGF appears to facilitate the entry of its receptor into coated pits. The selectivity of this process may imply recognition signals that are associated with receptors that will facilitate their entry to and association with coated pits. Since clathrin and other proteins that are associated with coated pits are located in the cytoplasmic side of the plasma membrane, it is possible that the putative recognition sites of receptors reside in their cytoplasmic domains. Indeed, mutations in the cytoplasmic domain of LDL receptor impair its ability to cluster in coated pits and abolish receptor endocytosis.[26,27,28] However, EGF-receptor does not share sequence similarity with the LDL receptor or with other receptors that enter the cells via the coated pits pathway. Hence, it is possible that the clustering in coated pits is mediated by chemical modifications added posttranslationally to the receptor protein or by some other as yet uncharacterized processes.

It is clear from the results presented so far that EGF-receptor is a multifunctional allosteric protein with several regulatory sites. These regulatory sites modulate various receptor functions such as: ligand affinity, receptor clustering and endocytosis, protein tyrosine kinase activity, and the state of receptor phosphorylation. An approach that affords the analysis of the various structural domains of the EGF-receptor is the use of site-directed in vitro mutagenesis of EGF-receptor cDNA combined with transfection experiments into heterologous animal cells. This approach was used to generate various EGF-receptor mutants that were used for addressing questions of mechanism concerning the action and regulation of EGF-receptor.

DNA sequences encoding the human EGF-receptor and various EGF-receptor deletion mutants were transfected into Chinese hamster ovary cells (CHO) that were devoid of EGF-receptor. A functional EGF-receptor was expressed in these cells.[29,30] Like the native receptor, the EGF-receptor expressed in CHO cells has an apparent molecular mass of 170 kD. Moreover, it possesses an intrinsic protein tyrosine kinase activity that is stimulated by EGF leading to increased receptor self-phosphorylation and to the phosphorylation of exogenous substrates. The reconstituted EGF-receptor exhibits typical two-affinity states toward EGF. Approximately 10% of the receptors exhibit a K_d of $\simeq 3 \times 10^{-10}$ M (high affinity receptors) while the rest of the receptors have a K_d of $\simeq 2 \times 10^{-9}$ M (low affinity receptors). It appears that both affinity states of the receptor are encoded by a single gene and that the 170-kD EGF-receptor expressed on the cell surface has the capacity to acquire two affinity states toward EGF. This may be mediated by different oligomerization states of EGF-receptor on the cell surface or by interactions with other unknown membrane or cytoskeletal molecules.

The reconstituted EGF-receptor undergoes efficient EGF-induced endocytosis. Moreover, EGF stimulates the CHO cells expressing the EGF-receptor to undergo DNA synthesis. Deletion of 63 amino acids from the COOH-terminal tail of EGF-receptor, which removes two autophosphorylation sites, abolishes the high affinity state of the EGF-receptor. Nevertheless, this receptor mutant is able to undergo endocytosis and to respond mitogenically to the growth factor. Hence, the high affinity state of EGF-receptor is not required for efficient endocytosis and DNA synthesis. Further deletions from the cytoplasmic domain give rise to low affinity endocytosis-defective receptor mutants. One of these mutants possesses 98 cytoplasmic amino acids and is devoid of most of the protein tyrosine kinase domain. A larger deletion from the carboxy-terminal tail yields a receptor mutant with a short cytoplasmic tail of nine amino acids without Thr654, which is the major kinase C phosphorylation site. Finally, deletion of the transmembrane region of the human EGF-receptor yields an EGF receptor ligand binding domain that is secreted from the CHO cells.[30]

A murine retrovirus vector was used to express the human EGF-receptor cDNA and two insertional mutants of the EGF-receptor in mouse 3T3 cells.[31] Synthetic SalI linkers encoding four amino acid residues were inserted in the kinase domain of EGF-receptor to explore the role of this region on other known biological properties of the EGF-receptor.[31] A four amino acid insertion after residue 888 of the EGF-receptor abolished protein tyrosine kinase activity, high affinity binding, internalization, and mitogenic responsiveness. Another four amino mutation at residue 708 abolished the protein tyrosine kinase activity of the immunoprecipitated receptor. However, this receptor mutant exhibited both the high and low affinity binding affinities towards EGF and was able to internalize ^{125}I-EGF efficiently into these cells. Surprisingly, EGF was able to stimulate DNA synthesis in these cells as efficiently as in NIH-3T3 cells expressing the wild type receptor construct.[31] The properties of this mutant may suggest that the kinase activity is not required for internalization and mitogenesis. However, an alternative and more plausible interpretation is that the insertional mutation at the kinase domain renders the immunoprecipitated kinase unstable and therefore defective in the in vitro assays for autophosphorylation and exogenous substrate phosphorylation, while in intact cells, this receptor mutant may possess kinase activity capable of phosphorylating the physiologically important substrates.

As shown earlier, Thr654 is the major kinase C phosphorylation site of EGF-receptor. To probe the function of this amino acid residue, Thr654 was replaced either by an alanine residue[32] or by a tyrosine[33] residue utilizing in vitro site-directed mutagenesis. The Ala654 mutant of EGF-receptor was expressed in mouse L cells (devoid of endogenous EGF-receptor) and in RAT1 cells that possess endogenous EGF-receptor. It appears that the Ala654 mutant of EGF-receptor expressed in the heterologous cells did not internalize in response to TPA, although EGF was still able to induce endocytosis and degradation of the mutated receptor. On the basis of these results it was proposed that two independent mechanisms, one using Thr654 phosphorylation and a second mechanism that is EGF-dependent, play a role in the internalization of EGF-receptor.[32] The second mutant of EGF-receptor contained a tyrosine residue instead of Thr654.[33] In this receptor mutant Thr654 is replaced by an amino acid that is a potential phosphate acceptor site but cannot be phosphorylated by kinase C. The tyr654 mutant of EGF-receptor was expressed in NIH-3T3 (<1,000 endogenous EGF-receptors/cell) and in CHO cells that are devoid of EGF-receptor. Like the wild type receptor expressed on the surface of the NIH-3T3 cells, the tyr654 receptor mutant exhibited both the high and the low affinity states toward EGF. The addition of TPA abolished the high affinity state of both the wild type and the tyr654 receptor mutant in a similar manner.[34] Interestingly, the CHO cells expressing the tyr654 receptor mutant exhibited a single binding affinity toward the EGF.

The mitogenic capacity of EGF can either be enhanced[34] or inhibited[35] by TPA, depending on cell type. It appears that in the NIH-3T3 cells expressing the wild type EGF-receptor, TPA blocks the mitogenic capacity of EGF.[33] Interestingly, the inhibition mediated by TPA in cells expressing the wild type receptor does not occur in cells expressing the tyr4654 EGF-receptor mutant. Namely, EGF is able to stimulate DNA synthesis in these cells even in the presence of TPA, which inhibits the mitogenic capacity of EGF in the 3T3 cells expressing the wild type receptor.[23] On the basis of these results, we propose that the phosphorylation of Thr654 by kinase C may provide a negative control mechanism for EGF-induced mitogenesis.

References

1. G. Carpenter and S. Cohen, Epidermal growth factor, Annu. Rev. Biochem. 48:193-216 (1979).
2. J. Schlessinger, A.B. Schreiber, A. Levi, I. Lax, T. Libermann and Y. Yarden, Regulation of cell proliferation by epidermal growth factor, CRC Crit. Rev. Biochem. 14:93-111, (1983).
3. Y. Yarden, I. Harari and J. Schlessinger, Purification of an active EGF-receptor kinase with monoclonal anti-receptor antibodies, J. Biol. Chem. 260:315-319 (1985).
4. J. Downward, Y. Yarden, E. Mayes, G. Scrace, N. Totty, P. Stockwell, A. Ullrich, J. Schlessinger and M.D. Waterfield, Close similarity of epidermal growth factor receptor and V-erb-B oncogene protein sequences, Nature (Lond.) 307:521-527 (1984).
5. A. Ullrich, L. Coussens, J.S. Hayflick, T.J. Dull, A. Gray, A.W. Tam, J. Lee, Y. Yarden, T.A. Libermann, J. Schlessinger, J. Downward, E.L.V. Mayes, N. Whittle, M.D. Waterfield and P.H. Seeburg, Human epidermal growth factor receptor cDNA sequence and aberrant expression of the amplified gene in A-431 epidermoid carcinoma cells, Nature (Lond.) 309:418-425 (1984).
6. T. Hunter and J.A. Cooper, Protein-tyrosine kinases, Annu. Rev. Biochem. 54:897-930 (1985).
7. P.J. Betrics and G.N. Gill, Self-phosphorylation enhances the protein-kinase activity of the epidermal growth factor receptor, J. Biol. Chem. 260:14642-14647 (1985).
8. Y. Ebina, L. Ellis, K. Jarnagin, M. Edery, L. Graf, E. Clausner, J.-H. Ou, F. Masiarz, X.W. Kan, I.D. Goldfine, R.A. Roth and N.J. Rutter, The human insulin receptor cDNA: the structural basis for hormone activated transmembrane signalling, Cell 40:747-758.
9. R. Ullrich, J.R. Bell, E.Y. Chen, R. Herrera, L.M. Petruzzelli, T.J. Dull, A. Gray, L. Coussens, X.-C. Liao, M. Tsubokawa, A. Mason, P.H. Seeburg, C. Grunfeld, D.M. Rosenfeld and J. Ramachandran, Human insulin receptor and its relationship to the tyrosine kinase family of oncogenes, Nature (Lond.) 313:756-761 (1985).
10. E. Livneh, L. Glaser, D. Segal, J. Schlessinger and B.Z. Shilo, The drosphila EGF-receptor gene homolog conservation of both hormone binding and kinase domains, Cell 40:599-607 (1985).
11. T. Yamamoto, T. Nishida, M. Miyajima, S. Kawai, T. Ooi and K. Toyoshima, The erb-B gene of avian erythroblastosis virus is a member of the src gene family, Cell 35:71-78 (1983b).
12. R. Kris, I. Lax, M. Gullick, M. Waterfield, A. Ullrich, M. Fridkin and J. Schlessinger, Antibodies against a synthetic peptide as a probe for the kinase activity of the avian EGF-receptor and V-erb-B proteins, Cell 40:619-625 (1985).

13. I. Lax, R. Kris, I. Sasson, A. Ullrich, M.J. Hayman, H. Beug and J. Schlessinger, Activation of C-erb-B in avian leukosis virus-induced erythroblastosis leads to the expression of a truncated EGF-receptor kinase, EMBO J. (Eur. Mol. Biol. Organ.) 4:3179-3182 (1985).
14. T.W. Nilsen, P.A. Maroney, R.G. Goodwin, F.M. Rottman, L.B. Crittenden, M.A. Raines and H.-J. Kung, C-erb-B activation in ALV-induced erythroblastosis: novel RNA processing and promoter insertion result in expression of an amino-truncated EGF-receptor, Cell 41:719-726 (1985).
15. G.J. Todaro, C. Fryling and J.E. DeLarco, Transforming growth factors produced by certain human tumor cells: polypeptides that interact with human EGF-receptors, Proc. Natl. Acad. Sci USA 77:5258-5262 (1980).
16. T.A. Libermann, H.R. Nussbaum, N. Razon, R. Kris, i. Lax, M. Soreq, N. Whittle, M.D. Waterfield, A. Ullrich and J. Schlessinger, Amplification, enhanced expression and possible rearrangement of the EGF-receptor gene in primary human brain tumors of glial origin, Nature (Lond.) 313:144-147 (1985).
17. A.C. King, and P. Cuatrecasas, Resolution of high and low affinity epidermal growth factor receptors: inhibition of high affinity component by low temperature, cycloheximide and phorbol esters, J. Biol. Chem. 257:3053-3060 (1982).
18. K.D. Brown, P. Dicker and E. Rozengurt, Inhibition of epidermal growth factor binding to surface receptors by tumor promoters, Biochem. Biophys. Res. Commun. 86:1037-1043 (1979).
19. M. Shoyab, J.E. DeLarco and G.J. Todaro, Biologically active phorbol esters specifically alter affinity of epidermal growth factor receptors, Nature (Lond.) 279:387-391 (1979).
20. C. Cochet, G.N. Gill, J. Meisenheider, J.A. Cooper and T. hunter, C-kinase phosphorylates the epidermal growth factor-receptor and reduces its epidermal growth factor stimulated tyrosine protein kinase activity, J. Biol. Chem. 259:2553-2558 (1984).
21. S. Iwashita and C.F. Fox, Epidermal growth factor and potent tumor promoters induce epidermal growth factor receptor phosphorylation in a similar but distinctively different manner in human epidermal carcinoma A-431 cells, J. Biol. Chem. 259:2559-2567 (1984).
22. T. Hunter, N. Ling and N.A. Cooper, Protein Kinase-C phosphorylation of the EGF-receptor at a threonine residue close to the cytoplasmic face of the plasma membrane, Nature (Lond.) 314:480-483 (1984).
23. R.J. Davis and M.P. Czech, Tumor-promoting phorbol diesters cause the phosphorylation of epidermal growth factor receptors in normal human fibroblasts at threonine-654, Proc. Natl. Acad. Sci. USA 82:1974-1978 (1985).

24. Y. Yarden and J. Schlessinger, Self-phosphorylation of EGF-receptor:evidence for a model of intermolecular allosteric activation, Biochemistry 26:1434-1442 (1987).
25. Y. Yarden and J. Schlessinger, EGF induces rapid, reversible aggregation of the purified EGF-receptor, Biochemistry 26: 1443-1451 (1987).
26. M.S. Brown, R.G.W. Anderson and J.R. Goldstein, Recycling receptors: the round trip itinerary of migrant membrane proteins, Cell 32:663-667 (1983).
27. M.A. Lehrman, J.L. Goldstein, M.S. Brown, D.W. Russell and W.J. Schneider, Internalization defective LDL-receptors produced by genes with nonsense and frameshift mutations that truncate the cytoplasmic domain, Cell 41:735-743 (1985).
28. M.A. Lehrman, W.J. Schneider, T.C. Sudhof, M.S. Brown, J.L. Goldstein and D.W. Russell, Mutation in LDL-receptor: Alu-Alu-recombination deletes exons encoding transmembrane and cytoplasmic domains, Science (Wash. DC) 227:140-146 (1985).
29. E. Livneh, M. Benveniste, R. Prywes, S. Felder, Z. Kam and J. Schlessinger, Large deletions in the cytoplasmic kinase domain of epidermal growth factor receptor do not affect its lateral mobility, J. Cell Biol. 103:327-331 (1986).
30. E. Livneh, R. Prywes, O. Kashles, N. Reiss, I. Sasson, y. Mory, A. Ullrich and J. Schlessinger, Reconstitution of human EGF-receptors and its deletion mutants in cultured hamster cells, J. Biol. Chem. 261:12490-12497 (1986).
31. R. Prywes, E. Livneh, A. Ullrich and J. Schlessinger, Mutations in the cytoplasmic domain of EGF-receptors affect EGF-binding and receptor internalization, EMBO J. (Eur. Mol. Biol. Organ.) 5:2179-2190 (1986).
32. C.-R. Lin, S.-W. Chen, C.S. Lazar, D.C. Carpenter, G.N. Gill, R.N. Evans and M.G. Rosenfeld, Protein kinase-C phosphorylation at Thr654 of the unoccupied EGF-receptor and EGF binding regulate functional receptor loss by independent mechanisms, Cell 44:839-848 (1986).
33. E. Livneh, T.J. Dull, R. Prywess, A. Ullrich and J. Schlessinger, Release of phorbol ester-induced mitogenic block by mutation at Thr654 of EGF-receptor, J. Biol. Chem., submitted.
34. P. Dicker and E. Rozengurt, Stimulation of DNA synthesis by tumor promoters and pure mitogenic factors, Nature (Lond.) 279:723-726 (1978).
35. J.F. Lechner and M.E. Kaighn, EGF-growth promoting activity is neutralized by phorbol esters, Cell Biol. Int. Rep. 4:23-28 (1980).

COLONY STIMULATING FACTOR-1 STIMULATED MACROPHAGE MEMBRANE PROTEIN PHOSPHORYLATION

Paul T. Jubinsky, Yee-Guide Yeung, Rosalba Sacca, Wei Li and E. Richard Stanley

Department of Microbiology and Immunology and Department of Cell Biology
Albert Einstein College of Medicine
Bronx, N.Y. 10461

INTRODUCTION

Colony Stimulating Factor-1 (CSF-1) is required for the _in vitro_ survival, proliferation and differentiation of mononuclear phagocytes (precursor cell → monoblast → promonocyte → monocyte → macrophage) (reviewed in 1). Purified murine or human CSF-1 has been shown to be a homodimer consisting of two identical ~14,000 M_r polypeptide chains that are maintained as a dimer by disulfide bonds. It is heavily glycosylated with N-linked oligosaccharides of the acidic complex type. The variable molecular weight (M_r) of the native molecule within the range of 45,000-70,000 is entirely ascribable to variation in glycosylation. While the carbohydrate does not appear to be necessary for biological, antibody-binding, or receptor-binding activities, all of these activities are lost when the molecule is gently reduced to yield monomeric subunits[2].

Human CSF-1 is encoded by a single gene that is 18kb in length and comprised of 10 exons[3-5]. The gene has been mapped to the long arm of chromosome 5 (5q33.1), proximal to the CSF-1 receptor gene and the genes for several other growth factors and growth factor receptors[6]. cDNA clones predict a protein with a 32 amino acid leader sequence, whose amino acid sequence extends beyond the carboxyterminal amino acid of the mature CSF-1 subunit and includes a hydrophobic stretch of 23 amino acids. Recent studies indicate that CSF-1 is synthesized as a transmembrane glycoprotein which rapidly dimerizes and subsequently appears at the cell surface with its amino terminal portion exteriorized. The mature dimeric form of the growth factor appears to be released by proteolytic cleavage from the homodimeric transmembrane precursor[7]. Studies of the nature and distribution of CSF-1 producing cells are incomplete. However, fibroblasts are known to release CSF-1[8] and work is being carried out to identify other possible CSF-1 producing cell types.

The _in vitro_ actions of CSF-1 are mediated via a specific receptor that is selectively expressed on all mononuclear phagocytic cell types[1]. Murine CSF-1 is found in the circulation at a concentration of 1nM (25pg/ml). Changes in serum CSF-1 are correlated with changes in circulating monocytes and are elevated in response to agents that stimulate the mononuclear phagocytic system[9,10]. Intravenous injection of CSF-1 causes an increase in the number of mononuclear phagocyte precursors in the spleen[11]. Thus CSF-1 appears to play a major role in mononuclear phagocyte production in vivo. Recent experiments indicate that tissue macrophages (particularly Kupffer cells) are responsible for clearance of the bulk of circulating CSF-1, via CSF-1 receptor mediated endocytosis and intracellular destruction[12]. Thus the CSF-1 receptor, which mediates the proliferative response of immature precursor cells, also, via its presence on macrophages, negatively regulates macrophage production by lowering the CSF-1 concentration. This feedback control is specific because the CSF-1 receptor is selectively expressed on mononuclear phagocytes, of which the overwhelming majority are macrophages. It appears to be the basic coarse control for the regulation of macrophage production. Mechanisms responsible for regulating the release of CSF-1 by producing cells have not yet been described.

During pregnancy, the concentration of mouse uterine CSF-1 increases 10^3-fold whereas increases in other tissues and serum are ~2-fold. The increase in uterine CSF-1 is regulated by chorionic gonadotrophin via its action on the ovary[9]. As CSF-1 is elevated early in pregnancy (~5-fold by day 5 of gestation[9]), human trophoblastic cell lines express the CSF-1 receptor[13] and murine placental trophoblasts proliferate in response to CSF-1[14], it is likely that CSF-1 not only regulates mononuclear phagocyte production but also plays an important role in the formation and maintenance of the placenta.

The pleiotropic response to CSF-1 addition to quiescent cells includes very rapid morphological changes of membrane ruffling (at 1-2 min.) and vacuole formation (within 15 min.)[8], stimulation of the protein synthetic rate and inhibition of the rate of protein degradation[15]. The response is mediated via a single class of high affinity ^{125}I-CSF-1 binding sites[16]. Utilizing ^{125}I-CSF-1 binding assays, the murine cell surface CSF-1 receptor has been identified, purified and characterized[17-19]. The purified receptor is a single polypeptide chain of 165kD that is not covalently associated with any other protein or as a homopolymer. It exhibits ligand stimulated autophosphorylation in tyrosine as well as tyrosine kinase activity for exogenous substrates[19]. Evidence from several sources indicates that the CSF-1 receptor is the c-fms proto-oncogene product; 1)both molecules are tyrosine kinases of Mr ~165kD[19,20], 2) both molecules are selectively expressed in mononuclear phagocytes and in human choriocarcinoma cell lines of fetal origin[1,13,20], 3) antiserum to the v-fms product selectively precipitates the receptor CSF-1 complex from solubilized membranes but does not react with CSF-1[20], 4) ^{125}I-CSF-1 specifically binds to the v-fms product on cells transformed with a v-fms containing retrovirus[21] and 5) fibroblasts transfected with the c-fms gene, but not control fibroblasts, can be stimulated to proliferate in response to human CSF-1[22]. While the ligand binding domain of the v-fms oncogene is not truncated[23],v-fms induces growth factor independence in a CSF-1 dependent cell line by a non-autocrine mechanism[24] suggesting that the v-fms product provides growth stimulating signals in the absence of ligand. In the v-fms product the C-terminal 40 amino acids of the c-fms protein have been replaced by 11 unrelated residues[23]. This truncation results in the removal of a tyrosine, at position 969, which, in c-fms, negatively regulates cell proliferation[22]. However, the truncation only potentiates transformation. It does not account for the transforming activity of the v-fms product[22].

The mechanism by which CSF-1 receptor occupation is coupled to the many biological effects of CSF-1 is not known. However, a mechanism involving the covalent modification of proteins by phosphorylation is strongly suggested by the ability of CSF-1 to increase the tyrosine kinase activity of purified receptor preparations[19]. Studies with other polypeptide hormones, such as epidermal growth factor[25], insulin[26], somatomedin C[27] and platelet-derived growth factor[28] also implicate a role for phosphorylation in growth factor action. In the present paper, we demonstrate that CSF-1 selectively stimulates the phosphorylation of proteins in purified macrophage membranes. These observations, support the involvement of protein phosphorylation in CSF-1 regulated cell proliferation and provide the basis for further investigation into the role of phosphorylation in signal transduction.

MATERIALS AND METHODS

Reagents

Murine L cell CSF-1 was purified as described.[29] Purity was checked by SDS-PAGE under reducing and non-reducing conditions and by complexing with rabbit anti-CSF-1 antibody. CSF-1 concentration was determined by CSF-1 specific radioimmuno or radioreceptor assays[29]. Aprotinin, soybean trypsin inhibitor (Type I-S), leupeptin, glutamine, asparagine, 2-mercaptoethanol and Tris-base were purchased from Sigma (St. Louis, Missouri). Bis-acrylamide was obtained from Bio-Rad, and ultrapure sucrose, sodium dodecylsulfate and urea were purchased from Schwarz-Mann. ATP was brought from Calbiochem-Behring (La Jolla, Ca.). [γ-^{32}P] ATP (3000 Ci/mmole) and carrier free Na^{125}I(13-17 mCI/μgI) were obtained from Amersham. Fetal calf serum (FCS), horse serum (HS), penicillin, streptomycin and HEPES buffer were bought from Gibco and alpha medium purchased from Kansas City Biologicals (Lenaxs, Kan.). Anti-phosphotyrosine antiserum was a gift from Dr. C.H. Heldin (Institute for Medical and Physiological Chemistry, Uppsala, Sweden). Protein A Sepharose CL-4B was obtained from Sigma.

Cell cultures

The BAC1[30] and J774.2[31] murine macrophage cell lines were obtained from Dr. Betty Diamond (Albert Einstein College of Medicine). The BAC1.2F5 clone[32] was developed from the BAC1 line for studies of CSF-1 action and the Swiss mouse 3T3.L1 fibroblast line was a gift from Dr. C.S. Rubin (Albert Einstein College of Medicine). Primary bone marrow derived macrophages (BMM) were obtained from endotoxin resistant C3H/HeJ mice as described[8]. J774.2 cells were cultured at 37°C in suspension cultures containing 10% heat inactivated (56°C, 30min) HS, 2mg/ml penicillin and 2 mg/ml streptomycin in alpha medium. BAC1, BAC1.2F5 and BMM cells were grown in petri dishes (Falcon) in 15% FCS, 0.292 mg/ml glutamine, 0.02 mg/ml asparagine, 5 x 10^{-5}M 2-mercaptoethanol, 2 mg/ml penicillin, 2 mg/ml streptomycin and 3,000 units/ml Stage 1 L cell CSF-1[29] at 37°C. Cells were incubated for a further 16 h in the absence of CSF-1 in order to upregulate CSF-1 receptors. They were then harvested by scraping in 4 mM iodoacetic acid, 1 mM EGTA, 10 μg/ml leupeptin, 0.1 unit/ml aprotinin, 1,000 units/ml soybean trypsin inhibitor in PBS pH 7.4 at 0°C. 3T3-L1 fibroblasts were cultured on tissue culture dishes (Falcon) in 15% FCS in alpha medium and harvested by scraping at 0°C as described above.

Membrane phosphorylation

Membranes were prepared by homogenization and fractionation and membrane protein determined as described previously[18]. The standard phosphorylation assay was carried out in triplicate at 0°C under the following conditions unless otherwise indicated. The reaction mixture (final incubation volume, 50 μl) contained membranes (30-60 μg of membrane protein; binding capacity, 0.5-15 fmole of CSF-1) in 12mM

$MnCl_2$, 8mM $MgCl_2$, 20 mM NaCl, 25 mM HEPES pH 7.3. Membranes were preincubated with and without 10^5 U/ml of purified murine CSF-1 (8 x 10^7 U/mg protein) for 40 min. The phosphorylation reaction was initiated by addition of 12 μl of 83 μM ATP containing 20 μCi [γ-^{32}P] ATP (final ATP concentration, 20 μM). Eight minutes later the reaction was stopped by the addition of 18 μl of 4 times concentrated Laemmli[33] electrophoretic sample buffer containing mercaptoethanol and immediately heating to 100 C for 4 min. The resulting mixtures were subjected to sodium dodecyl sulfate polyacrylamide gel electrophoresis (SDS-PAGE)[33] and the gels dried and autoradiographed on Kodak XAR-5 film using an intensifying screen (Picker Corp.). ^{32}P incorporation in individual bands or in whole lanes was determined by Cerenkof counting in a liquid scintillation counter.

Precipitation with anti-phosphotyrosine antiserum

The phosphorylated membrane preparations were solubilized with 1 ml of 50 mM Tris-HCl pH7.4, 150 mM NaCl, 20 mM EDTA, 1% Triton X-100, 1% sodium deoxycholate, 1 mM phenylmethylsulfonyl fluoride and 0.5 units/ml aprotinin. Insoluble debris was removed by centrifugation and sodium dodecyl sulfate added to a final concentration of 0.01% Lysates were then incubated overnight with 2 μl of rabbit antiphosphotyrosine serum[28] alone or in the presence of 50 μg of either phosphoserine, phosphothreonine or phosphotyrosine, at 2°C. Protein A Sepharose (50 μl, 1:1 slurry) was added and the incubation continued for 1 hr. with continuous agitation. The pellet was collected by centrifugation and washed five times with 50mM Tris-HCl pH7.4, 150mM NaCl, 2mM EDTA, 1% Triton X-100, 0.01% sodium dodecyl sulfate (SDS), 1% sodium deoxycholate, 1mM phenylmethylsulfonyl fluoride, 0.5 units/ml aprotinin and twice with 50 mM tris-HCl pH7.4. The pellet was solubilized by addition of 50 μl of SDS sample buffer and subjected to SDS-PAGE.

RESULTS

Most of the experiments were carried out with membranes prepared from BAC1 cells and cells of the BAC1 clone, BAC1.2F5[32]. Both cell line require CSF-1 for growth and survival, although BAC1.2F5 cells posses a larger average number of CSF-1 receptors per cell (~120,000)[17]. BAC1 or BAC1.2F5 membranes were phosphorylated in triplicate under the standard conditions and subjected to SDS-PAGE as described. Autoradiography showed that at least 14 bands were phosphorylated in the absence of CSF-1 (see Fig. 5). Preincubation of the membranes with CSF-1 increased the total amount of ^{32}P incorporated into most of the proteins phosphorylated in the absence of CSF-1. These proteins represented only a fraction of the protein bands resolved on Coomassie Blue stained gels (data not shown). A variety of other hormones and growth factors, including epidermal growth factor, nerve growth factor, insulin, glucagon and dexamethasone failed to significantly stimulate the phosphorylation of BAC1 membrane proteins (data not shown).

Conditions for Optimum Membrane Protein Phosphorylation

Divalent cations are required for the activity of protein kinases and many phosphatases. They are believed to function either by forming a complex with ATP or by binding to regulatory sites on these enzymes. Their effect on the phosphorylation of total BAC1 membrane protein in the presence or absence of CSF-1 is shown in Table 1. Slight stimulation of phosphorylation was observed in the absence of divalent cations (5mM EGTA) or the presence of Ca^{++}. However, greater stimulation was observed with either Mg^{++} (maximum at 4mM) or Mn^{++} (maximum at 2mM) in the presence or absence of CSF-1. Mn^{++} was more

Table 1. Influence of Divalent Cations on the Incorporation of Phosphate into BAC1 Membrane Proteins

Cation	^{32}P incorporated (cpm x 10^{-3})	
	-CSF-1	+CSF-1
Ca^{++}	7.9	13.7
Mg^{++}	14.5	28.6
Mn^{++}	44.8	153.0
Mn^{++} + Mg^{++}	41.3	173.3

Following preincubation of BAC1 membranes in the presence or absence of CSF-1, incubations were continued for 10 min at 0°C with 0.4 μM [γ-^{32}P]ATP in 5mM EGTA, 12 mM $CaCl_2$, $MgCl_2$ or $MnCl_2$, or 8 mM $MnCl_2$ plus 12mM $MgCl_2$ (final volume 50 μl.) The reactions were terminated by addition of 50 μl of 20% tricholoracetic acid in 2% SDS, the precipitates collected by centrifugation at 12,000 g, washed with 10% TCA (2 x 5 ml) on glass fiber filters (Whatman, GFC), dried and counted for radioactivity. Means of duplicate determinations.

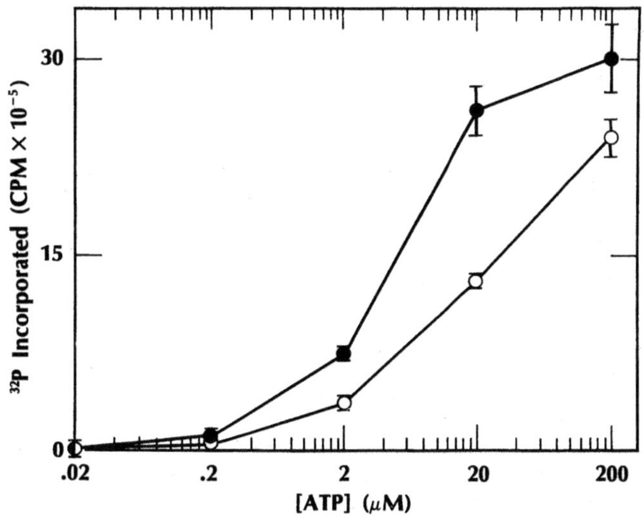

Fig. 1. Effect of ATP concentration on ^{32}P incorporation into BAC1 membrane protein. Conditions for phosphorylation and processing of the samples were as described in the Methods. Each point represents the mean of triplicate samples ± S.D. (●) with CSF-1; (O) without CSF-1.

effective than Mg^{++} and the combination (Mn^{++} + Mg^{++}) was not additive. Similar results were obtained with BAC1.2F5. In case of possible additive effects of Mg^{++} and Mn^{++} on the phosphorylation of individual protein bands, these cations were used in combination for subsequent experiments in which other conditions of the phosphorylation reaction were varied.

To further characterize BAC1 membrane protein phosphorylation, the ATP concentration was varied between 0.02 and 200 μM. The amount of ^{32}P incorporated into total protein increased with increasing ATP concentration (Figure 1). The ATP concentration yielding half-maximum ^{32}P incorporation in the presence of CSF-1 was 6.5 μM, which is comparable to the K_m reported for the EGF receptor (2.9 μM)(41). In the absence of CSF-1, the ATP concentration yielding half-maximum ^{32}P incorporation was ≥ 29.2nM. CSF-1 therefore increased the affinity of membranes for ATP. At 20 μM ATP, CSF-1 induced phosphorylation was at a maximum, while phosphorylation in the absence of CSF-1 continued to rise. In order to obtain the maximum incorporation in the presence of CSF-1, a concentration of 20 μM ATP was used in subsequent experiments.

The time course of total ^{32}P incorporation into BAC1 membrane protein at 0°C is shown in Figure 2. A large increase in ^{32}P incorporation in the presence or absence of CSF-1 occurred within 30 seconds of the addition of [γ-^{32}P] ATP. Phosphorylation in the presence of CSF-1 proceeded at a faster rate than phosphorylation in its

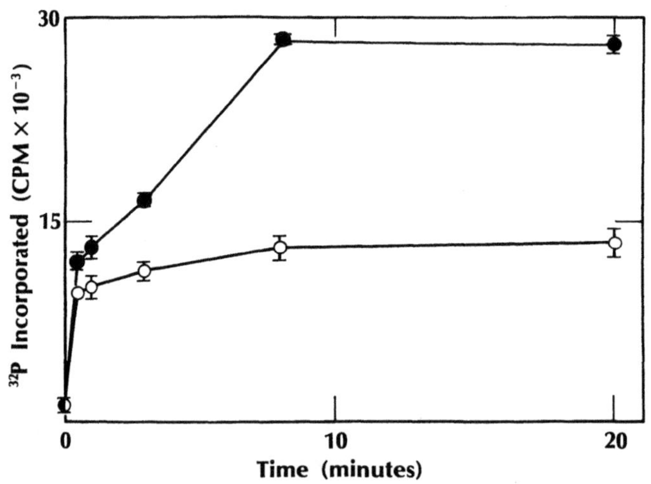

Fig. 2. Time course of ^{32}P incorporation into BAC1 membrane protein. Conditions for phosphorylation and processing of the samples were as described in the Methods. Time points represent the means of triplicate samples ± S.D. (●) with CSF-1. (O) without CSF-1.

absence. Maximum incorporation was achieved by 8 min. Autoradiographs of the gels indicated that there was a uniform increase in ^{32}P incorporation among the individual phosphorylated proteins. In addition, excision and counting of three individual protein bands (M_rs 46,000; 35,000; 21,000) did not demonstrate a significantly different time course from the time course observed for the ^{32}P incorporation into total membrane protein, either in the presence or absence of CSF-1. In similar experiments carried out with BAC1.2F5 membranes, there was a more marked stimulation of CSF-1 receptor phosphorylation and maximum phosphorylation of all protein bands had occurred by 4 min.

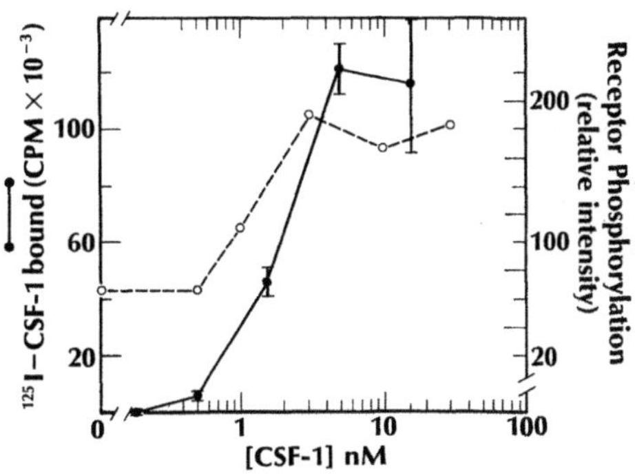

Fig. 3. Dose-response of ^{125}I-CSF-1 binding and CSF-1-stimulated protein phosphorylation in BAC1.2F5 membranes. For ^{125}I-CSF-1 binding, membranes (40 μg protein) were incubated in 25 mM HEPES pH 7.4, 50 mM NaCl with the indicated concentrations of ^{125}I-CSF-1 at 0°C for 60 min. Control incubations were preincubated with 22 nM CSF-1 for 60 min prior to addition of ^{125}I-CSF-1. Aliquots of each reaction mixture were filtered through a 0.22 μm membrane filter which was washed extensively prior to determination of the membrane associated radioactivity. Results represent the difference between the averages of triplicate test and control incubations, ± S.D. For phosphorylation, membranes (40 μg protein) were incubated in 25 mM HEPES pH7.4, 50 mM NaCl with the indicated concentrations of CSF-1 for 60 minutes. The phosphorylation reaction was initiated by the addition of [γ-^{32}P] ATP, Mn^{++} and Mg^{++} and terminated after 4 minutes by the addition of SDS sample buffer. Aliquots of the membrane preparations were then fractionated by SDS-PAGE and autoradiographed. The relative intensities of the receptor (165,000 M_r) band, were obtained by densitometric scanning of the lanes.

CSF-1 Dose-response Curve

Preliminary experiments with BMM and BAC1 membrane preparations indicated that maximum phosphorylation was often not achieved at CSF-1 concentrations as high as 30,000 U/ml. Since saturation of binding sites on intact cells occurs at 500-1000 U/ml[16], binding and phosphorylation experiments were carried out simultaneously on BAC1.2F5 membrane preparations. These experiments showed that the binding of CSF-1 was inhibited by the divalent cations. Preincubation of membrane preparations with different concentrations of CSF-1 in the absence of divalent cations, followed by incubation in the presence of the divalent cations and [γ-^{32}P] ATP resulted in the simultaneous saturation of both the CSF-1 binding sites and the CSF-1 stimulated phosphorylation (Figure 3). Approximately 5000 U/ml (2nM) was required for the maximum response. This concentration of CSF-1 is close to the concentrations in mouse serum (1,000-2,000 U/ml[12]) and in culture media in which maximum proliferation of BAC1.2F5 cells occurs (3,000 U/ml)[32]. There were no CSF-1 concentration dependent changes in the pattern of phosphorylation of individual proteins.

Turnover of Incorporated Phosphate

Almost all of the BAC1 membrane proteins exhibiting CSF-1 stimulated phosphorylation in the presence of Mn^{++} plus Mg^{++} were also phosphorylated in the absence of CSF-1. To determine whether this effect was due to CSF-1 induced activation of a phosphatase(s) rather than activation of kinases, experiments were performed to determine the

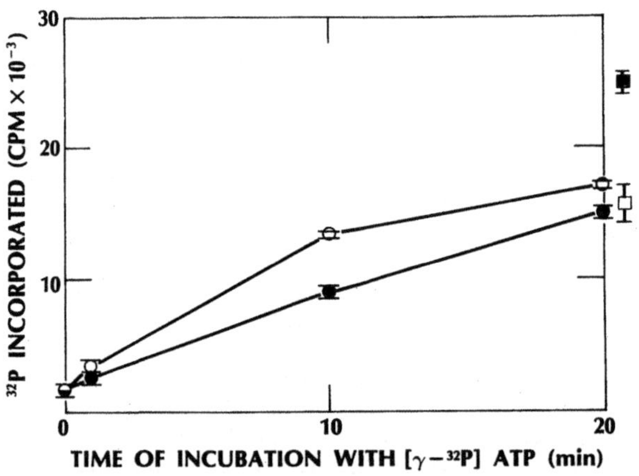

Fig. 4. The stability of phosphate incorporated into BAC1 membrane proteins. Membranes were phosphorylated with unlabeled ATP (20 μM) for 20 min in the presence or absence of CSF-1 under the standard conditions. A trace amount of [γ-^{32}P] ATP (20 μCi) was added and the amount of ^{32}P incorporated at the indicated times determined by SDS-PAGE and counting of excised lanes as described in the Methods. ●, with CSF-1; ○, without CSF-1. Controls, in which the [γ-^{32}P] ATP was simultaneously added with the unlabeled ATP, were incubated for 20 min. ■, with CSF-1; □ without CSF-1.

relative turnover of incorporated phosphate (Figure 4). A BAC1 membrane preparation was preincubated with or without CSF-1 (15,000 U/ml) for 40 min. Unlabeled ATP (20 μM) was added and the incubation continued for another 20 min. A trace amount of [γ-^{32}P] ATP (20 μCi) was then added for the times indicated in Figure 4A, and the reaction stopped and incorporation determined as described previously. Since after a 20 min incubation with 20 μM ATP, the number of sites occupied with unlabeled phosphate is constant (fig 2), any incorporation of ^{32}P reflects phosphate turnover. Controls, in which unlabeled ATP and [γ-^{32}P] ATP were added simultaneously for a 20 minute incubation, exhibited a CSF-1 increased ^{32}P incorporation of 1.6-fold.

In the absence of CSF-1, the total amount of ^{32}P incorporated by 20 min. attained the level of the corresponding control, indicating that there was rapid turnover of the incorporated phosphate (fig. 4A). By contrast, in the presence of CSF-1, the turnover of incorporated phosphate was lower than in its absence at all times points, indicating that not only was there insignificant turnover of phosphate at the CSF-1 dependent sites but also that CSF-1 inhibits the turnover of phosphate at the CSF-1 independent sites. Three phosphorylated protein bands (M_rs 150,000; 45,000; 39,000) were examined individually. They exhibited turnover kinetics in the presence and absence of CSF-1 which were similar to the turnover kinetics of the total membrane protein (data not shown).

CSF-1 Stimulated Protein Phosphorylation in Membranes from Different Cells

The patterns of membrane phosphorylation in the membranes from five different murine cell populations are presented in Figure 5. The cell populations examined were primary macrophages (BMM), the BAC1, BAC1.2F5 and J774.2 macrophage lines, and the 3T3-L1 fibroblast line. With the exception of the 3T3-L1 line, cells from all of these populaions express the CSF-1 receptor[17]. However, the J774.2 cells in contrast to BMM, BAC1 and BAC1.2F5 cells, do not require CSF-1 for proliferation.

The patterns of CSF-1 stimulated phosphorylation of membranes from BAC1 cells and the BAC1.2F5 clone were similar, except for a more pronounced phosphorylation of the 165,000 M_r CSF-1 receptor band in the case of BAC1.2F5 membranes. The patterns of protein phosphorylation and CSF-1 response in these CSF-1 dependent cell lines were also very similar to the response in membranes from the CSF-1 dependent primary macrophage population, BMM. Only minor differences were apparent i.e., the absence of M_r 56,000 and 60,000 bands in the presence and absence of CSF-1. In contrast, the pattern of protein phosphorylation in membranes from the CSF-1 independent (autonomous) J774.2 macrophage line was markedly different. Despite a relatively similar pattern observed in the absence of CSF-1, only three J774.2 membrane proteins (M_rs 165,000, 95,000, 49,000) exhibited a CSF-1 dependent increase in ^{32}P incorporation. CSF-1 had no effect on phosphorylation of 3T3 L1 membrane proteins and the pattern of protein phosphorylation was different from the pattern observed with membranes obtained from the macrophage populations. All macrophage cell lines examined exhibited CSF-1 dependent phosphorylation of a membrane protein with the M_r ($\sim$165,000) of the purified CSF-1 receptor. These results indicate that the CSF-1 induced membrane phosphorylation is selective for macrophages and that the pattern of CSF-1 dependent membrane protein phosphrylation is similar for primary macrophages and a macrophage line which required CSF-1 for growth, but very different in a macrophage cell line that is CSF-1 independent for growth.

Tyrosine Phosphorylation of Macrophage Membrane Proteins

The CSF-1 receptor has been shown to be a tyrosine kinase[19]. In order to identify candidate substrates for this kinase, solubilized, phosphorylated BAC1.2F5 membrane preparations were reacted with an anti-phosphotyrosine antiserum, complexes precipitated with protein A-Sepharose, and subjected to SDS-PAGE. Control incubations were carried out with the antiserum in the presence of phosphotyrosine, phosphoserine

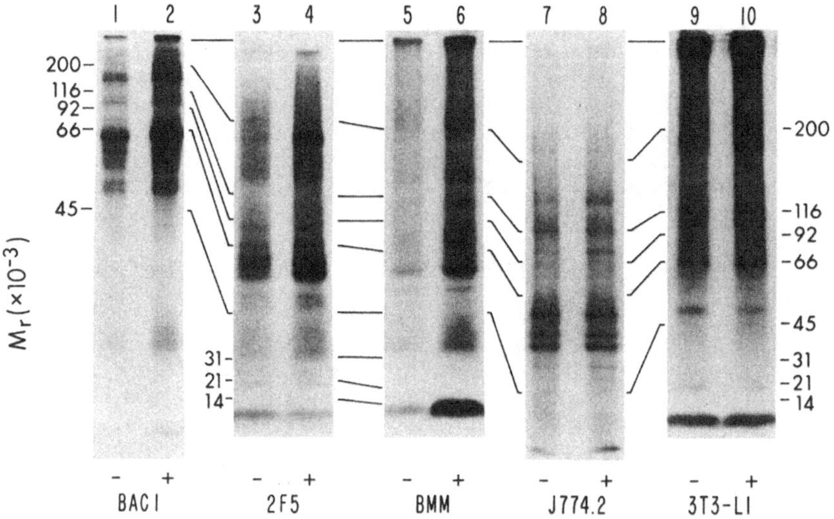

Fig. 5. Autoradiography of SDS gel electrophoresis of membranes from 4 different cell lines phosphorylated in the presence (+) and absence (-) of CSF-1. Membrane preparations were phosphorylated and samples processed as described in the Methods. Marker proteins used were Myosin (M_r 200,000), α-galactosidase (M_r 116,500), phosphorylase B (M_r 92,5000), bovine serum albumin (M_r 66,200), ovalbumin (M_r 45,000), carbonic anhydrase (M_r 31,000), soybean trypsin inhibitor (M_r 21,500) and lysozyme (M_r 14,400).

and phosphothreonine. As shown in figure 6, at least 6 phosphorylated protein bands (M_rs, 165,000 (CSF-1 receptor); 120,000; 95,000; 80,000; 72,000 and 59,000) were specifically complexed by the anti-phosphotyrosine antibodies and the incorporation of ^{32}P into all of these bands was stimulated by CSF-1. In a separate experiment, immunoprecipitated bands exhibiting increased ^{32}P incorporation with CSF-1 were also shown to exhibit CSF-1 stimulated ^{32}P incorporation into tyrosine by phosphoaminoacid analysis[34] (data not shown).

DISCUSSION

CSF-1 has been shown to selectively stimulate phosphate incorporation into macrophage membrane proteins at physiological concentrations. As with platelet derived growth factor and epidermal

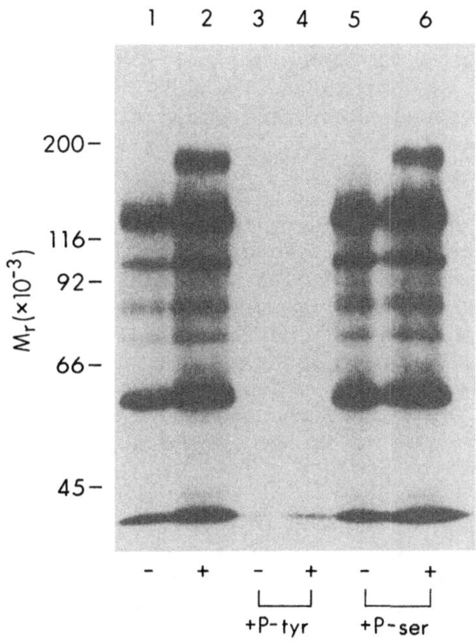

Fig. 6. Precipitation of phosphotyrosine containing BAC1.2F5 membrane proteins with anti-phosphotyrosine antiserum. Membranes incubated with $[\gamma\text{-}^{32}P]$ATP in the presence (+) or absence (-) of 4.4 nM CSF-1. Equal aliquots of the membrane preparations were disrupted with detergent and subject to immunoprecipitation by antiphosphotyrosine antiserum alone (lanes 1 and 2) or antiserum in the presence of either phosphotyrosine (lanes 3 and 4), phosphoserine (lanes 5 and 6) or phosphothreonine (data not shown). Results with the phosphothreonine group were indistinguishable from those for the phosphoserine group.

growth factor stimulated membrane phosphorylation, as well as insulin receptor autophosphorylation[35-38], Mn^{++} was shown to be the preferred cation for CSF-1 stimulated phosphorylation. The stimulation by CSF-1 was dependent on the CSF-1 responsiveness of the cells. Primary macrophages (BMM) and macrophage cell lines (BAC1 and BAC1.2F5) which require CSF-1 for growth and survival, exhibited a CSF-1 dependent phosphorylation in $\sim$14 bands. In contrast, membranes from J774.2 cells, which are CSF-1 independent for survival and growth, responded to

CSF-1 with increased phosphorylation in only three bands, one of which was the CSF-1 receptor. Since the other two proteins still exhibit CSF-1 stimulated phosphorylation they are likely to be tightly coupled to the CSF-1 receptor. The less marked response of these membranes could be caused by the constitutive activation of a kinase(s) which is normally regulated by CSF-1. This could lead to an apparent decrease in the CSF-1 stimulated phosphorylation as well as relieving the CSF-1 requirement for growth. All of the CSF-1 receptor bearing cells exhibited CSF-1 dependent phosphorylation of the 165,000 dalton band which corresponds to the CSF-1 receptor. The intensity of the CSF-1 stimulated phosphorylation of the 165,000 dalton band reflected the relative amounts of the CSF-1 receptor present (BAC1.2F5, 120,000 receptors/cell; BMM, 50,000 receptors/cell; J774.2, 20,000 receptors/cell[17]). In contrast to the CSF-1 dependent phosphorylation exhibited in all macrophage membrane preparations, no response was observed in membranes from 3T3-L1 fibroblasts, which do not posses CSF-1 receptors.

The complexity of the CSF-1 induced membrane protein phosphorylation suggests that several protein kinases are activated by CSF-1-receptor binding. To uncover differential protein phosphorylation, suggestive of a series of coupled kinase reactions, the time course of phosphorylation and the effect of ATP concentration were examined. In the time course experiment with membranes exhibiting the slowest kinetics (BAC1), total phosphorylation was shown to be saturable within 8 min in the absence of phosphatase inhibitors. Over the 20 min. period studied, there was no change in the degree of phosphorylation of any band with respect to another. This lack of differential protein phosphorylation may be explained by its absence, the absence in membrane preparations of essential kinases or substrates or our inability to detect small or rapid (within 30 sec.) changes. In contrast to the failure to detect differential protein phosphorylation with time, changes were observed when the ATP concentration was varied between 0.2 and 2 μM (data not shown). These changes were probably due to the concentration dependent activation or inhibition of protein kinases or phosphatases. Localized changes in the ATP concentration within cells could therefore be regulatory. Total incorporation of ^{32}P into membrane proteins increases with increasing ATP. However, the percentage stimulation of ^{32}P incorporation by CSF-1 was significantly higher at lower ATP concentrations ($\sim$1μM) (compare Table 1, figure 1), as expected from the CSF-1 induced increase in affinity for ATP.

As all of the phosphorylation experiments described were carried out in the absence of phosphatase inhibitors, the observed CSF-1 stimulated ^{32}P incorporation into membrane proteins could have resulted from activation of kinases, or alternatively, from alterations in phosphatase activity. In order to distinguish between these possibilities, the turnover of incorporated ^{32}P was measured. In the absence of CSF-1, there was rapid turnover of the incorporated phosphate which, by 20 minutes, had turned over at least once. In contrast, in the presence of CSF-1, the turnover of incorporated phosphate was no greater than the turnover in the absence of CSF-1, indicating that the CSF-1 stimulated increase in ^{32}P incorporation is due to activation of kinase(s). The fact that the turnover of incorporated phosphate in the presence of CSF-1 was even less than in its absence, indicates that CSF-1 actually inhibits turnover. Possible mechanisms include the inhibition of phosphatases which act at phosphorylation sites which would normally turnover in the absence of CSF-1.

In membranes from CSF-1 responsive cells, many proteins exhibit a CSF-1 dependent increase in phosphate incorporation. A number of these protein bands have molecular weights similar to proteins whose phosphorylation is stimulated by other growth factors and transforming oncogene products[39-41]. The significant degree of overlap suggests that some of the phosphorylated proteins may be common to pathways for sveral different growth factors. Several protein bands (M_r 165,000; 120,000; 95,000; 80,000; 72,000; 59,000) exhibited increased phosphorylation in tyrosine in response to CSF-1 and are candidate substrates for the CSF-1 receptor-kinase. Further analysis of this system in conjunction with whole cell experiments should provide additional insights into the mechanism of growth factor signal transduction.

ACKNOWLEDGEMENTS

We thank Michelle Hankins, Suzanne Saltzman and Fran Pellerito for technical assistance and Dr. W-K. Liu for phosphoaminoacid determinations. This work was supported by National Institutes of Health Grant Ca26504 and the Albert Einstein Cancer Center Core Grant. P.T.J. was supported by National Institutes of Health Training Grant GM7288 from the National Institute of General medical Sciences. E.R.S. is the recipient of an Irma T. Hirschl Career Scientist Award.

REFERENCES

1. E. R. Stanley, L. J. Guilbert, R. J. Tushinski, S. H. Bartelmez, CSF-1 - A mononuclear phagocyte lineage-specific hemopoietic growth factor. J. Cell Biochem. 21:151-159 (1983).
2. S. K. Das and E. R. Stanley, Structure-function studies of a colony stimulating factor (CSF-1). J. Biol. Chem. 257:13679-13684 (1982).
3. E. S. Kawasaki, M. B. Ladner, A. M. Wang, J. N. Van Arsdell, M. K. Warren, M. Y. Coyne, V. L. Schweickart, M-T. Lee, K. J. Wilson, A. Boosman, E. R. Stanley, P. Ralph, and D. F. Mark, Molecular cloning of a complementary DNA encoding human macrophage-specific colony-stimulating factor (CSF-1). Science 230:291-296 (1985).
4. E. S. Kawasaki, M. B. Ladner, A. M. Wang, J. Van Arsdell, M. K. Warren, M. Y. Coyne, V. L. Schweickart, M-T. Lee, K.J. Wilson, A. Boosman, E. R. Stanley, P. Ralph, and D. Mark, Human macrophage colony stimulating factor (CSF-1): Isolation of genomic and complementary DNA clones, in: Immune Regulation by Characterized Polypeptides. UCLA Symposia on Molecular and Cellular Biology, New Series, Vol. 41 Editors: A. Goldstein, J.F. Bach and H. Wigzell. Alan R. Liss, Inc. New York, NY, pp. 387-396.
5. P. Ralph, M. B. Ladner, A. M. Wang, E. S. Kawasaki, L. McConlogue, J. F. Weaver, S. A. Weiss, P. Shadley, K. Kothsy, M. K. Warren, E. R. Stanley and H. E. Broxmeyer, The Molecular and Biological Properties of the Human and Murine Members of the CSF-1 Family, in: "Fifth International Lymphokine Workshop". In Press. (1987).
6. M. J. Pettenati, M. M. Lebeau, R. S. Lemons, E. A. Shima, E. S. Kawasaki, R. A. Larson, C. J. Sherr, M. O. Diaz and J. D. Rowley, Assignment of CSF-1 to 5q 33.1: Evidence for clustering of genes regulating hematopoiesis and for their involvement in the deletion of the long arm of chromosome 5 in myeloid disorders. Proc. Natl. Acad. Sci. USA 84:2970-2974 (1987).

7. C. W. Rettenmier, M. F. Roussel, R. A. Ashmun, P. Ralph, L. K. Price and C. J. Sherr, Synthesis of membrane-bound CSF-1 and down modulation of CSF-1 receptors in NIH-3T3 cells transformed by cotransfection of human CSF-1 and c-fms (CSF-1 receptor) genes. Molec. Cell Biol. 7:2378-2387 (1987).
8. R. J. Tushinski, I. T. Oliver, L. J. Guilbert, P. W. Tynan, J. R. Warner and E. R. Stanley, Survival of mononuclear phagocytes depends on a lineage-specific growth factor that the differentiated cells selectively destroy. Cell 28:71-81 (1982).
9. A. Bartocci, J. W. Pollard, E. R. Stanley, Regulation of colony-stimulating factor 1 during pregnancy. J. Exp. Med. 164:956-961 (1986).
10. J. W. Sheridan and D. Metcalf, A low molecular weight factor in lung conditioned medium stimulating granulocyte and monocyte colony formation in vitro. J. Cell. Physiol., 81:11-23 (1973).
11. H. E. Broxmeyer, D. E. Williams, S. Cooper, R. K. Shadduck, S. Gillis, A. Waheed, D. L. Urdal and D. C. Bricknell, Comparative effects in vivo of recombinant murine interleukin 3, natural murine colony-stimulating factor-1 and recombinant murine granulocyte-macrophage colony-stimulating factor on myelopoiesis in mice. J. Clin. Invest., 79:721-730.
12. A. Bartocci, D. S. Mastrogiannis, A. Migliorati, R. J. Stockert, A. W. Wolkoff and E. R. Stanley, Macrophages specifically regulate the concentration of their own growth factor in the circulation. Proc. Natl. Acad. Sci. USA 84:6179-6183 (1987).
13. C. W. Rettenmier, R. Sacca, W. L. Furman, M. F. Roussel, J. Holt, A. W. Nienhuis, E. R. Stanley and C. J. Sherr, Expression of the human c-fms proto-oncogene product (CSF-1 receptor) on peripheral blood mononuclear cells and choriocarcinoma cell lines. J. Clin. Invest. 77:1740-1746 (1986).
14. I. Athanassakis, R. C. Bleackley, V. Paetkau, L. J. Guilbert, P. J. Barr, and T. G. Wegmann, The immunostimulatory effect of T cells and T cell lymphokines on murine fetally-derived placental cells. J. Immunol. 138:37-44 (1987).
15. R. J. Tushinski and E. R. Stanley, The regulation of macrophage protein turnover by a colony stimulating factor (CSF-1). J. Cell Physiol. 116:67-75 (1983).
16. L. J. Guilbert and E. R. Stanley, The interaction ^{125}I-CSF-1 with bone marrow derived macrophages. J. Biol. Chem. 261:4024-4032 (1986).
17. C. J. Morgan and E. R. Stanley, Chemical cross-linking of the mononuclear phagocyte specific growth factor CSF-1 to its receptor at the cell surface. BBRC. 119:35-41 (1984)
18. Y. G. Yeung, P. T. Jubinsky and E. R. Stanley, Solubilization and assay of a colony stimulating factor receptor from murine macrophages. J. Cell. Biochem. 31:259-269 (1986)
19. Y. G. Yeung, P. T. Jubinsky, A. Sengupta, D. C. Y. Yeung and E. R. Stanley, Purification of the colony-stimulating factor 1 receptor and demonstration of its tyrosine kinase activity. Proc. Natl. Acad. Sci. USA 84:1268-1271 (1987).

20. C. J. Sherr, C. W. Rettenmier, R. Sacca, M. F. Roussel, A. T. Look, and E. R. Stanley, The c-fms proto-oncogene product is related to the receptor for the mononuclear phagocyte growth factor, CSF-1. Cell 41:665-676 (1985).
21. R. Sacca, E. R. Stanley, C. J. Sherr, C. W. Rettenmier, Specific binding of the mononuclear phagocyte colony stimulating factor, CSF-1, to the product of the v-fms oncogene. Proc. Natl. Acad. Sci. USA 83:3331-3335 (1986).
22. M. F. Roussel, T. J. Dull, C. W. Rettenmier, P. Ralph, A. Ullrich and C. J. Sherr, Transforming potential of the c-fms proto-oncogene (CSF-1 receptor). Nature 325:549-552 (1987).
23. L Coussens, C. Van Beveren, D. Smith, E. Chen, R. L. Mitchell, C. M. Isacke, I. M. Verma and A. Ullrich. Structural alteration of viral homologue of receptor proto-oncogene fms at carboxy terminus. Nature 320:277-280 (1986)
24. E. F. Wheeler, C. W. Rettenmier, A. T. Look and C. J.Sherr, The v-fms oncogene induces factor independence and tumorigenicity in CSF-1 dependent macrophage cell line. Nature 324:377-380 (1986).
25. G. Carpenter, L. King, and S. Cohen, Rapid enhancement of protein phosphorylation in A431 cell membrane preparations by epidermal growth factor. J. Biol. Chem. 254:4884-4891 (1979).
26. R. W. Rees-Jones and S. I. Taylor, An endogenous substrate for the insulin receptor-associated tyrosine kinase. J. Biol. Chem. 260:4461-4467 (1985)
27. S. Jacobs, F. C. Kull, H. S. Earp, M. E. Svoboda, J. J. Van Wyk and P. Cuatrecasas, Somatomedin-C stimulates the phosphorylation of the β-subunit of its own receptor. J. Biol.Chem. 258:9581-9584 (1983).
28. B. Ek and C. H. Heldin, Use of an antiserum against phosphotyrosine for the identification of phosphorylated components in human fibroblasts stimulated by platelet-derived growth factor. J.Biol. Chem. 259:11145-11152 (1984).
29. E. R. Stanley, The Macrophage colony stimulating factor, CSF-1, in: "Methods in Enzymology-Immunochemical Techniques" Vol. 116, S. P. Colowick and N. O. Kaplan, eds., Harcourt Brace Jovanovich, Publishers, Academic Press. pp. 564-587 (1985).
30. S. Schwarzbaum, R. Halpern and B. Diamond, The generation of macrophage-like cell lines by transfection with SV40 origin defective DNA. J. Immunol. 132:1158-1162 (1984).
31. P. Ralph, J. Prichard and M. Cohn, Reticulum cell sarcoma: An effector cell in antibody-dependent cell-mediated immunity. J. Immunol. 114:898-905 (1975).
32. C. J. Morgan, J. W. Pollard, and E. R. Stanley, Isolation and characterization of a cloned growth factor dependent macrophage cell line, BAC1.2F5. J. Cell. Physiol. 130:420-427.
33. U. K. Laemmli, Cleavage of structural proteins during the assembly of the head of bacteriophage T4. Nature 227:680-685 (1970).
34. A. Golden, S. P. Nemeth and J. S. Brugge, Blood platelets express high levels of the $pp60^{c-src}$ -

specific tyrosine kinase activity. Proc. Natl. Acad. Sci. USA 83:852-856 (1986)

35. O. Rosen, R. Herrera, R. Howe, L. A. Petruzzelli and M. Cobb, The rate-determining step in cAMP-mediated exocytosis in the rat parotid and submandibular glands appears to involve analogous 26-kDa integral membrane phosphoproteins. Proc. Natl. Acad. Sci. USA 80:3237-3240 (1983)

36. K-T Yu and M. P. Czech, Tyrosine phosphorylation of the insulin receptor β-subunit activates the receptor-associated Tyrosine kinase activity. J. Biol.Chem. 259:5277-5286 (1984)

37. K-T Yu and D. K. Werth, I. H. Poston, and Czech, M.P. src kinase catalyzes the phosphorylation and activation of the insulin receptor kinase. J. Biol. Chem. 260:5838-3846 (1985)

38. C. Erneux, S. Cohen and D. L. Garbers, The kinetics of tyrosine phosphorylation by the purified epidermal growth factor receptor kinase of A-431 cells. J. Biol. Chem. 4137-4142 (1983)

39. J. C. Chambard, A. Franch, A. Lecam and J. Pouyssegur Growth factor-stimulated protein phosphorylation in G_0/G_1 -arrested fibroblasts. J. Biol. Chem. 258:1706-1713 (1983)

40. R. D. Fleischmann, and J. M. Pawelek, Evidence that a 90-kDa phosphoprotein, an associated kinase, and a specific phosphatase are involved in the regulation of Cloudman melanoma cell proliferation by insulin. Proc. Natl. Acad. Sci. USA 82:1007-1011 (1985)

41. E. Erickson, D. J. Shealy and R. L Erickson, Evidence that viral transforming gene products and epidermal growth factor stimulate phosphorylation of the same cellular protein with similar specificity. J. Biol. Chem. 256:11381-11384 (1981)

ANALYSIS OF FUNCTIONAL DOMAINS IN THE EPIDERMAL GROWTH FACTOR RECEPTOR USING SITE-DIRECTED MUTAGENESIS

Gordon N. Gill, Michael G. Rosenfeld, William S. Chen, Paul J. Bertics*, and Cheri S. Lazar

Division of Endocrinology and Metabolism and
Eukaryotic Regulatory Biology Program
Department of Medicine
University of California, San Diego, School of Medicine
La Jolla, California 92093

Protein tyrosine kinase activity is a central feature of one class of retroviral onc genes and of several growth factor receptors including those for epidermal growth factor receptor (EGF) (Ushiro and Cohen, 1980; Hunter and Cooper, 1985), platelet-derived growth factor (PDGF) (Yarden et al., 1986), insulin and insulin-like growth factors (IGF) (Ebina et al., 1985; Ullrich et al., 1985), colony-stimulating factor 1 (CSF1) (Sherr et al., 1985), and other ligands (neu) (Bargmann et al., 1986a; Yamamoto et al., 1986). The EGF receptor is the proto-oncogene for the retroviral oncogene erb B (Yamamoto et al., 1983; Downward et al., 1984a; Lin et al., 1984; Ullrich et al., 1984; Xu et al., 1984); the CSF1 receptor for v-fms (Sherr et al., 1985; Coussens et al., 1986); and the neu receptor for the transforming gene neu (Bargmann et al., 1986b). The sis oncogene of the simian sarcoma virus is homologous to the B chain of PDGF and presumably exerts its transforming effects through interaction with the normal PDGF receptor (Doolittle et al., 1983; Waterfield et al., 1983; Huang et al., 1984). Constitutive expression of TGF-α and of EGF, both of which bind to the EGF receptor, also results in cell transformation, presumably via activation of normal EGF receptors (Rosenthal et al., 1986; Stern et al., 1987). Constitutive activation of ligands may be analogous to gene amplification and quantitative over-expression of receptor protein as occurs in some human tumors (Thompson and Gill, 1985) because the initial ligand receptor interaction in a bimolecular event dependent on the concentration of both ligand and receptor (Gill et al., 1985). There is thus good evidence for qualitative, i.e. mutational activation of proto-oncogene growth factor receptors by removal of regulatory constraints and quantitative over-expression of either ligand or receptor to result in constitutive activation via formation of ligand receptor complexes. There is considerable importance to understanding the structural determinants of receptor activation and expression of both receptor and ligand.

The EGF receptor is 170 kd transmembrane glycoprotein with intrinsic ligand-dependent protein tyrosine kinase activity (Cohen et al., 1982; Ullrich et al., 1984). EGF binding to the extracellular domain stimulates

---

*Present Address: Department of Physiological Chemistry, University of Wisconsin-Madison, Madison, Wisconsin 53706

its protein tyrosine kinase activity, and leads to increased intracellular phosphorylations as well as self-phosphorylation of the receptor (Hunter and Cooper, 1981; Hunter and Cooper, 1985; Gill et al., 1987). The binding of EGF to the receptor activates several immediate biochemical processes such as alterations of intracellular calcium (Sawyer and Cohen, 1981), sodium/hydrogen exchange with increased intracellular pH (Moolenaar et al., 1982), and increased transcription of a group of responsive genes (Murdoch et al., 1982; Greenberg and Ziff, 1984; Kruijer et al., 1984; Muller et al., 1984; these events culminate many hours later in DNA replication and cell division (Fig. 1).

Fig. 2 shows a schematic representation of the structure of the EGF receptor deduced from the cDNA sequence (Ullrich et al., 1984). The amino terminal cell surface domain is composed of about 621 amino acids, and is glycosylated at 11-12 asparagine residues (Mayes and Waterfield, 1984; Weber et al., 1984a). The extracellular region contains 51 cysteine residues which are preferentially localized in two regions, i.e. residues 134-313 and 446-612. The extracellular domain contains a single ligand binding site (Weber et al., 1984b). The extracellular binding domain is separated from the intracellular protein tyrosine kinase domain by a hydrophobic membrane spanning domain of about 23 amino acids. The inner boundary of the transmembrane domain is defined by a stop-transfer sequence of 3 dibasic amino acid residues. The ATP binding site begins 50 amino acids carboxy terminal from the inner membrane boundary in the EGF and all of the other transmembrane growth factor protein tyrosine kinase receptors. The carboxyl terminal domain which appears to function as a regulatory sequence contains the major sites of self-phosphorylation in the EGF receptor (Downward et al., 1984b).

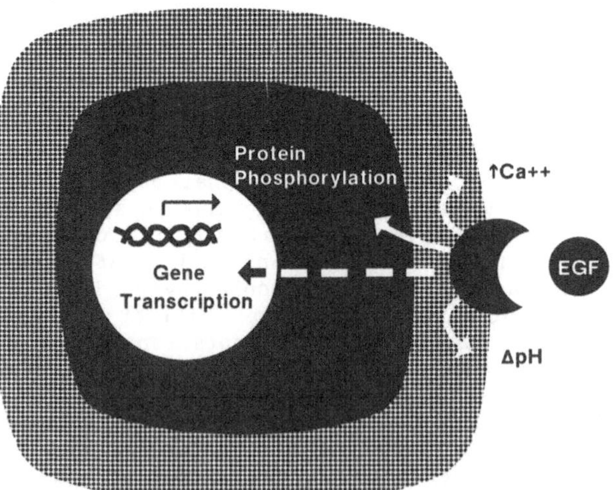

FIG. 1. Some effects of EGF binding to its cell surface receptor.

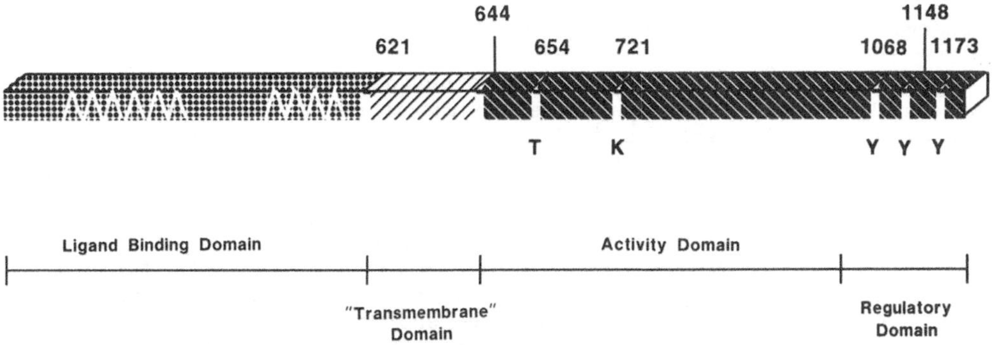

Fig. 2. Schematic representation of the EGF receptor. The cystein-rich domains in the extracellular ligand binding domain are indicated by wavy lines. The transmembrane domain is defined by hydropathy profiles using the program of Engleman et al. (1987). Numbered amino acid residues are discussed in the text.

Site-directed mutations which abolish the kinase activity of the v-src (Snyder et al., 1985), v-mos (Hannink and Donoghue, 1985), and v-fps (Weinmaster et al., 1985) oncogenes abolishes their capacity to transform cells. Because many transmembrane receptors interact with intrinsic membrane proteins to activate second messenger systems, it was not certain whether the protein tyrosine kinase activity of the EGF receptor mediated all the effects of ligand binding shown in Fig. 1. Several reports suggested that specific aspects of EGF receptor function, including changes in intracellular calcium and increases in DNA synthesis were independent of the intrinsic protein tyrosine kinase activity of the receptor (Defize et al., 1985; Prywes et al., 1985). To characterize the role of protein tyrosine kinase activity in EGF-activated, receptor-mediated processes, we prepared a mutant EGF receptor containing a methionine substitution for lysine 721 in the EGF receptor (Chen et al., 1987). Computer modeling of the ATP binding site based on the nucleotide binding site of glyceraldehyde 3-phosphate dehydrogenase indicated that this lysine provided the major energy for ATP binding via formation of a salt bridge to the β/γ phosphates of ATP (Sternberg and Taylor, 1984). Methionine was substituted for lysine because of similarities of bond lengths and angles which should optimally preserve the overall structure of the EGF receptor.

Fig. 3 shows the strategy for construction and expression of the mutant EGF receptor containing a methionine at residue 721. A synthetic oligonucleotide (21 mer) was used to alter a single nucleotide (A-T) to generate the methionine substitution (Zoller and Smith, 1983). The EGF receptor transcription unit was placed under the control of SV40 early promoter enhancer and the SV40 late splice polyadenylation regions (Lin et al., 1986). A mutant dihydrofolate reductase (DHFR) gene provided a dominant selectable marker (Simonsen and Levinson, 1984). The parental EGF receptor used for construction of mutants contained an alanine replacement for threonine at residue 654 to remove heterologous regulatory effects on the EGF receptor (Lin et al., 1986). This A654 receptor has been shown to function normally in response to EGF (Lin et al., 1986). Two recipient cells: mouse L cells (B82) and Chinese hamster ovary cells (CHO) which lack endogenous EGF receptor mRNA and protein were used as recipients. Transfections were carried out using the calcium phosphate precipitation procedure (Wigler et al., 1979), and clonal lines were isolated and expanded. Fig. 3B shows binding parameters derived from Scatchard analysis of ^{125}I EGF binding to derived lines. Both K721 and M721 EGF receptor-containing cells exhibited similar equilibrium dissociation constants for EGF binding. Greater expression of EGF receptors was obtained in permanently transfected B82 cell lines maintained in 1-10 μM methotrexate. As shown in Fig. 3C, both EGF receptors were expressed as 170 kd glycoproteins. These data indicate that the mutation did not affect EGF receptor synthesis and maturation into the mature cell surface binding species.

Whereas the K721 EGF receptor exhibited EGF-stimulated self-phosphorylation at tyrosine residues and exhibited EGF-stimulated protein tyrosine kinase activity assayed with exogenous peptide substrates such as angiotensin II, the methionine 721 containing EGF receptor was completely devoid of all protein tyrosine kinase activity whether assayed as self- or as exogenous substrate phosphorylation (Chen and et al., 1987).

Several EGF-stimulated activities were examined in cells expressing the K721 and M721 EGF receptors (Table I). One of the most rapid cellular responses induced by EGF is an increase in free intracellular calcium concentrations ($[Ca^{2+}]i$). Quantitation of rapid alterations in $[Ca^{2+}]i$ in individual cells was measured by fluorescence imaging using the dye fura 2 as an indicator (Poenie et al., 1985; Poenie et al., 1986). Cells expressing the K721 receptor exhibited a rapid response with a 4- to 5-fold increase of intracellular calcium with a plateau by 30 sec and a subsequent decline. B82 cells expressing the M721 EGF receptor exhibited no EGF-dependent alterations in basal calcium ($[Ca^{2+}]i$). These cells, however, did demonstrate a rapid, dramatic increase in $[Ca^{2+}]i$ in response to the ionophore ionomycin.

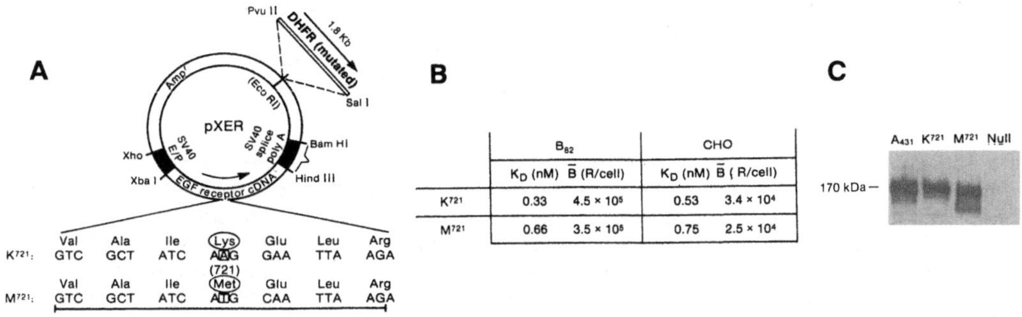

Fig. 3. Construction and expression of a mutant EGF receptor containing a lysine to methionine (K to M) substitution at amino acid residue 721. **Panel A:** Diagram of the chimeric plasmid used to transfect the K^{721} and M^{721} EGF receptor genes. The synthetic oligonucleotide complementary to the region indicated by a bar was used to generate the K to M substitution at amino acid 721 is shown. A mutant dihydrofolate reductase gene (DHFR) under SV40 early enhancer control provided the dominant selectable marker (Simonsen and Levinson, 1983). **Panel B:** EGF binding to permanent transfectants of B82 and CHO cells. Binding parameters were derived from Scatchard analysis of ^{125}I-EGF binding to representative selected clonal transfected cell lines as previously described (Gill and Lazar, 1981). **Panel C:** Quantitation of EGF receptors using polyclonal anti-human EGF receptor antibody. Following separation of 80 µg of membrane protein using SDS polyacrylamide gel electrophoresis, proteins were transferred to nitrocellulose, and EGF receptors were identified using a polyclonal anti-human EGF receptor antibody (Weber et al., 1984a).

TABLE I

Comparison of EGF-stimulated activities of K^{721} and M^{721} EGF receptors

	K^{721}	M^{721}
Self-phosphorylation	↑	-
Protein tyrosine kinase activity (peptide substrate)	↑	-
Intracellular $[Ca^{2+}]$	↑	-
Gene induction	↑	-
Cell proliferation	↑	-
Receptor metabolism (down-regulation)	↑	-

Binding of EGF to its receptor rapidly stimulates transcription of a set of genes, including proto-oncogenes such as c*fos* and c*myc* with accumulation of their mRNAs. The effect of EGF on gene expression was evaluated in the cell lines expressing the K721 and M721 EGF receptors using a reporter gene prepared by placing two copies of the c-*fos* 5' regulated enhancer element (Treisman, 1986) in front of the herpes thymidine kinase promoter. This was linked to the luciferase gene to report expression (de Wet et al., 1987). EGF stimulated approximately a 5-fold increase in luciferase activity in cells containing the K721 EGF receptor in a transient transfection assay, whereas cells expressing the M721 EGF receptor failed to exhibit any EGF-induced increase in luciferase activity.

Cell proliferation was measured in serum-free medium to avoid any growth effects caused by other mitogenic agents. Untransfected or cells expressing the M721 EGF receptor exhibited no response to EGF at any concentration, whereas transfected cell lines expressing the K721 EGF receptor exhibited EGF-dependent growth with CHO cells decreasing their population doubling time from 66 to 36 hrs in medium consisting of only DME/F-12 medium containing BSA. Similar effects were observed with B82 cells carrying the two EGF receptors. These results indicate that the protein tyrosine kinase activity of the EGF receptor is required for effects on calcium translocation, gene induction, and cell proliferation.

The EGF receptor is rapidly internalized and degraded after binding its ligand (homologous down-regulation) (Schlessinger et al., 1978). Although the kinetics of binding, the equilibrium dissociation constants for binding, and the number of receptors were similar, the M721 receptors exhibited no EGF-induced down-regulation while the K721 EGF receptors exhibited the expected homologous down-regulation as assessed by iodinated EGF binding to the cell surface and by decreased receptor protein as measured by quantitative immunoprecipitation of [^{35}S]methionine labeled cells (Lin et al., 1986). These results indicate that the EGF-induced effects on receptor metabolism also require intact intrinsic protein kinase activity. These results are consistent with the findings of decreased internalization in EGF receptors containing carboxy terminal truncations (Livneh et al., 1986). These data overall indicate that the intrinsic protein tyrosine kinase activity is essential to signal the complex series of events which culminate in DNA replication and cell division. The findings that insulin-dependent glucose transport also requires intrinsic protein tyrosine kinase activity in that receptor (Chou et al., 1987; Ebina et al., 1987) suggests that this may be a general and essential feature of all of the transmembrane receptor tyrosine kinase proteins.

Because the protein tyrosine kinase activity appears essential to the biological effects of EGF, factors regulating this activity are important. The binding of the single EGF ligand is with high affinity (calculated $\Delta G = -14$ kilocalories mol^{-1} (Knauer et al., 1984)). The carboxyl terminus contains the major site of self-phosphorylation, tyrosine 1173 (Y1173) which is phosphorylated in vivo and in vitro in response to EGF (Hunter and Cooper, 1981; Downward et al., 1984b). The EGF receptor can be self-phosphorylated at 5-7 additional tyrosine residues and is especially prominent tyrosine at residues 1068 and 1148 also located at the carboxyl terminus (Downward et al., 1984b; Gill et al., 1985). Self-phosphorylation occurs via an intramolecular mechanism which exhibits a Km for ATP that is approximately 10-fold less than that required for phosphorylation of exogenous substrates (Weber et al., 1984b). Self-phosphorylation is competitive with exogenous peptide substrates with a half-maximal inhibitory concentration of peptide equal to the peptide substrates steady-state affinity (Bertics and Gill, 1985). Comparison of exogenous peptide phosphorylation as a function of EGF receptor phosphorylation indicated that self-phosphorylation enhanced tyrosine protein kinase activity assayed at lower concentrations of substrate without affecting the Vmax. These studies suggested that the phosphorylation rate of a given substrate would depend on its concentration and relative affinity for the enzyme compared to that of the self-phosphorylation site. At low substrate levels, the phosphorylation rate would be slowed by competition by self-phosphorylation. At high substrate levels, the phosphorylation rate would appear maximal because the substrate could now compete with the self-phosphorylation site. The fact that phosphorylation occurs in vivo in response to EGF indicates that the receptor is not saturated with high affinity substrate in the cell and suggests that this reaction may play a role in receptor regulation in vivo. Phosphorylation of the carboxyl terminal domain was presumed to decrease its affinity for the enzyme active site and result in additional confirmational changes which permit exogenous substrates greater access to the substrate binding domain and enhanced tyrosine phosphorylation of these substrates (Bertics et al., 1985).

In contrast to these studies (Bertics and Gill, 1985) was the report of Downward et al. (1985) who found no effect of self-phosphorylation on the activity of purified EGF receptor/kinase. Both that study and our own agree that there is no influence of self-phosphorylation on the Vmax of the reaction where high saturating levels of exogenous substrates effectively compete with the self-phosphorylation sites. These differing conclusions made it important to study further the role of self-phosphorylation in regulating EGF receptor activity. This was accomplished by preparing a mutant EGF receptor in which tyrosine 1173 was converted to a phenylalanine residue (Bertics et al., 1987). The rationale was that the substitution of a structurally similar amino acid at the major kinetically preferred self-phosphorylation site in the carboxyl terminus would allow this region to still serve as a competitive inhibitor versus exogenous substrates, but would prevent its "removal" via phosphorylation. If the proposed model of action of self-phosphorylation is correct, one would predict that the mutant receptor would exhibit impaired kinase activity at low substrate levels, but essentially normal kinase activity at saturating substrate levels. Such results would indicate also that the mutation did not generate an inactive enzyme. Site-directed mutagenesis, cell transfections, and selections were carried out as described above. In addition to kinetic studies utilizing membranes prepared from the cells containing the two receptors, F1173 receptor was purified by immunoaffinity chromatography (Weber et al., 1984b), and the kinetic properties of the purified enzyme were compared to those of the Y1173 EGF receptor. Table II shows the characteristics of the two receptors. Mutant EGF receptors containing F1173 are equally and fully active at high substrate concentrations where effective competition occurs, but exhibit markedly reduced protein tyrosine kinase activity at low exogenous substrate concentrations. These results support the idea that the carboxyl terminus functions as a tight binding inhibitor and that substitution of F for Y at residue 1173 diminishes the capacity to activate the protein via self-phosphorylation. The F1173 mutation does not affect EGF binding nor receptor internalization nor does it affect the Km for ATP for self-phosphorylation. Interestingly, the kinetics of self-phosphorylation remain intramolecular, suggesting that there is redundancy in the regulatory constraints present in the carboxyl terminus such that other residues such as tyrosines 1068 and 1148 may function as competitive inhibitors in a fashion similar to that of the kinetically preferred Y1173 (Bertics et al., 1987).

When assayed in vivo, EGF stimulated growth of CHO cells containing both the Y1173 and F1173 EGF receptors with similar ED50s for EGF. However, the capability of EGF to stimulate growth of the mutant F1173 receptor was reduced by more than 50%. These results suggest that the kinetic effects observed in vitro are also important in vivo. These results are also compatible with there being residual inhibitory effects of the remaining tyrosine residues in the carboxyl terminus, i.e. redundancy in the regulatory function of the carboxyl terminal region.

TABLE II

Characteristics and Comparison of the Normal EGF Receptor (Tyr 1173) and a Mutant EGF Receptor (Phe 1173) Lacking the Major Self-phosphorylation Site

Receptor Property	Tyr 1173 EGF Receptor	Phe 1173 EGF Receptor
Major sites of self-phosphorylation	tyrosines 1173, 1148, 1068	tyrosines 1148, 1068
Km for ATP with self-phosphorylation	200 nM	200 nM
Mechanics of self-phosphorylation	intramolecular	intramolecular
Kinetics of protein-tyrosine kinase activity	exhibits linear kinetics in a double reciprocal plot with peptide substrates	same Vmax as normal receptor but reduced activity at low substrate levels--competitively inhibited
In vitro stimulation of kinase activity by EGF	yes ED(50) = 20-30 nM	yes ED(50) = 10-30 nM
Ability to mediate cell growth in response to EGF	yes ED(50) = 0.01 nM	reduced >50% compared to control ED(50) = 0.01 nM

Reprinted with permission from Bertics et al. (1987).

These results are summarized in a model for allosteric activation of the EGF receptor in Fig. 4. We assume that high affinity EGF binding occurs in a cleft with resulting closing and extensive conformational changes, including motion of entire helices as is well described by crystallographic studies of proteins such as hexokinase. Binding of ligand EGF is the major activator of the receptor tyrosine kinase activity which is essential for the biological activity of the receptor. A rapid secondary response is self-phosphorylation of the carboxyl terminus which removes an alternate substrate/competitive inhibitory constraint allowing access of exogenous substrates to the active site of the enzyme and full phosphotransferase activity. Self-phosphorylation also appears essential for activity of the insulin receptor (Rosen et al., 1983', Ellis et al., 1986) although the mechanism through which this occurs has not yet been defined.

Studies of recovered viruses following infection with viruses which cause insertional mutagenesis in the EGF receptor indicate that the acquisition of erb B deletes the extracellular ligand binding domain (Fung et al., 1983', Nilsen et al., 1985). Further studies also indicate that the precise boundaries of transduced fragments of erb B have a determining effect on the nature of produced transformation. Those containing an intact carboxyl terminus were strain restricted and caused erythroblastosis only (Gamett et al., 1986). Angiosarcomas and fibrosarcoma-inducing viruses contained either a truncated carboxyl terminus and/or an internal deletion in this region (Gamett et al., 1986). These studies with erb B also point to the importance of the two major regulatory domains defined by mutational analysis of the EGF receptor located at the amino and carboxyl terminus. Because erb B possesses protein tyrosine kinase activity (Kris et al., 1985; Gilmore et al., 1985), it is likely that constitutive expression of this activity is essential for transformation as it is for the other effects of the activated EGF receptor described.

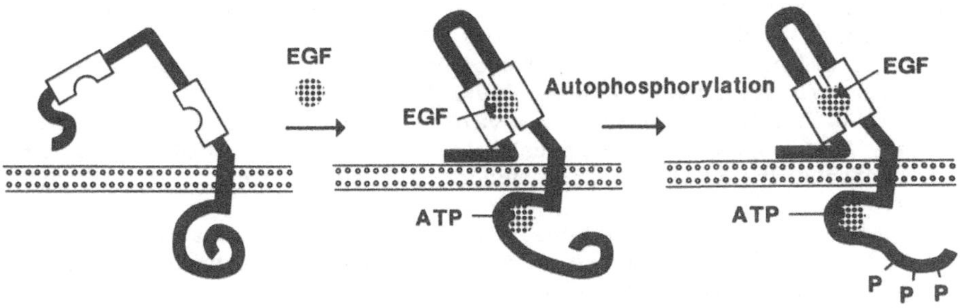

Fig. 4. Model of allosteric activation of the EGF receptor. See text for details.

AKNOWLEDGEMENTS

We gratefully acknowledge the contributions of Cliff D. Carpenter, Laura Hubler, and Charles Nelson. These studies were supported by grants from the National Institutes of Health, American Cancer Society, and Council for Tobacco Research - U.S.A., Inc. MGR is a Howard Hughes Medical Institute Investigator.

REFERENCES

Bargmann, C. I., Hung, M-C., and Weinberg, R. A., 1986a, The neu oncogene encodes an epidermal growth factor receptor-related protein. Nature 319:226-230.

Bargmann, C. I., Hung, M-C., and Weinberg, R. A., 1986b, Multiple independent activations of the neu oncogene by a point mutation altering the transmembrane domain of p185. Cell 45:649-657.

Bertics, P. J., Chen, W. S., Lazar, C. S., Hubler, L., Rosenfeld, M. G., and Gill, G. N., 1987. Submitted for publication.

Bertics, P. J., Hubler, L., Chen, W. S., Carpenter, C. D., Rosenfeld, M. G., and Gill, G. N., 1987, Epidermal growth factor receptor: Structure and regulation by self-phosphorylation. in: "Growth Regulation of Cancer," Marc E. Lippman, ed., Alan R. Liss, Inc., New York. In press.

Bertics, P. J. and Gill, G. N., 1985, Self-phosphorylation enhances the protein-tyrosine kinase activity of the epidermal growth factor receptor. J. Biol. Chem. 260:14642-14627.

Bertics, P. J., Weber, W., Cochet, C., and Gill, G. N., 1985, Regulation of the epidermal growth factor receptor by phosphorylation. J. Cell. Biochem. 29:195-208.
Chen, W. S., Lazar, C. S., Poenie, M., Tsien, R. Y., Gill, G. N., and Rosenfeld, M. G., 1987, Intrinsic protein tyrosine kinase activity mediates the immediate and late actions of the EGF receptor. Nature. In press.
Chou, C. K., Dull, T. J., Russell, D. S., Gherzi, R., Lebwohl, D., Ullrich, A., and Rosen, O. M., 1987, Human insulin receptors mutated at the ATP-binding site lack protein tyrosine kinase activity and fail to mediate postreceptor effects of insulin. J. Biol. Chem. 262:1842-1847.
Cohen, S., Ushiro, H., Stoscheck, C., and Chinkers, M., 1982, A native 170,000 epidermal growth factor receptor-kinase complex from shed plasma membrane vesicles. J. Biol. Chem. 257:1523-1531.
Coussens, L., Van Beveren, C., Smith, D., Chen, E., Mitchell, R. L., Isacke, C. M., Verma, I. M., and Ullrich, A., 1986, Structural alteration of viral homologue of receptor proto-oncogene fms at carboxyl terminus. Nature 320:277-280.
Defize, L. H. K., Moolenaar, W. H., van der Saag, P. T., and de Laat, S. W., 1986, Dissociation of cellular responses to epidermal growth factor using anti-receptor monoclonal antibodies. EMBO J. 5:1187-1192.
de Wet, J. R., Wood, K. V., DeLuca, M., Helinski, D. R., and Subramani, S., 1987, Firefly luciferase gene: Structure and expression in mammalian cells. Mol. Cell. Biol. 7:725-737.
Doolittle, R. F., Hunkapiller, M. W., Hood, L. E., Devare, S. G., Robbins, K. C., Aaronson, S. A., and Antoniades, H. N., 1983, Simian sarcoma virus onc gene, v-sis, is derived from the gene (or genes) encoding a platelet-derived growth factor. Science 221:275-277.
Downward, J., Yarden, Y., Mayes, E., Scrace, G., Totty, N., Stockwell, P., Ullrich, A., Schlessinger, J., and Waterfield, M. D., 1984a, Close similarity of epidermal growth factor receptor and v-erb-B oncogene protein sequences. Nature 307:521-527.
Downward, J., Parker, P., and Waterfield, M. D., 1984b, Autophosphorylation sites on the epidermal growth factor receptor. Nature 311:483-485.
Downward, J., Waterfield, M. D., and Parker, P. J., 1985, Autophosphorylation and protein kinase C phosphorylation of the epidermal growth factor receptor. J. Biol. Chem. 260:14538-14546.
Ebina, Y., Araki, E., Taira, M., Shimada, F., Mori, M., Craik, C. S., Siddle, K., Pierce, S. B., Roth, R. A., and Rutter, W. J., 1987, Replacement of lysine residue 1030 in the putative ATP-binding region of the insulin receptor abolishes insulin- and antibody-stimulated glucose uptake and receptor kinase activity. Proc. Natl. Acad. Sci. U.S.A. 84:704-708.
Ebina, Y., Ellis, L., Jarnagin, K., Edery, M., Graf, L., Clauser, E., Ou, J., Masiarz, F., Kan, Y. W., Goldfine, I. D., Roth, R. A., and Rutter, W. J., 1985, The human insulin receptor cDNA: The structural basis for hormone-activated transmembrane signalling. Cell 40:747-758.
Ellis, L., Clauser, E., Morgan, D. O., Edery, M., Roth, R. A., and Rutter, W. J., 1986, Replacement of insulin receptor tyrosine residues 1162 and 1163 compromises insulin-stimulated kinase activity and uptake of 2-deoxyglucose. Cell 45:721-732.
Engelman, D. M., Steitz, T. A., and Goldman, A., 1987, Identifying non-polar transbilayer helices in amino acid sequences of membrane proteins. In press.
Fung, Y-K. T., Lewis, W. G., Crittenden, L. B., and Kung, H-J., 1983, Activation of the cellular oncogene c-erbB by LTR insertion: Molecular basis for induction of erythroblastosis by avian leukosis virus. Cell 33:357-368.

Gamett, D. C., Tracy, S. E., and Robinson, H. L., 1986, Differences in sequences encoding the carboxyl-terminal domain of the epidermal growth factor receptor correlate with differences in the disease potential of viral erbB genes. Proc. Natl. Acad. Sci. U.S.A. 83: 6053-6057.

Gill, G. N., Bertics, P. J., and Santon, J. B., 1987, Epidermal growth factor and its receptor. Mol. Cell. Endocr. 51:169-186.

Gill, G. N., Bertics, P. J., Thompson, D. M., Weber, W., and Cochet, C., 1985, Structure and regulation of epidermal growth factor receptor. Cancer Cells 3:11-18.

Gill, G. N., and Lazar, C. S., 1981, Increased phosphotyrosine content and inhibition of proliferation in EGF-treated A431 cells. Nature 293: 305-308.

Gill, G. N., Weber, W., Thompson, D. M., Lin, C., Evans, R. M., Rosenfeld, M. G., Gamou, S., and Shimizu, N., 1985, The relation between production of epidermal growth factor receptors, gene amplification, and chromosome 7 translocation in variant A431 cells. Somatic Cell Mol. Genet. 11:309-318.

Gilmore, T., DeClue, J. E., and Martin, G. S., 1985, Protein phosphorylation at tyrosine is induced by the v-erbB gene product in vivo and in vitro. Cell 40:609-618.

Greenberg, M. E., and Ziff, E. B., 1984, Stimulation of 3T3 cells induces transcription of the c-fos proto-oncogene. Nature 311:433-438.

Hannink, M., and Donoghue, D. J., 1985, Lysine residue 121 in the proposed ATP-binding site of the v-mos protein is required for transformation. Proc. Natl. Acad. Sci. U.S.A. 82:7894-7898.

Huang, J. S., Huang, S. S., and Deuel, T. F., 1984, Transforming protein of simian sarcoma virus stimulates autocrine growth of SSV-transformed cells through PDGF cell-surface receptors. Cell 39:79-87.

Hunter, T., and Cooper, J. A., 1981, Epidermal growth factor induces rapid tyrosine phosphorylation of proteins in A431 human tumor cells. Cell 24:741-752.

Hunter, T., and Cooper, J. A., 1985, Protein-tyrosine kinases. Annu. Rev. Biochem. 54:897-930.

Knauer, D. J., Wiley, H. S., and Cunningham, D. D., 1984, Relationship between epidermal growth factor receptor occupancy and mitogenic response. J. Biol. Chem. 259:5623-5631.

Kris, R. M., Lax, I., Gullick, W., Waterfield, M. D., Ullrich, A., Fridkin, M., and Schlessinger, J., 1985, Antibodies against a synthetic peptide as a probe for the kinase activity of the avian EGF receptor and v-erbB protein. Cell 40:619-625.

Kruijer, W., Cooper, J. A., Hunter, T., and Verma, I., 1984, Platelet-derived growth factor induces rapid but transient expression of the c-fos gene and protein. Nature 312:711-716.

Lin, C. R., Chen, W. S., Kruijer, W., Stolarsky, L. S., Weber, W., Evans, R. M., Verma, I. M., Gill, G. N., and Rosenfeld, M. G., 1984, Expression cloning of human EGF receptor complementary DNA: Gene amplification and three related messenger RNA products in A431 cells. Science 224:843-848.

Lin, C. R., Chen, W. S., Lazar, C. S., Carpenter, C. D., Gill, G. N., Evans, R. M., and Rosenfeld, M. G, 1986, Protein kinase C phosphorylation at Thr 654 of the unoccupied EGF receptor and EGF binding regulate functional receptor loss by independent mechanisms. Cell 44:839-848.

Livneh, E., Prywes, R., Kashles, O., Reiss, N., Sasson, I., Mory, Y., Ullrich, A., and Schlessinger, J., 1986, Reconstitution of human epidermal growth factor receptors and its deletion mutants in cultured hamster cells. J. Biol. Chem. 261:12490-12497.

Mayes, E. L. V., and Waterfield, M. D., 1984, Biosynthesis of the epidermal growth factor receptor in A431 cells. EMBO J. 3: 531-537.

Moolenaar, W. H., Yarden, Y., de Laat, S. W., and Schlessinger, J., 1982,

Epidermal growth factor induces electrically silent Na^+ influx in human fibroblasts. J. Biol. Chem. 257:8502-8506.
Muller, R., Bravo, R., Burckhardt, J., and Curran, T., 1984, Induction of c-fos gene and protein by growth factors precedes activation of c-myc. Nature 312:716-720.
Murdoch, G. H., Potter, E., Nicolaisen, A. K., Evans, R. M., and Rosenfeld, M. G., 1982, Epidermal growth factor rapidly stimulates prolactin gene transcription. Nature 300:192-194.
Nilsen, T. W., Maroney, P. A., Goodwin, R. G., Rottman, F. M., Crittenden, L. B., Raines, M. A., and Kung, H-J., 1985, c-erbB activation in ALV-induced erythroblastosis: Novel RNA processing and promoter insertion result in expression of an amino-truncated EGF receptor. Cell 41:719-726.
Poenie, M., Alderton, J., Steinhardt, R., and Tsien, R., 1986, Calcium rises abruptly and briefly throughout the cell at the onset of anaphase. Science 233:886-889.
Poenie, M., Alderton, J., Tsien, R. Y., and Steinhardt, R. A., 1985, Changes of free calcium levels with stages of the cell division cycle. Nature 315:147-149.
Prywes, R., Livneh, E., Ullrich, A., and Schlessinger, J., 1986, Mutations in the cytoplasmic domain of EGF receptor affect EGF binding and receptor internalization. EMBO J. 5:2179-2190.
Rosen, O. M., Herrera, R., Olowe, Y., Petruzzelli, L. M., and Cobb, M. H., 1983, Phosphorylation activates the insulin receptor tyrosine protein kinase. Proc. Natl. Acad. Sci. U.S.A. 80:3237-3240.
Rosenthal, A., Lindquist, P. B., Bringman, T. S., Goeddel, D. V., and Derynck, R., 1986, Expression in rat fibroblasts of a human transforming growth factor-α cDNA results in transformation. Cell 46:301-309.
Sawyer, S. T., and Cohen, S., 1981, Enhancement of calcium uptake and phosphatidylinositol turnover by epidermal growth factor in A-431 cells. Biochemistry 20:6280-6286.
Schlessinger, J., Shechter, Y., Willingham, M. C., and Pastan, I., 1978, Direct visualization of binding, aggregation, and internalization of insulin and epidermal growth factor on living fibroblastic cells. Proc. Natl. Acad. Sci. U.S.A. 75:2659-2663.
Sherr, C. J., Rettenmier, C. W., Sacca, R., Roussel, M. F., Look, A. T., and Stanley, E. R., 1985, The c-fms proto-oncogene product is related to the receptor for the mononuclear phagocyte growth factor, CSF-1. Cell 41:665-676.
Simonsen, C. C., and Levinson, A. D., 1983, Isolation and expression of an altered mouse dihydrofolate reductase cDNA. Proc. Natl. Acad. Sci. U.S.A. 80:2495-2499.
Snyder, M. A., Bishop, J. M., McGrath, J. P., and Levinson, A. D., 1985, A mutation at the ATP-binding site of $pp60^{v-src}$ abolishes kinase activity, transformation, and tumorigenicity. Mol. Cell. Biol. 5: 1772-1779.
Stern, D. F., Hare, D. L., Cecchini, M. A., and Weinberg, R. A., 1987, Construction of a novel oncogene based on synthetic sequences encoding epidermal growth factor. Science 235:321-324.
Sternberg, M. J. E., and Taylor, W. R., 1984, Modelling the ATP-binding site of oncogene products, the epidermal growth factor receptor and related proteins. FEBS Lett. 175:387-392.
Thompson, D. M., and Gill, G. N., 1985, The EGF receptor: structure, regulation and potential role in malignancy. Cancer Surveys 4:767-788.
Treisman, R., 1986, Identification of a protein-binding site that mediates transcriptional response of the c-fos gene to serum factors. Cell 46:567-574.
Ullrich, A., Bell, J. R., Chen, E. Y., Herrera, R., Petruzzelli, L. M., Dull, T. J., Gray, A., Coussens, L., Liao, Y-C., Tsubokawa, M.,

Mason, A., Seeburg, P. H., Grunfeld, C., Rosen, O. M., and Ramachandran, J., 1985, Human insulin receptor and its relationship to the tyrosine kinase family of oncogenes. Nature 313:756-761.

Ullrich, A., Coussens, L., Hayflick, J. S., Dull, T. J., Gray, A., Tam, A. W., Lee, J., Yarden, Y., Libermann, T. A., Schlessinger, J., Downward, J., Mayes, E. L. V., Whittle, N., Waterfield, M. D., and Seeburg, P. H., 1984, Human epidermal growth factor receptor cDNA sequence and aberrant expression of the amplified gene in A431 epidermoid carcinoma cells. Nature 309:418-425.

Ushiro, H., and Cohen, S., 1980, Identification of phosphotyrosine as a product of epidermal growth factor-activated protein kinase in A-431 cell membranes. J. Biol. Chem. 2558363-8365.

Waterfield, M. D., Scrace, G. T., Whittle, N., Stroobant, P., Johnsson, A., Wasteson, A., Westermark, B., Heldin, C-H., Huang, J. S., and Deuel, T. F., 1983, Platelet-derived growth factor is structurally related to the putative transforming protein $p28^{sis}$ of simian sarcoma virus. Nature 304:35-39.

Weber, W., Gill, G. N., and Spiess, J., 1984a, Production of an epidermal growth factor receptor-related protein. Science 224:294-297.

Weber, W., Bertics, P. J., and Gill, G. N., 1984b, Immunoaffinity purification of the epidermal growth factor receptor: Stoichiometry of EGF binding and kinetics of self-phosphorylation. J. Biol. Chem. 259:14631-14636.

Weinmaster, G., Zoller, M. J., and Pawson, T., 1985, A lysine in the ATP-binding site of $P130^{gag-fps}$ is essential for protein-tyrosine kinase activity. EMBO J. 5:69-76.

Wigler, M., Pellicer, A., Silverstein, S., Alex, R., Urlaub, G., and Chasio, L., 1979, DNA mediated transfer of the adenosine phosphoribosyltransferase locus into mammalian cells. Proc. Natl. Acad. Sci. U.S.A. 76:1373-1376.

Xu, Y-H., Ishii, S., Clark, A. J. L., Sullivan, M., Wilson, R. K., Ma, D. P., Roe, B. A., Merlino, G. T., and Pastan, I., 1984, Human epidermal growth factor receptor cDNA is homologous to a variety of RNAs overproduced in A431 carcinoma cells. Nature 309:806-810.

Yamamoto, T., Ikawa, S., Akiyama, T., Semba, K., Nomura, N., Miyajima, N., Saito, T., and Toyoshima. K., 1986, Similarity of protein encoded by the human c-erb-B-2 gene to epidermal growth factor receptor. Nature 319:230-234.

Yarden, Y., Escobedo, J. A., Kuang, W-J., Yang-Feng, T. L., Daniel, T. O., Tremble, P. M., Chen, E. Y., Ando, M. E., Harkins, R. N., Francke, U., Fried, V. A., Ullrich, A., and Williams, L. T., 1986, Structure of the receptor for platelet-derived growth factor helps define a family of closely related growth factor receptors. Nature 323:226-232.

Zoller, M. J., and Smith, M., 1983, Oligonucleotide-directed mutagenesis of DNA fragments cloned into M13 derived vectors. Methods Enzymol. 100:468-501.

REGULATION OF EGF RECEPTOR AND

TRANSFORMING GROWTH FACTOR-ALPHA EXPRESSION

Jeffrey E. Kudlow, Jeffrey D. Bjorge, Michael S. Kobrin and Andrew J. Paterson

Departments of Clinical Biochemistry and Medicine
Banting and Best Diabetes Centre
University of Toronto
Toronto, Ontario, Canada M5G 1L5

Introduction

In 1957, Salmon and Daughaday demonstrated the presence of a sulfation factor in the serum, whose activity was stimulated by growth hormone[1] It is now known that this activity is primarily composed of insulin-like growth factor-I (IGF-I)[2] This finding demonstrated a principle upon which our laboratory has focused; that is, that a wide variety of hormones whose initial mechanisms of action are diverse, can stimulate growth through pathways involving growth factors as intermediates. This concept arose out of the finding that protein tyrosine kinases are intimately associated with growth regulation. The first example of a relationship between tyrosine kinases and growth regulation was the product of the src oncogene of the Rous sarcoma virus, $pp60^{src}$. This oncogene is responsible for the aberrant growth of cells transformed by the virus. Since that time, several oncogenes have been shown to code for highly homologous tyrosine kinases but in addition, several growth factor receptors have been shown by direct sequence analysis to be src-related tyrosine specific protein kinases including the receptors for epidermal growth factor (EGF), platelet derived growth factor (PDGF), IGF-I, insulin, and colony stimulating factor-I[3-6]. The cDNA's for all of these receptors have been cloned and many structure-function relationships within these receptors have been established using techniques such as site directed mutagenesis. These studies have strengthened the general concept that the growth response to growth factors is mediated by growth factor dependent, tyrosine specific protein kinases.

Growth also occurs in response to hormones whose actions do not appear to be directly mediated through protein tyrosine kinases. There has been the long known clinical observation that hormones capable of stimulating differentiated function in their target tissues, such as hormone synthesis, can after chronic stimulation, result in tissue growth through both hypertrophy and hyperplasia. What has not been established is how these hormones whose action does not directly involve the tyrosine kinases act to stimulate growth. One mechanism is demonstrated by the example of growth hormone; that is, the primary hormone stimulates the expression of the growth factor(s) in the target tissue, resulting in a local (autocrine or paracrine) increase in the growth factor concentration, or the secretion of the growth factor into the blood (endocrine). Another mechanism involves alteration in tissue responsiveness to growth factor via alterations in (a) receptor number or affinity or (b) pathways within the cell upon which the receptors impinge (post-receptor). Our laboratory and others have made progress in understanding the factors involved in controlling the expression of the EGF receptor and one of its ligands, transforming growth factor-α (TGF-α).

Regulation of EGF Receptor Expression by EGF

EGF, in addition to stimulating cellular proliferation, stimulates the transcription of a number of genes including c-fos[7], c-myc[7], prolactin[8], and actin[9]. In particular, c-myc and c-fos are thought to play important roles in the cell growth cycle and in differentiation[10-12]. Using monoclonal antibodies developed in our own laboratory[13], we measured the rate of synthesis of nascent EGF receptor protein and showed that EGF stimulates the synthesis of its own receptor in human breast cancer cells[14] (MDA468 cell line). Half-maximal stimulation of receptor biosynthesis occurred at an EGF concentration of 10^{-10} M (Fig. 1a and 1b), with peak stimulation seen 6-8 hours after EGF addition (Fig. 2).

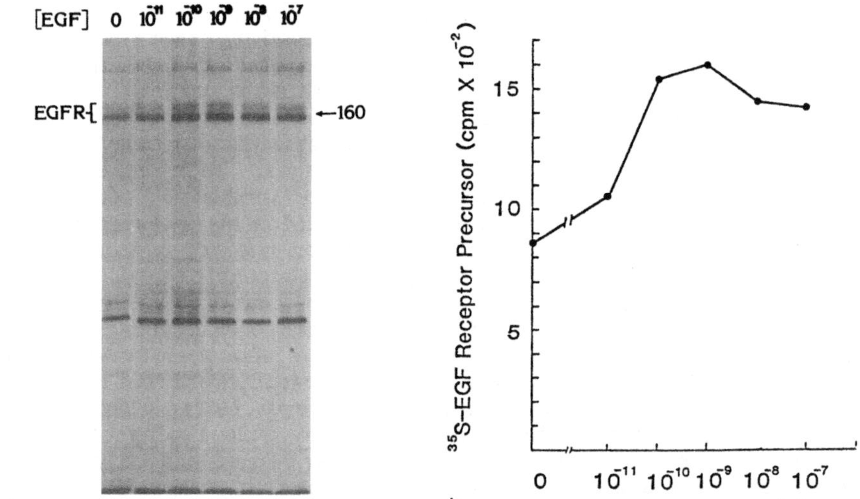

Fig. 1. Effect of EGF concentration on the EGF receptor precursor synthetic rate. MDA 468 cells grown in Dulbecco's modified Eagles medium containing 10% calf serum were incubated for 6 hours in the presence of various concentrations of EGF. The cells were then washed and metabolically labeled with L-[^{35}S]methionine for 1 hour under the same EGF concentrations, after which the cells were lysed and the receptor immunoprecipitated using monoclonal antibody B1D8 and fixed Staphylococcus aureus. The immunoprecipitates were run on SDS-polyacrylamide gel electrophoresis and the labeled bands were detected by autoradiography (Fig. 1a). Labeled bands corresponding to the EGF receptor precursor were cut out and the radioactivity was quantitated by scintillation counting (Fig. 1b).

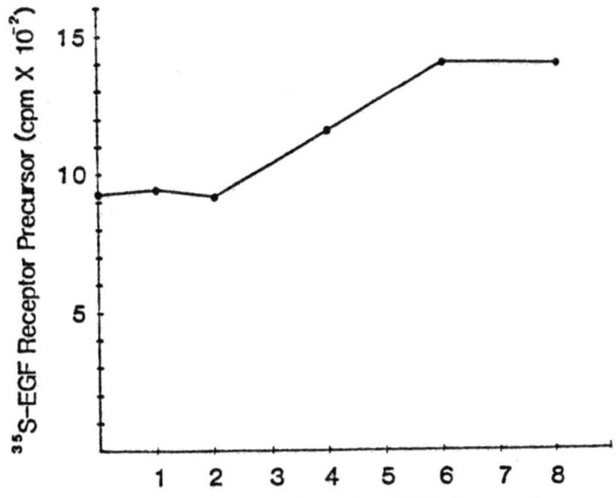

2

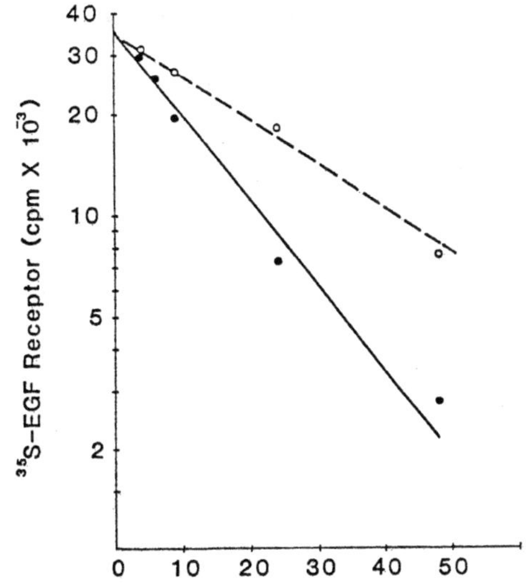

3

Fig. 2. Effect of incubation time with EGF on the rate of synthesis of EGF receptor precursor. MDA 468 cells were incubated for varying times in the presence of 10^{-9}M EGF. The cells were washed, metabolically labeled, and the immunoprecipitates processed as in Fig. 1. Quantitation of the 160-kDa precursor is shown. (Figure on left)

Fig. 3. Effect of EGF on the degradation rate of the receptor. MDA 468 cells were metabolically labeled overnight with L-[^{35}S] methionine in the absence of EGF. The cell monolayers were washed and the label was chased in regular growth medium without label in the absence or the presence of 10^{-9}M EGF. At the indicated times following exposure to EGF, the cells were lysed and the receptor immunoprecipitated. The immunoprecipitates were run on SDS-PAGE, and the labeled proteins were detected by autoradiography. Labeled bands corresponding to the mature EGF receptor were cut out and the radioactivity quantitated by scintillation counting is shown. EGF absent (o). EGF present (•). (Figure on right)

EGF simultaneously accelerates the rate of receptor degradation (Fig. 3) resulting in a net loss of cell surface receptors. This stimulation of receptor synthesis by EGF would work in opposition to down-regulation, which has been thought to desensitize the cell to EGF.

We have since shown that EGF stimulates the accumulation of EGF receptor mRNA (Fig. 4) with kinetics virtually identical to that shown for the stimulation of synthesis of the receptor protein. This similarity of kinetics and our measurement of EGF receptor mRNA halflife suggest that the EGF effect on receptor expression is regulated transcriptionally. Similar observations were made almost simultaneously in other laboratories in other cell lines[15,16] suggesting that this effect of EGF on receptor expression is general.

Involvement of Protein Kinase C in the Regulation of EGF Receptor Expression

To further characterize the mechanism through which EGF stimulates receptor synthesis, we have examined the role of protein kinase C. Protein kinase C is a phospholipid-dependent, calcium activated kinase that phosphorylates proteins on serine and threonine residues[17]. In

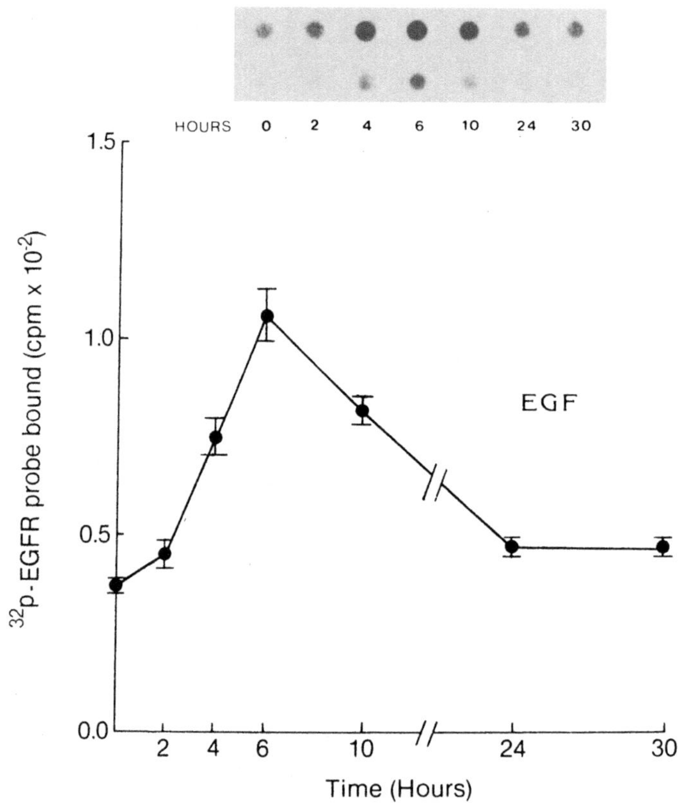

Fig. 4. Effects of EGF on EGF receptor mRNA levels in MDA468 cells. MDA468 cells were treated with 10^{-9}M EGF for varying times. After the various incubations, the mRNA was extracted using guanidine-HCl. Following precipitation of the RNA and resolubilization, equal amounts of total RNA were dotted onto nitrocellulose filters (5 ug/dot for the upper row of dots, 1.25 ug/dot for the lower row). The filters were dried, baked, prehybridized, and probed with [^{32}P]-labeled EGF receptor cDNA probe. The EGF receptor probe consists of a 760 base pair sequence corresponding to the intracellular domain of the human EGF receptor (kindly provided by Dr. G.N. Gill). Filters were then air dried and exposed to Kodak X-Omat AR film with Dupont Cronex intensifying screens for 1-2 days. Quantitation was performed by localizing the dot on the nitrocellulose using radioactive ink alignment markers, excising the dot, and counting the radioactivity using liquid scintillation counting. Each point represents the mean of triplicate dots +/- 1 S.D. and a representative dot is shown for each time point.

vivo, protein kinase C is thought to be activated by diacylglycerol derived from the hydrolysis of phosphoinositides by the action of phospholipase C. A number of hormones and growth factors are known to cause the hydrolysis of phosphoinositides[18] through the activation of phospholipase C by a membrane GTP-binding protein. These hormone could in turn make use of this step as a common entry point for activation of protein kinase C. In A431 cells EGF has also been shown to stimulate phosphoinositide turnover[19-21] and in both A431 and MDA468 cells, has been shown to stimulate the accumulation of inositol triphosphate[22]. In the cells expressing lower numbers of EGF receptors, these effects on second messenger generation have not been demonstrated[23,24].

Evidence has been obtained that tumor-promoting phorbol esters such as TPA can substitute for diacylglycerol in both in vivo and in vitro systems and activate protein kinase C in a manner similar to that of diacylglycerol[25,26]. It is currently accepted that protein kinase C may, in fact, be the receptor for TPA[27] and the numerous effects one sees when cells are treated with TPA may simply reflect activation of this enzyme. These findings have led to the widespread use of TPA as an activator of protein kinase C.

Activation of protein kinase C is known to have a variety of effects on cells[28] including interactions with the EGF receptor that result in lowered EGF receptor affinity and reduced tyrosine kinase activity[29-33], lowered EGF receptor number[34,35], altered receptor

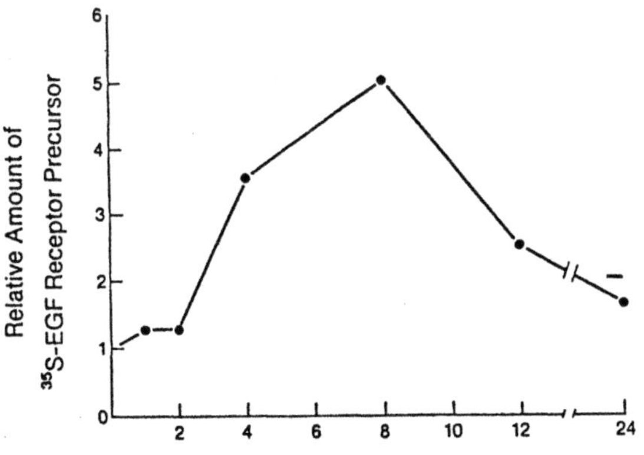

5

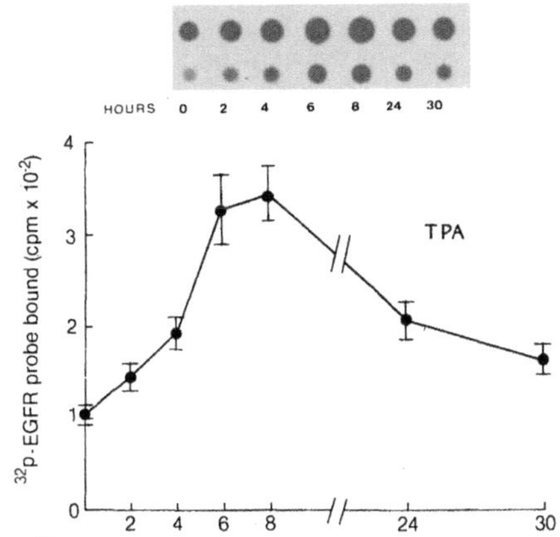

6

Fig. 5. Effect of the TPA incubation time on the rate of synthesis of the EGF receptor precursor. MDA468 cells were incubated for varying times in the presence of 100ng/ml TPA. The cells were then labeled for 1 hour, and processed as in Fig. 1. The receptor synthesis rate is expressed as a fold increment relative to the basal rate. Control cells treated with the TPA diluent (0.01% methanol in growth medium) showed no effect on receptor synthesis. (figure on left)

Fig. 6. Effects of TPA on EGF receptor mRNA levels in MDA468 cells. MDA468 cells were treated with 100ng/ml TPA for varying times. After the incubations, the mRNA was extracted, processed, and quantitated as in Figure 4. Equal amounts of total RNA were dotted onto nitrocellulose filters (5 ug/dot for the upper row of dots, 1.25 ug/dot for the lower row) which were probed with the EGF receptor cDNA probe. Each point represents the mean of triplicate dots +/- 1 S.D. and a representative dot is shown for each time point. (figure on right)

metabolism[33], and internalization of the receptor[37]. Phosphorylation of the EGF receptor by protein kinase C on an intracellular threonine residue[33,38,39] is thought to be responsible for several of these effects.

We tested the hypothesis that protein kinase C may be involved in the transduction of an EGF binding signal to the more distal effect of stimulating EGF receptor synthesis. To determine whether activation of protein kinase C could stimulate EGF receptor synthesis, MDA468 cells were treated with TPA. A detailed time course showed that stimulation with 100 ng/ml TPA resulted in an increased receptor precursor synthesis rate with a maximum at 8 hours, and a decline thereafter until the rate had reached a value just above the basal level at 24 hours (Fig. 5). The maximum increase observed at 8 hours was 5 fold over the basal rate of synthesis. This type of temporary response elicited by TPA is typical, and thought to represent the initial effects of TPA, followed by a downregulation of protein kinase C[40-42] and loss of effect after longer incubations[43-46]. This effect of TPA on the synthesis of EGF receptor protein corresponded remarkably with the effect of TPA on the EGF receptor mRNA, both in time course of induction and fold stimulation (Fig. 6).

Treatment of the MDA468 cells with increasing amounts of TPA resulted in a dose dependent stimulation of EGF receptor precursor synthesis (Fig. 7). The maximal stimulation, after 8 hours of treatment, occurred between 25ng/ml and 100ng/ml TPA, with no further changes in receptor synthesis rate with TPA doses as high as 1000ng/ml. Half-maximal stimulation occured at a dose of 5 ng/ml. This increase in receptor synthesis rate was specific to TPA, since the parent compound β-phorbol, which lacks tumor promoting activity, had no effect on receptor synthesis at all concentrations tested.

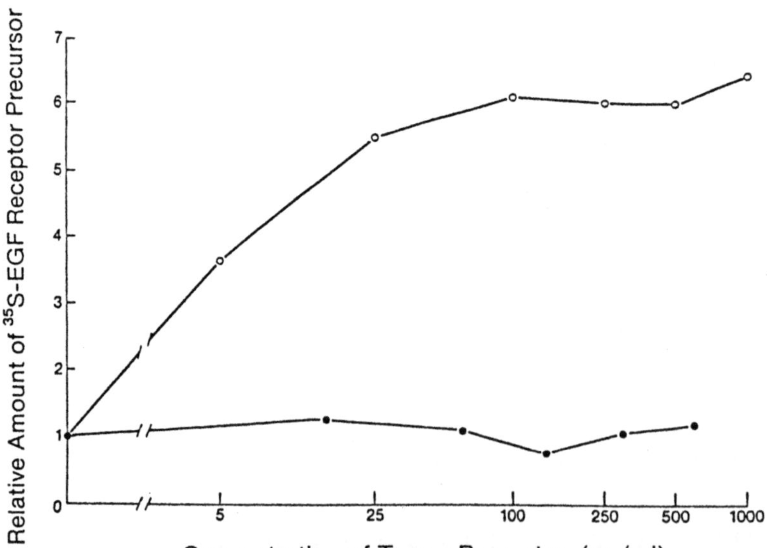

Fig. 7. Effect of TPA dose on the EGF receptor precursor synthetic rate. MDA468 cells were incubated with different doses of TPA or β-phorbol in growth medium for 8 hours. The cells were then labeled and processed as in Fig. 1. The receptor synthesis rate is expressed as a fold increment relative to the basal rate. TPA (O) ; B-phorbol (●).

The effects of TPA and EGF on EGF receptor synthesis were relatively specific since both EGF and TPA caused small decreases in general protein synthesis in the MDA468 cells. Also, the amount of labelled protein applied to the immunoprecipitation reactions was kept constant so that the increases in EGF receptor labelling were relative to other cellular proteins. The changes in EGF receptor mRNA were also specific. When equal amounts of total RNA from cells treated with nothing, β-phorbol, EGF and TPA were analyzed by northern blotting, an increase in EGF receptor hybridization signal was detected (Fig. 8a). When the EGF receptor probe was eluted from this blot and reprobed with the cDNA for the alpha chain of hexosaminidase A[47], a "housekeeping", lysosomal enzyme, no change in hybridization signal was observed (Fig. 8b).

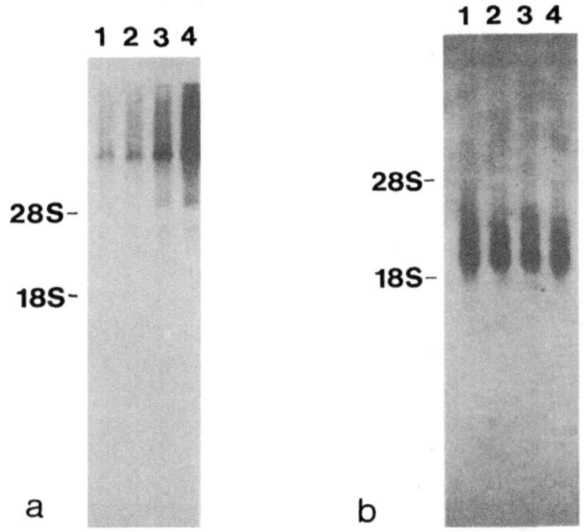

Fig. 8. Effects of EGF and TPA on EGF receptor and hexosaminidase A mRNA levels in MDA468 cell. MDA468 cells were left untreated (lane 1) or treated for 6 hours with 60 ng/ml 4β- phorbol (an inactive TPA analogue)(lane 2), 10^{-9}M EGF (lane 3), or 100ng/ml TPA (lane 4). The mRNA was extracted, equal amounts of total RNA run on a formaldehyde agarose gel, transferred to nitrocellulose, and hybridized with the cDNA probe for the EGF receptor (Fig. 8a). Following autoradiography of the blot, the EGF receptor probe was stripped from the nitrocellulose and the blot was then reprobed using a cDNA probe for the α-chain of hexosaminidase A (kindly provided by Dr. R.G Korneluk) (Fig. 8b). Autoradiograms at each step are shown.

Because TPA stimulated the rate of EGF receptor synthesis, we investigated the possibility that the total receptor number per cell would also be increased. MDA468 cells were treated with TPA for various times prior to harvesting. All cell harvesting was done at the same time to minimize differences in cell numbers. A previously optimized amount of the total cell homogenate was run on a 5% polyacrylamide gel, transferred to nitrocellulose, and probed with monoclonal antibody H9B4 against the receptor. Receptor-bound antibody was detected with ^{125}I-Protein A, which could be quantitated by gamma counting. Optimization was performed to ensure that the amount of ^{125}I-Protein A bound would be directly proportional to the amount of receptor present within a range established on a standard curve. Figure 9a shows that the amount of EGF receptor protein per cell, as determined by immunoblotting, does increase after treatment with TPA. The amount of receptor protein detected reached a maximum between 12 and 24 hours after TPA addition, 1.6 fold greater than detected in untreated cell (Fig. 9b).

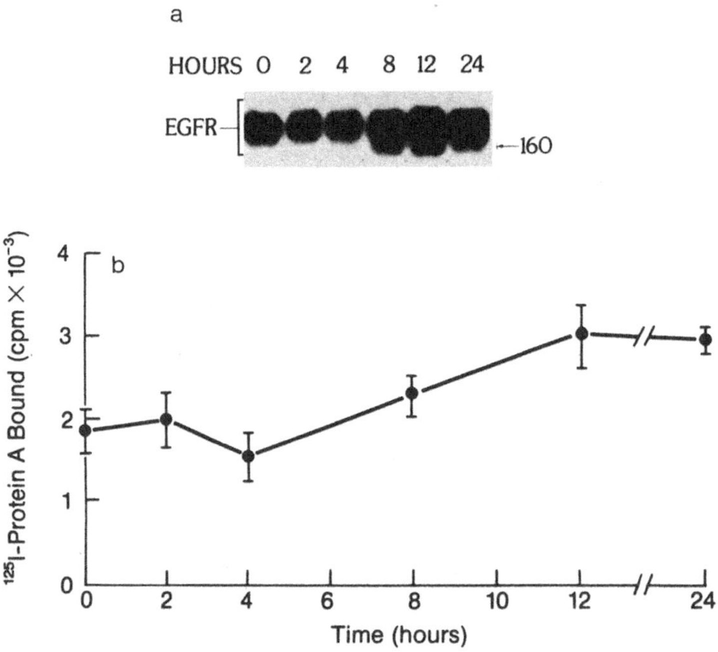

Fig. 9. Effect of TPA incubation on total cellular receptor protein. MDA468 cells were incubated for various times with 100 ng/ml TPA in growth medium. TPA treatment was initiated such that cell harvesting was carried out at the same time to minimize differences in cell numbers. The cells were lysed, and an aliquot was run on a 5% SDS-polyacrylamide gel. After transfer to nitrocellulose, the EGF receptor was detected with a monoclonal antibody against the receptor (H9B4), followed by rabbit anti-mouse IgG antibody, and ^{125}I-Protein-A. Results of autoradiography are shown in (a). The radioactive receptor bands were cut from the nitrocellulose and quantitated by gamma counting (b). The points represent the mean of triplicate determinations +/- s.e. The 12 and 24 hour points are significantly different than the 0 time point with p=.013 and p=.0041 respectively.

Because the amount of receptor protein detected on immunoblots increased with TPA treatment, we also examined the effect of TPA treatment on EGF binding. EGF binding studies were carried out in the presence of varying EGF concentrations after 4 and 24 hour TPA treatments. Scatchard plots of these binding data (Fig. 10) shows that after short incubations with TPA there is a loss of a small number of high affinity sites with no observed change in the low affinity receptors. This loss of high affinity sites has been attributed to protein kinase C mediated phosphorylation of the EGF receptor. After the 24 hour incubation, these high affinity sites have reappeared, and the number of both high and low affinity sites per cell increased by approximately 30%. This increase in receptor binding activity is in keeping with data presented above showing an increased receptor synthesis rate and an increased receptor protein as determined by immunoblotting.

From the Scatchard plot, we can estimate that unstimulated cells have 10^6 receptors. From Figure 3, we determined that the receptor half-life is 24 hours in these cells. Assuming exponential decay during half-life, and a synthesis rate matching degradation (steady state), then basal receptor synthesis rate at t=0 in the half-life experiment is (receptor number x ln 2)/half-life = 30,000 receptors per hour, agreeing with the determination in A431 cells[48]. After TPA stimulation, the synthesis rate rose to an average of 120,000 receptors per hour (4-fold) over a period of about 7 hours. Assuming that TPA had relatively small effects on degradation (30,000 degraded per hour), the net excess number of receptors synthesized each hour would be 120,000 minus 30,000 leading to an accumulation in 7 hours of 7 x 90,000=6.3 x 10^5 receptors. The resulting increment in receptors would be about 60% of the starting amount, agreeing with our experimental observations. This calculation of the fractional change in receptor mass is independent of absolute receptor number but dependent on there being no change in degradation and knowledge of the half-life and fold-increment in the synthesis rate. In cells with a shorter receptor half-life, the increment in synthesis rate as measured in the MDA468 cells would result in a greater effect on total receptor number.

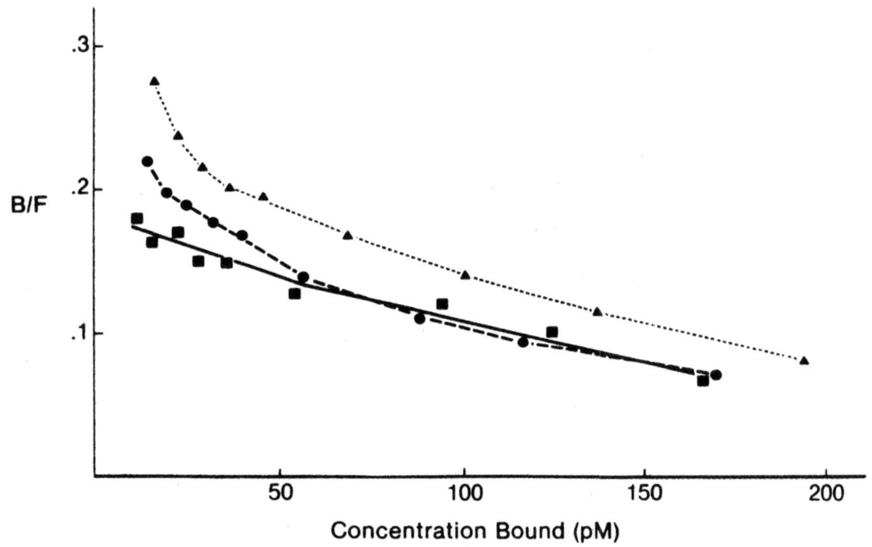

Fig. 10. Scatchard plot of EGF binding data following TPA treatment of MDA468 cells. 6.2 X 10^4 cells per well were treated for 24, 2, and 0 hours with TPA. The cells were then washed with ice cold binding medium and incubated with varying concentrations of unlabeled EGF with a constant amount of ^{125}I-EGF (20,000cpm per well; 4.9 X 10^5 cpm/pmol) in .5 ml of binding medium for 6 hours at 4^oC. Duplicate counts bound were averaged prior to analysis by the method of Scatchard. 24 hours TPA (▲); 2 hours TPA (■); 0 hours TPA (●).

Since both EGF and TPA stimulation result in increased receptor expression, experiments were performed which might indicate whether there was a common mechanism for stimulation by both these agents. EGF and TPA had no additive effect on receptor synthesis. However, TPA can influence EGF binding to cells through a variety of mechanisms[29-39]. Therefore, showing a lack of effect of EGF in the presence of TPA may simply reflect the decreased binding of EGF to its receptors. To circumvent this problem, we have pretreated the MDA468 cells with TPA for 48 hours, at which time we demonstrated more receptors per cell of both the high and low affinity types. Such pretreatment has been shown by other investigators to "downregulate" protein kinase C[40-42] and result in the loss of some of its putative actions[43-46]. Figure 11 shows that cells pretreated with the inactive agent β-phorbol respond in a typical fashion to EGF and TPA treatments. Pretreatment with TPA for 48 hours resulted in a return of the receptor synthesis rate to the untreated level. Depletion of TPA from the medium was ruled out because we could demonstrate that the TPA containing culture medium, after being on the cells for 48 hours could both decrease binding of EGF and stimulate receptor synthesis in previously untreated cells. In addition, the TPA pretreated cells became resistant to both the stimulatory effects of TPA and EGF.

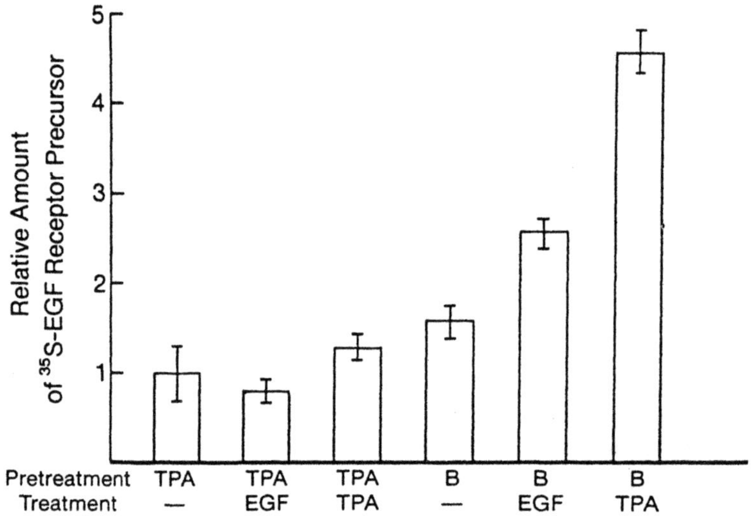

Fig. 11. Effect of TPA pretreatment of MDA468 cells on the stimulation of EGF receptor precursor synthesis caused by TPA and EGF. MDA468 cells were pretreated with either beta phorbol (B) (60ng/ml) or TPA (100ng/ml) for 2 days in DME medium with 10% calf serum. Subsequently, the cells were left untreated, or treated with EGF (10^{-9}M) or TPA (100ng/ml) for 8 hours to allow for potential stimulation of receptor synthesis. Labeling, immunoprecipitation, and quantitation were performed as in Fig. 1. The results represent the mean +/- 1 S.D. of triplicate wells. The receptor synthesis rate is expressed relative to the basal rate.

The observations that long term TPA treatment resulted in a return of the high affinity EGF binding sites, a lowering of EGF receptor synthesis rate to basal levels and an inability of further TPA treatment to stimulate receptor synthesis are compatable with the concept that such treatment depletes the cell of protein kinase C. Assuming that the only effect of chronic TPA treatment is to deplete cells of protein kinase C, the failure of EGF to stimulate receptor synthesis in such cells is compatable with the notion that protein kinase C plays an important role in signal transmission between the activated EGF receptor and the final response of increased EGF receptor synthesis.

However, chronic TPA treatment may have had other effects on the cells resulting in impaired EGF action. These effects could have been on the EGF receptor or at points distal to the receptor in the transducing pathway. While we could not rule out the latter, we demonstrated that the EGF receptor was intact. After long term TPA treatment high affinity EGF binding sites reappeared. But it was still possible that the EGF receptor was made biologically inactive by TPA treatment. To further confirm the integrity of the EGF receptor, we showed that when MDA468 cells were pretreated with TPA, they retained their known sensitivity to the growth inhibitory effects of EGF[49] (Fig 12). Therefore, chronic TPA treatment does not inactivate the EGF receptor nor the distal effectors of EGF action involved in growth inhibition. Furthermore, protein kinase C does not appear to be involved in the growth inhibitory effect of EGF on these cells.

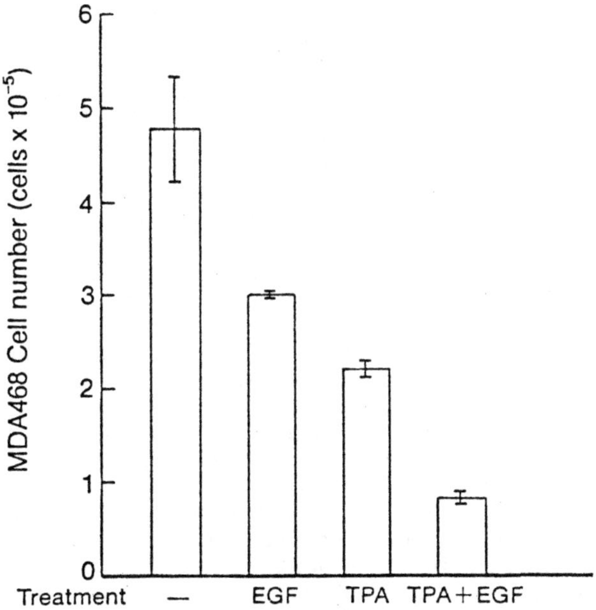

Fig. 12. The effects of EGF on MDA468 cell growth after pretreatment with TPA. MDA468 cells in 3.5 cm tissue culture dishes were pretreated for 2 days with or without TPA followed with or without treatment for a further 6 days with 10^{-9}M EGF. After these treatments, the cells were trypsinized and counted. The mean cell counts are shown +/- 1 S.D. of duplicate wells.

The ability of protein kinase C to stimulate EGF receptor synthesis is yet another means of this regulatory system interacting with the EGF receptor. Thus, protein kinase C has a complex role in the autoregulation of the EGF receptor, affecting its activity, cellular distribution and now, synthesis rate. A number of other receptor systems including platelet-derived growth factor[50], thyrotropin releasing hormone[51], and vasopressin[52] have been shown to stimulate phosphoinositide turnover, thereby presumably utilizing protein kinase C as a signal transduction mechanism, raising the potential for cross-regulatory pathways between the EGF receptor and other receptors. The synergism of growth factors[53-55] and the ability of a variety of hormones to stimulate proliferation may be a result of their ability to activate protein kinase C which in turn stimulates EGF receptor synthesis, sensitizing the cell to this growth factor.

Regulation of Transforming Growth Factor-alpha Expression

Transforming growth factor-alpha (TGF-α) was first discovered in the conditioned medium of retrovirally transformed fibroblasts[56,57] and subsequently in spontaneous tumors[58]. Its primary sequence was determined[59] and its cDNA cloned subsequently[60,61]. The growth factor is only 20% homologous with mouse EGF although the positions of the disulfide bounds is exactly conserved between EGF and TGF-α. Despite the relatively low primary sequence homology between EGF and TGF-α, both compete for binding to the same EGF receptor[62,63]. Blocking monoclonal antibodies that we developed to the EGF receptor, block both the mitogenic activity of TGF-α and EGF, suggesting both growth factors use the same receptor[64].

In investigtions designed to determine the effect of EGF on primary bovine pituitary cells in culture, we observed that these cells proliferated in the absence of added growth factors[65]. We postulated that the cells could grow because they could make their own growth factors. Indeed, the conditioned medium from these cells was mitogenic on fibroblasts and adrenal cells. Furthermore, the medium contained an activity that competed with EGF for binding to the EGF receptor[66] and an activity that stimulated soft agar colony formation by NRK cells[66,64]. Both the adrenal[67] and EGF-like growth factors were purified from vast quantities of conditioned medium. The EGF-like factor was purified to homogeneity[64,68] and compared to TGF-α purified from the conditioned medium of a human melanoma cell line. Using amino acid analysis and HPLC peptide mapping, we showed that the pituitary derived material was similar to the melanoma TGF-α. To further characterize the growth factor, we synthesized a synthetic peptide representing the 17-amino acid carboxyl terminus of rat TGF-α, conjugated this peptide to limpet hemocyanin and immunized mice. Hybridomas were made and the monoclonal antibodies screened against iodinated synthetic peptide. Several positive clones were detected but in secondary screens against purified human melanoma derived TGF-α, only two monoclonal antibodies could be shown to recognize the authentic growth factor. One of these antibodies, MF9, showed high affinity and was used in further studies. MF9 recognized human and rat TGF-α in a solid phase radioimmune assay developed in our laboratory but did not crossreact with high concentrations of EGF[68]. Furthermore, MF9 recognized the TGF-α secreted by the cultured pituitary cells strongly suggesting identity of the pituitary EGF-like growth factor as TGF-α. The purified growth factor was subjected to N-terminal sequence analysis, establishing its identity[69]. This work on the pituitary demonstrated that normal adult cells can express TGF-α.

Early in our studies, we noticed that the production rate of growth factor by subcultured pituitary cells was 10-fold greater than the production by primary cultures[65,66]. We were concerned that artifacts related to culture conditions resulted in the expression of the growth factor. Using monoclonal antibody, MF9, we asked whether normal pituitary contains TGF-α immunoreactivity. The indirect immunoperoxidase technique with anti- TGF-α monoclonal antibody MF9 to localize TGF-α immunoreactivity was performed on tissue sections of bovine pituitary glands. Twelve male bovine calves were studied and identical results obtained with each pituitary gland. TGF-α immunoreactivity was localized in many adenohypophysial cells in a scattered distribution but concentrated predominately in the posterior and antero-lateral regions; staining was conspicuously absent in the anterior median wedge. There was no TGF-α immunoreactivity in fibroblasts, nor in cells in the intermediate and posterior lobes (Fig. 13). Positivity in the anterior pituitary cells was diffuse in the cytoplasm and no nuclear staining was identified. Occasional cells showed weaker staining, scattered focally in the cytoplasm. The immunostaining was completely inhibited if the MF9 antibody was preincubated with saturating concentrations of the synthetic peptide to which it was raised. Furthermore, a human specific anti-EGF receptor monoclonal IgG_1 used at similar concentrations to MF9, did not stain the bovine tissue examined.

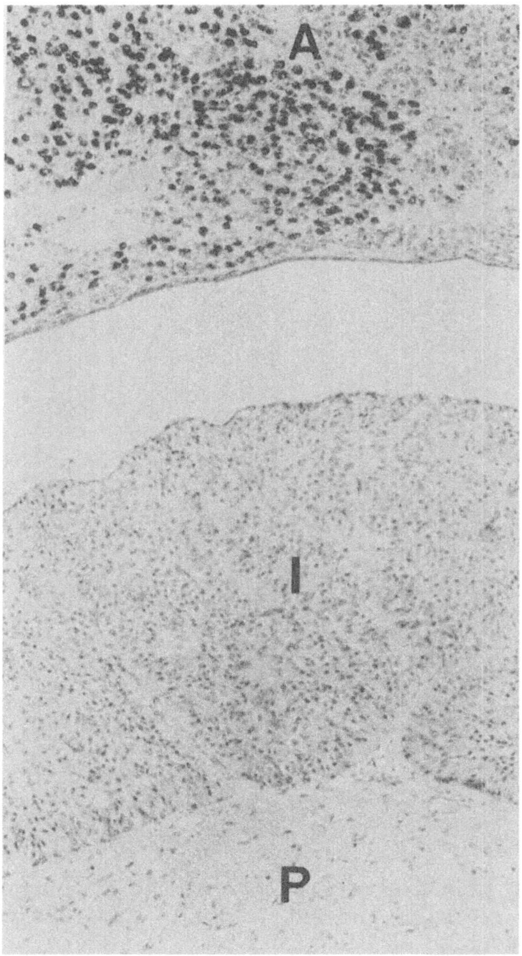

Fig. 13. Cells in the anterior lobe (A) of the bovine pituitary contain TGF-α immunoreactivity. The intermediate (I) and posterior (P) lobes do not contain detectable immunoreactivity. The immunoperoxidase technique was used for TGF-α with hematoxylin counterstain. Original magnification, x 115.

To determine whether TGF-α immunoreactivity was localized in specific hormone-secreting cell types, consecutive mirror sections of bovine pituitaries were examined using antibody MF9 and antibodies against each anterior pituitary hormone. The primary antisera were directed against ovine prolactin, human adrenocorticotropin (1-39), and the β-chains of human thyrotrophin, follicle-stimulating hormone and luteinizing hormone. Examination of mirror sections showed that the TGF-α immunoreactivity did not localize in cells containing adrenocorticotropin, thyrotropin, follicle-stimulating hormone and luteinizing hormone. A significant proportion of the cells containing

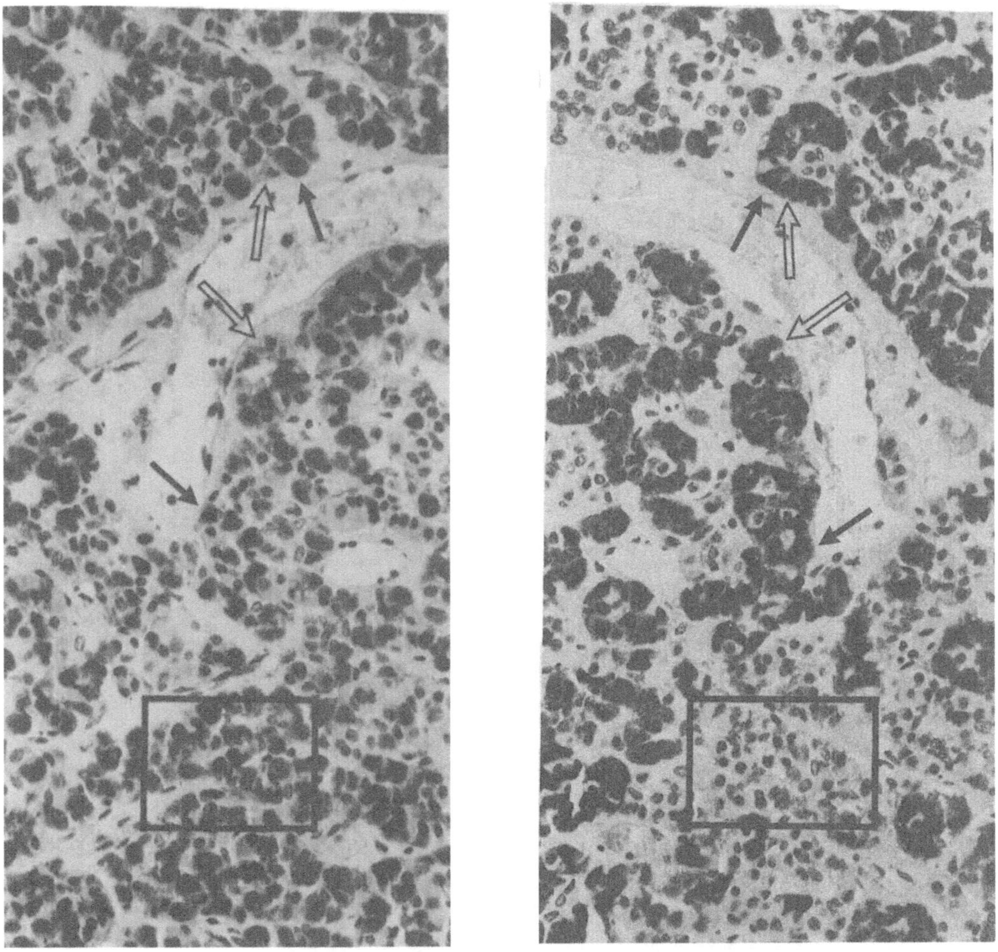

Fig. 14. Mirror sections of the anterior lobe of the bovine pituitary stained for TGF-α (left) and PRL (right). Numerous cells of the posterolateral portion of the gland contain TGF-α immunoreactivity diffusely in the cytoplasm. Many of the cells containing TGF-α immunoreactivity also contain PRL (arrows); some PRL-containing cells do not contain detectable TGF-α immunoreactivity (arrow outline), and many cells containing TGF-α immunoreactivity do not contain PRL (box). The immunoperoxidase technique for TGF-α (left) and PRL (right) with hematoxylin counterstain. Magnification, x285.

immunoreactive prolactin also contained TGF-α immunoreactivity, however the number of cells containing TGF-α immunoreactivity greatly exceeded the number of prolactin-containing cells (Fig. 14). Similar observations were made in the rat and human pituitary. In these species, for which we also had anti-growth hormone serum, we also showed colocalization of TGF-α with growth hormone. The TGF-α did not colocalize with any of the other pituitary hormones.

The colocalization of TGF-α with prolactin in the pituitary suggested that regulation of prolactin and TGF-α expression might be similar. Lactotrophs have been shown to have EGF receptors[70], implying an autocrine mode of action of TGF-α. Indeed we have now shown that TGF-α secretion by bovine pituitary cells into the culture medium is stimulated by TPA (Fig. 15).

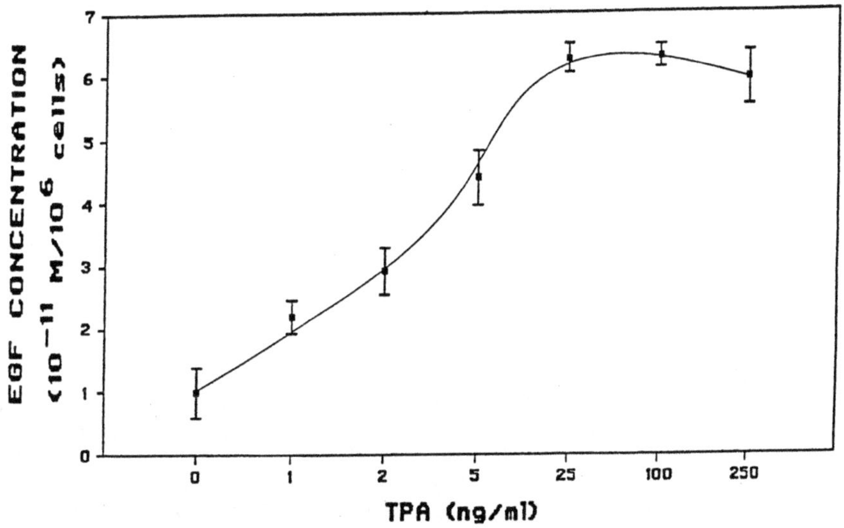

Fig. 15. Effect of TPA on the level of TGF-α secreted into the conditioned medium of bovine pituitary cells. Bovine pituitary cells in culture were treated with various doses of TPA for 24 hours. After the incubation period, the conditioned medium was collected and assayed for the presence of EGF displacing activity using an EGF radioreceptor assay. The EGF concentration is expressed per 1 X 10^6 cells to correct for changes in cell number during the incubation period. Each point represents the mean result of triplicate determinations +/- 1 S.D.

When we examined the MDA468 cells which had just recently been shown in out laboratory to contain TGF-α mRNA in addition to EGF receptors, we found both TPA and EGF could stimulate TGF-α mRNA accumulation (Fig 16). Since EGF can stimulate TGF-α expression, the TGF-α produced by the MDA468 cells could presumably stimulate its own expression (EGF and TGF-α bind to the same receptor).

Furthermore, EGF and therefore TGF-α can stimulate EGF receptor expression. Thus, when a stimulus such as TPA or a hormone that activates protein kinase C is received by the cell, the cell increases its expression of both growth factor and receptor which in turn stimulate, by a positive feedback autocrine loop, the continued expression of receptor and ligand. We believe the 10-fold permanent induction of TGF-α secretion in the pituitary cells by subculturing results from such a positive feedback loop. Indeed, if the pituitary cultures are washed frequently to prevent

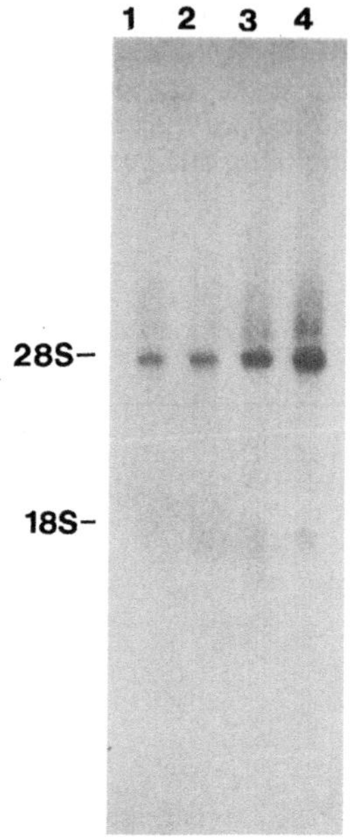

Fig. 16. Effects of EGF and TPA on TGF-α mRNA levels in MDA468 cell. MDA468 cells were left untreated (lane 1) or treated for 6 hours with 60 ng/ml 4β- phorbol (an inactive TPA analogue)(lane 2), 10^{-9}M EGF (lane 3), or 100ng/ml TPA (lane 4). After the treatments, the mRNA was extracted and processed as in Figure 8. Hybridization was with a 670 base pair cDNA probe for human TGF-α (kindly provided by Dr. R. Derynck). Results from the autoradiogram are shown.

the accumulation of TGF-α in the medium, the TGF-α mRNA levels fall. Of note, the pituitary cells are not transformed and cease growing at confluence. Cancer of the pituitary is exceedingly rare and yet we have demonstrated a positive feedback loop involving a growth factor and its receptor. Clearly, normal tissues produce inhibitors of this positive feedback loop capable of imposing negative growth control on the system. Nevertheless, we hypothesize that hormones such as TRH can ultimately stimulate the proliferation of target cells by favouring a growth stimulatory environment through up-regulation of growth factor and its receptor. This concept may explain the clinical observations such as the hyperplasia seen in the pituitary resulting from the elaboration of releasing factor by hypothalamic lesions[71,72] or the shifts in relative populations of pituitary cells in conditions such as hypothyroidism[73] and pregnancy[74].

REFERENCES

1. Daughaday, W.H., Hall, K., Raben, M.S., Salmon Jr, W.D., Van den Brande, J.L., Van Wyk, J.J., Somatomedin: proposed designation for sulphation factor, Nature 235:107-110 (1972)

2. Klapper, D.G., Svoboda, M.E., Van Wyk, J.J., Sequence analysis of somatomedin-C: Confirmation of identity with insulin-like growth factor I. Endocrinology 112:2215-2223 (1983)

3. Salomon, D.S., Perroteau, I., Growth factors in cancer and their relationship to oncogenes, Cancer Investigation 4:43-60 (1986)

4. Goustin, A.S., Leof, E.B., Shipley, G.D., Moses, H.L., Growth factors in cancer. Cancer Research 46:1015-1029

5. Weinstein, B.I., Growth factors, oncogenes and multistage carcinogenesis. J. Cell. Biochem. 33:213-224 (1987)

6. Hunter, T., A thousand and one protein kinases, Cell 50:821-822 (1987)

7. Bravo, R., Burckhardt, J., Curran, T., and Muller, R. Stimulation and inhibition of growth by EGF in different A431 cell clones is accompanied by the rapid induction of c-fos and c-myc proto-oncogenes, EMBO J. 4:1193-1197 (1985).

8. Murdoch, G.H., Waterman, M., Evans, R.M., and Rosenfeld, M.G. Molecular mechanisms of phorbol ester, thyrotropin-releasing hormone, and growth factor stimulation of prolactin gene transcription, J. Biol. Chem. 260:11852-11858 (1985).

9. Elder, P.K., Schmidt, L.J., Ono, T., and Getz, M.J. Specific stimulation of actin gene transcription by epidermal growth factor and cycloheximide, Proc. Natl. Acad. Sci. U.S.A. 81:7476-7480 (1984).

10. Kelly, K., Cochran, B.H., Stiles, C.D., and Leder, P. Cell-specific regulation of the c-myc gene by lymphocyte mitogens and platelet-derived growth factor, Cell 35:603-610 (1983).

11. Greenberg, M.E., and Ziff, E.B. Stimulation of 3T3 cells induces transcription of the c-fos proto-oncogene, Nature 311:433-438 (1984).

12. Muller, R., Bravo, R., Burckhardt, J., and Curran, T. Induction of c-fos gene and protein by growth factors precedes activation of c-myc, Nature 312:716-720 (1984).

13. Kudlow, J.E., Khosravi, M.J., Kobrin, M.S., Mak, W.W. Inability of anti-epidermal growth factor receptor monoclonal antibody to block "autocrine" growth stimulation in transforming growth factor-secreting melanoma cells, J. Biol. Chem. 259:11895-11900 (1984)

14. Kudlow, J.E., Cheung, C.-Y. M., and Bjorge, J.D. Epidermal growth factor stimulates the synthesis of its own receptor in a human breast cancer cell line, J. Biol. Chem. 261:4134-4138 (1986).

15. Clark, A.J.L., Ishii, S., Richert, N., Merlino, G.T., and Pastan, I. Epidermal growth factor regulates the expression of its own receptor, Proc. Natl. Acad. Sci. U.S.A. 82:8374-8378 (1985).

16. Earp, H.S., Austin, K.S., Blaisdell, J., Rubin, R.A., Nelson, K.G., Lee, L.W., and Grisham, J.W. Epidermal growth factor (EGF) stimulates EGF receptor synthesis, J. Biol. Chem. 261:4777-4780 (1986).

17. Takai, Y., Kikkawa, U., Kaibuchi, K., and Nishizuka, Y. Membrane phospholipid metabolism and signal transduction for protein phosphorylation, Advances in Cyclic Nucl. Phosph. Res. 18:119-158 (1984).

18. Berridge, M.J. Inositol trisphosphate and diacylglycerol as second messengers, Biochem. J. 220:345-360 (1984).

19. Sawyer, S.T., and Cohen, S. Enhancement of calcium uptake and phosphatidylinositol turnover by epidermal growth factor in A-431 cells, Biochemistry 20:6280-6286 (1981).

20. Smith, K.B., Losonczy, I., Sahai, A., Panneerselvam, M., Fehnel, P., and Salomon, D.S. Effect of 12-O-tetradecanoylphorbol-13-acetate (TPA) on the growth inhibitory and increased phosphatidylinositol (PI) responses induced by epidermal growth factor (EGF) in A431 cell, J. Cell. Physiol. 117:91-100 (1983).

21. Macara, I.G. Activation of $^{45}Ca^{2+}$ influx and $^{22}Na^{+}/H^{+}$ exchange by epidermal growth factor and vanadate in A431 cells is independent of phosphatidylinositol turnover and is inhibited by phorbol ester and diacylglycerol, J. Biol. Chem. 261:9321-9327 (1986).

22. Wahl, M.I., Sweatt, J.D., Carpenter, G. Epidermal growth factor stimulates inositol triphosphate formation in cells which overexpress the EGF receptor, Biochem. Biophys. Res. Comm. 142:688-695 (1987)

23. Taylor, D., Uhing, R.J., Blackmore, P.F., Prpic, V., Exton, J.H. Insulin and epidermal growth factor do not affect phosphoinositide metabolism in rat liver plasma membranes and hepatocytes, J. Biol. Chem. 260:2011-2014 (1985)

24. Besterman, J.M., Watson, S.P., Cuatrecasas, P. Lack of association of epidermal growth factor-, insulin-, and serum-induced mitogenesis with stimulation of phosphoinositide degradation in Balb/c 3T3 fibroblasts, J. Biol. Chem. 261:723-727 (1986)

25. Castagna, M., Takai, Y., Kaibuchi, K., Sano, K., Kikkawa, U., and Nishizuka, Y. Direct activation of calcium-activated, phospholipid-dependent protein kinase by tumor-promoting phorbol esters, J. Biol. Chem. 257:7847-7851 (1982).

26. Yamanishi, J., Takai, Y., Kaibuchi, K., Sano, K., Castagna, M., and Nishizuka, Y. Synergistic functions of phorbol ester and calcium in serotonin release from human platelets, Biochem. Biophys. Res. Commun. 112:778-786 (1983).

27. Niedel, J.E., Kuhn, L.J., and Vandenbark, G.R. Phorbol diester receptor copurifies with protein kinase C, Proc. Natl. Acad. Sci. U.S.A. 80:36-40 (1983).

28. Nishizuka, Y. Studies and perspectives of protein kinase C, Science 233:305-312 (1986).

29. Shoyab, M., DeLarco, J.E., and Todaro, G.J. Biologically active phorbol esters specifically alter affinity of epidermal growth factor membrane receptors, Nature 279:387-391 (1979).

30. Davis, R.J., and Czech, M.P. Tumor-promoting phorbol diesters mediate phosphorylation of the epidermal growth factor receptor, J. Biol. Chem. 259:8545-8549 (1984).

31. Friedman, B., Frackelton, Jr., A.R., Ross, A.H., Connors, J.M., Fujiki, H., Sugimura, T., and Rosner, M.R. Tumor promoters block tyrosine-specific phosphorylation of the epidermal growth factor receptor, Proc. Natl. Acad. Sci. U.S.A. 81:3034-3038 (1984).

32. Cochet, C., Gill, G.N., Meisenhelder, J., Cooper, J.A., and Hunter, T. C-kinase phosphorylates the epidermal growth factor receptor and reduces its epidermal growth factor-stimulated tyrosine protein kinase activity, J. Biol. Chem. 259:2553-2558 (1984).

33. Downward, J., Waterfield, M.D., and Parker, P.J. Autophosphorylation and protein kinase C phosphorylation of the epidermal growth factor receptor, J. Biol. Chem. 260:14538-14546 (1985).

34. Lee, L.S., and Weinstein, I.B. Tumor-promoting phorbol esters inhibit binding of epidermal growth factor to cellular receptors, Science 202:313-315 (1978).

35. Salomon, D.S. Inhibition of epidermal growth factor binding to mouse embryonal carcinoma cells by phorbol esters mediated by specific phorbol ester receptors, J. Biol. Chem. 256:7958-7966 (1981).

36. King, A.C., and Cuatrecasas, P. Resolution of high and low affinity epidermal growth factor receptors, J. Biol. Chem. 257:3053-3060 (1982).

37. Beguinot, L., Hanover, J.A., Ito, S., Richert, N.D., Willingham, M.C., and Pastan, I. Phorbol esters induce transient internalization without degradation of unoccupied epidermal growth factor receptors, Proc. Natl. Acad. Sci. U.S.A. 82:2774-2778 (1985).

38. Davis, R.J., and Czech, M.P. Tumor-promoting phorbol diesters cause the phosphorylation of epidermal growth factor receptors in normal human fibroblasts at threonine-654, Proc. Natl. Acad. Sci. U.S.A. 82:1974-1978 (1985).

39. Hunter, T., Ling, N., Cooper, J.A. Protein kinase C phosphorylation of the EGF receptor at a threonine residue close to the cytoplasmic face of the plasma membrane, Nature 311:480-483 (1984).

40. Phillips, M.A., and Jaken, S. Specific desensitization to tumor-promoting phorbol esters in mouse pituitary cells, J. Biol. Chem. 258:2875-2881 (1983).

41. Collins, M.K.L., and Rozengurt, E. Homologous and heterologous mitogenic desensitization of Swiss 3T3 cells to phorbol esters and vasopressin: Role of receptor and postreceptor steps, J. Cell. Physiol. 118:133-142 (1984).

42. Rodriguez-Pena, A., and Rozengurt, E. Disappearance of Ca^{2+} -sensitive, phospholipid-dependent protein kinase activity in phorbol ester-treated 3T3 cells, Biochem. Biophys. Res. Commun. 120:1053-1059 (1984).

43. Lee, L.S., and Weinstein, I.B. Mechanism of tumor promoter inhibition of cellular binding of epidermal growth factor, Proc. Natl. Acad. Sci. U.S.A. 76:5168-5172 (1979).

44. Collins, M.K.L., and Rozengurt, E. Binding of phorbol esters to high-affinity sites on murine fibroblastic cells elicits a mitogenic response, J. Cell. Physiol. 112:42-50 (1982).

45. Vara, F., and Rozengurt, E. Stimulation of Na^+/H^+ antiport activity by epidermal growth factor and insulin occurs without activation of protein kinase C, Biochem. Biophys. Res. Commun. 130:646-653 (1985).

46. Grinstein, S., Mack, E., and Mills, G.B. Osmotic activation of the Na^+/H^+ antiport in protein kinase C-depleted lymphocytes, Biochem. Biophys. Res. Commun. 134:8-13 (1986).

47. Korneluk, R.G., Mahuran, D.J., Neote, K., Klavins, M.H., O'Dowd, B.F., Tropak, M., Willard, H.F., Anderson, M-J., Lowden, J.A., Gravel, R.A. Isolation of cDNA clones coding for the alpha-subunit of human beta-hexosaminidase: extensive homology between the alpha- and beta-subunits and studies of Tay-Sachs disease, J. Biol. Chem. 261:8407-8413 (1986)

48. Krupp, M.N., Connolly, D.T., and Lane, M.D. Synthesis, turnover, and down-regulation of epidermal growth factor receptors in human A431 epidermoid carcinoma cells and skin fibroblasts, J. Biol. Chem. 257:11489-11496 (1982).

49. Filmus, J., Pollak, M.N., Cailleau, R., and Buick, R.N. MDA-468, a human breast cancer cell line with a high number of epidermal growth factor (EGF) receptors, has an amplified EGF receptor gene and is growth inhibited by EGF, Biochem. Biophys. Res. Commun. 128:898-905 (1985).

50. Davis, R.J., and Czech, M.P. Platelet-derived growth factor mimics phorbol diester action on epidermal growth factor receptor phosphorylation at threonine-654, Proc. Natl. Acad. Sci. U.S.A. 82:4080-4084 (1985).

51. Fearon, C.W., and Tashjian, Jr., A.H. Thyrotropin-releasing hormone induces redistribution of protein kinase C in GH_4C_1 rat pituitary cells, J. Biol. Chem. 260:8366-8371 (1985).

52. Thomas, A.P., Marks, J.S., Coll, K.E., and Williamson, J.R. Quantitation and early kinetics of inositol lipid changes induced by vasopressin in isolated and cultured hepatocytes, J. Biol. Chem. 258:5716-5725 (1983).

53. Pledger, W.J., Hart, C.A., and Wharton, W.R. Mammalian cell proliferation is regulated by the synergisitic actions of multiple growth factors, Adv. Exp. Med. Biol. 138:287-300 (1982).

54. Rozengurt, E. Growth factors, cell proliferation and cancer: An overview, Mol. Biol. Med. 1:169-181 (1983).

55. Assoian, R.K. Biphasic effects of type β transforming growth factor on epidermal growth factor receptors in NRK fibroblasts, J. Biol. Chem. 260:9613-9617 (1985).

56. DeLarco, J.E., Todaro, G.J., Growth factors from murine sarcoma virus-transformed cells. Proc. Natl. Acad. Sci. U.S.A. 75:4001-4005 (1978)

57. Todaro, G.J., Fryling, C., DeLarco, J.E., Transforming growth factors produced by certain human tumor cells: polypeptides that interact with epidermal growth factor receptors. Proc. Natl. Acad. Sci. U.S.A. 77:5258-5262 (1980)

58. Massague, J., Epidermal growth factor-like transforming growth factor I. isolation, chemical characterization, and potentiation by other transforming factors from feline sarcoma virus-transformed rat cells. J. Biol. Chem. 258:13606-13613 (1983)

59. Marquardt, H., Hunkapiller, M.W., Hood, L.E., Twardzik, D.R., DeLarco, J.E., Stephenson, J.R., Todaro, G.J., Transforming growth factors produced by retrovirus-transformed rodent fibroblasts and human melanoma cells: amino acid sequence homology with epidermal growth factor. Proc. Natl. Acad. Sci. U.S.A. 80:4684-4688 (1983)

60. Derynck R., Roberts A.B., Winkler M.E., Chen E.Y., Goeddel D.V., Human transforming growth factor-α: precursor structure and expression in E. coli. Cell 38:287 (1984)

61. Lee D.C., Rose T.M., Webb N.R., Todaro G.J. Cloning and sequence analysis of a cDNA for rat transforming growth factor-α. Nature 313:489-492 (1985)

62. Massague, J., Epidermal growth factor-like transforming growth factor II. J. Biol. Chem. 258:13614-13620 (1983)

63. Carpenter, G., Stoschek, C.M., Preston, Y.A., DeLarco, J.E., Antibodies to the epidermal growth factor receptor block the biological activities of sarcoma growth factor. Proc. Natl. Acad. Sci. U.S.A. 80:5627-5630 (1983)

64. Samsoondar, J, Kobrin, M.S., Kudlow, J.E. Alpha-transforming growth factor secreted by untransformed bovine anterior pituitary cells in culture: I. Purification from conditioned medium, J. Biol. Chem. 261:14408-14413 (1986)

65. Kudlow, J.E., Gerrie, B.M., Production of growth factor activity by cultured bovine calf anterior pituitary cells. Endocrinology 113:104-112 (1983)

66. Kudlow, J.E., Kobrin, M.S., Secretion of epidermal growth factor-like mitogens by cultured cells from bovine anterior pituitary glands. Endocrinology 115:911-921 (1984)

67. Samsoondar, J., Kudlow, J.E., Partial purification of an adrenal growth factor produced by normal bovine anterior pituitary cells in culture. Endocrinology 120:929-935 (1987)

68. Kobrin M.S., Samsoondar J., Kudlow J.E. Alpha-Transforming growth factor secreted by untransformed bovine anterior pituitary cells in culture. II. Identification using a sequence-specific monoclonal antibody. J Biol Chem 261:14414 (1986)

69. Kobrin, M.S., Asa, S.L., Samsoondar, J., Kudlow, J.E. Alpha-transforming growth factor in the bovine anterior pituitary gland: secretion by dispersed cells and immunohistochemical localization. Endocrinology 121:1412-1416 (1987)

70. Chabot J-G., Walker P., Pelletier G. Distribution of epidermal growth factor binding sites in the adult rat anterior pituitary gland. Peptides 7:45-50 (1986)

71. Asa S.L., Kovacs K., Tindall G.T., Barrow D.L., Horvath E., Vecsei P. Cushing's disease associated with an intracellar gangliocytoma producing corticotropin-releasing factor. Ann. Internal Med. 101:789-793 (1984)

72. Asa S.L., Scheithauer B.W., Bilbao J.M., Horvath E., Ryan N., Kovacs K., Randall R.V., Laws E.R., Singer W., Linfoot J.A., Thorner M.O., Vale W.W. A case for hypothalamic acromegaly: a clinicopathological study of six patients with hypothalamic gangliocytoma producing growth hormone-releasing factor. J. Clin. Endocrinol. Metab. 58:796-803 (1984)

73. Scheithauer B.W., Kovacs K., Randall R.V., Ryan N. Pituitary gland in hypothyroidism. Histologic and immunocytologic study. Arch. Pathol. Lab. Med. 109:499-507 (1985)

74. Asa S.C., Penz G., Kovacs K., Ezrin C. Prolactin cells in the human pituitary. A quantitative immunocytochemical analysis. Arch. Pathol. Lab. Med. 106:360-369 (1982)

STRUCTURAL AND FUNCTIONAL STUDIES OF PROTEIN KINASE C

Gerard M. Housey,[1,2] Mark D. Johnson,[1] W.L. Wendy Hsiao,[1]
Catherine A. O'Brian,[1], and I. Bernard Weinstein[1,3]

[1]Comprehensive Cancer Center and Institute of Cancer Research
[2]Department of Genetics and Development
[3]Department of Medicine
College of Physicians and Surgeons
Columbia University
New York, New York 10032

ABSTRACT

It has now become clear that protein kinase C (PKC) plays a fundamental role in cellular growth control in higher eukaryotes. In addition, since PKC has also been shown to be the high-affinity intracellular receptor for several classes of tumor promoters, the study of PKC has also become a central focus of current work in cancer research. We have isolated cDNA clones encoding several forms of this enzyme, and we have used these clones to begin to study in detail the role of PKC in growth control and tumor promotion. The complete primary structure of one of these forms, designated $PKC_{\beta 1}$, exhibits structural and functional characteristics which are shared among all of the currently identified forms of PKC. These include an amino terminal cysteine-rich domain which mediates Ca^{2+} and phospholipid binding, tumor promoter binding, and membrane association, and a carboxy terminal catalytic domain which possesses serine/threonine protein kinase activity. To further characterize the function of PKC, we have generated a series of rat fibroblast cell lines which stably overexpress a full-length cDNA encoding the β1 form of this enzyme. These cell lines contain a 20- to 53-fold increase in PKC activity, and also have an increase in high affinity phorbol ester receptors, relative to control cells. They also exhibit dramatically enhanced morphologic changes in response to treatment with the tumor promoter 12-0-tetradecanoyl phorbol-13-acetate (TPA). These cell lines grow to a higher saturation density in monolayer culture and, when maintained at post-confluence, develop small, dense foci. In contrast to the control cells, which display complete anchorage dependence, the cell lines that overproduce PKC form small colonies in soft agar in the absence of TPA, and larger colonies in the presence of TPA. Thus, the mere overproduction of a single form of PKC is sufficient to confer anchorage independent growth and other growth abnormalities in rat fibroblasts. Taken together, these results provide direct evidence that PKC plays a critical role in normal cellular growth control and that it mediates several, and perhaps all of the cellular effects of the phorbol ester tumor promoters.

INTRODUCTION

Protein kinase C (PKC) is a Ca^{2+}- and phospholipid-dependent serine/threonine protein kinase of fundamental importance in cellular growth control. PKC is activated endogenously by a wide variety of growth factors, hormones, and neurotransmitters, and has been shown to be a high affinity receptor for the phorbol ester tumor promoters as well as other agents which possess tumor promoting activity (for reviews see Nishizuka 1986; 1984; Ashendel, 1984). Once activated, PKC has been shown to phosphorylate several intracellular protein substrates, including the epidermal growth factor (EGF) receptor (Hunter et al., 1984), $pp60^{src}$ (Gould et al., 1985), the insulin receptor (Bollag et al., 1986), ras p21 (Jeng et al., 1987), and many others (Nishizuka, 1986). However, the functional role of the phosphorylations of these and other substrates by PKC is presently unknown.

Several laboratories, including our own, have recently isolated cDNA clones encoding distinct forms of PKC, thus demonstrating that PKC is encoded by a multigene family (Ono et al., 1986, Knopf et al., 1986, Parker et al., 1986; Coussens et al., 1986; Makowske et al., 1986; Ohno et al., 1987; Housey et al., 1987). The multiple forms of PKC exhibit considerable tissue specificity (Knopf, et. al., 1986; Brandt et al., 1987; Ohno, et al, 1987; Housey, et. al., 1987) which suggests that there may be subtle differences in the function(s) of each of the distinct forms. However, all of the cDNA clones which have been isolated thus far that encode distinct forms of PKC share at least 65% overall deduced homology at the amino acid level, and transient expression experiments with some of these cDNA clones have shown that they encode serine/threonine protein kinase activities which bind to, or are activated by, the phorbol ester tumor promoters (Knopf, et. al., 1986, Ono et. al., 1987).

In order to further elucidate the role of PKC in growth control and tumor promotion, it would be desirable to generate cell lines which stably produce elevated levels of each of the various forms of this enzyme. Such cell lines would provide model systems for studying the cellular events which occur subsequent to the activation of specific forms of PKC by tumor promoters and other agents. In addition, they would be useful to study the interactions between chemical carcinogens and tumor promoters with respect to the mechanism(s) underlying the multistage carcinogenic process (Weinstein et. al., 1987).

The present studies were undertaken, therefore, in order to generate such cell lines and to characterize their growth properties in culture. In these initial studies we have employed the Rat 6 embryo fibroblast cell line, a variant of the rat embryo fibroblast cell line established by Freeman et al., (1972), which was isolated in our laboratory (Hsiao et al., 1986). This cell line has an unusually flat morphology, even when maintained in culture at post-confluence for extended periods of time, displays anchorage dependent growth and, thus far, has not undergone spontaneous transformation. It was also ideal for these studies since it has a very low level of endogenous PKC activity and a low level of high affinity receptors for phorbol esters. Detailed aspects of the experimental procedures employed in these studies have been published elsewhere (Housey, et al., 1988).

RESULTS

Primary Structure of the β1 form of PKC

Figure 1 shows the full-length deduced amino acid sequence of $PKC_{\beta 1}$, of one of several forms of PKC which we have previously isolated (cDNA clone RP58), and whose sequence we have previously reported (Ono et al., 1986; Housey et al., 1987, 1988). The various PKC family members isolated thus far all exhibit a common primary structural organization consisting of two major domains. The amino terminal domain consists of amino acid residues 1-330 and appears to mediate the binding of all of the known co-factors and activators of PKC including phorbol esters, diacylglycerol, Ca^{2+} and phospholipid. This domain also exhibits a pattern of conserved cysteine residues (the R1 region in Figure 1) beginning at C_{50} and spaced as follows : $C-X_2-C-X_{13}-C-X_2-C-X_7-C-X_7$, where X designates any amino acid. This first repeated cysteine region is followed, 28 amino acids later, by an identical pattern of cysteine residues beginning at C_{115}. Similar cysteine repeat regions have been found both in proteins which chelate divalent cations and in proteins which possess DNA-binding activity (Berg, 1986). Thus, either of these properties may be shared by the amino terminal domains of the PKC family members as well.

The carboxyl terminal domain of PKC, from residues 331-671, possesses its serine/threonine protein kinase activity. Interestingly, two consensus ATP-binding sites are present in this domain, as shown in Figure 1. The first of these, beginning at G_{349}, is most likely to be the primary ATP-binding site since it is conserved in all of the PKC family members and in virtually all of the protein kinases as well (for review see Hunter and Cooper, 1985). However, we cannot rule out the possibility that the second consensus ATP binding site, beginning at G_{585}, also functions in some manner to regulate PKC activity. Site-directed mutagenesis studies of this domain will facilitate an understanding of the relative importance of each of these regions.

We used the $PKC_{\beta 1}$ cDNA clone for the present studies for the following reasons. With the exception of the brain, where its expression is very high, $PKC_{\beta 1}$ is expressed at very low levels in most tissues, and its expression is virtually undetectable in Rat 6 fibroblasts (see below). Thus, we reasoned that using this form would maximize the phenotypic differences observed between control cells and cells overexpressing the introduced form of PKC. The $PKC_{\beta 1}$ form is also of particular interest because within the PKC gene family its deduced carboxy terminal domain displays the highest overall homology to the catalytic subunit of the cyclic AMP-dependent protein kinase (PKA_c) and the cyclic GMP-dependent protein kinase (PKG) (Housey et al., 1987). The latter observation suggests that PKA_c, PKG, and the β1 form of PKC may share a common serine/threonine protein kinase progenitor, and that the additional PKC genes may have been derived through evolutionary divergence from the β1 form. Several of the most highly conserved amino acid residues found in these and other serine/threonine protein kinases are shown in Figure 1.

Stable Overproduction of $PKC_{\beta 1}$ in Rat-6 Cell Lines

The overall strategy used to generate cell lines stably overproducing PKC has been described in detail elsewhere (Housey et al., 1988). Briefly, a full-length $PKC_{\beta 1}$ cDNA sequence was inserted into the Eco RI site of of the cDNA expression vector pMV7 (Kirschmeier et al., 1988), using standard techniques, and the resulting plasmid, designated pMV7-$PKC_{\beta 1}$, was transfected onto Ψ-2 cells (Mann et al., 1983)

```
                            .                   .         <----------
1-60     MADPAAGPPPSEGEESTVRFARKGPLRQKNVHEVKNHKFTARFFKQPTFCSHCTDFIWGF
                                                          |--|-------
                                                            2

         -------------------.--------R1---------.--------------------
61-120   GKQGFQCQVCCFVVHKRCHEFVTFSCPGADKGPASDDPRSKHKFKIHTYSSPTFCDHCGS
         ------|--|-------|-------|----------------------------|--|--
          13     2    7       7                 28                2

         ------------------.---------->         .
121-180  LLYGLIHQGMKCDTCMMNVHKRCVMNVPSLCGTDHTERRGRIYIQAHIDREVLIVVVRDA
         -----------|--|-------|-------|
             13       2    7      7

                            .                   .
181-240  KNLVPMDPNGLSDPYVKLKLIPDPKSESKQKTKTIKCSLNPEWNETFRFQLKESDKDRRL
                           |--S2-|

                            .                   .
241-300  SVEIWDWDLTSRNDFMGSLSFGISELQKAGVDGWFKLLSQEEGEYFNVPVPPEESEGNEE

                            .                   .
301-360  LRQKFERAKIGQGTKAPEEKTANTISKFDNNGNRDRMKLTDFNFLMVLGKGSFGKVMLSE
                                                         + +  +

                            .                   .
361-420  RKGTDELYAVKILKKDVVIQDDDVECTMVEKRVLALPGKPPFLTQLHSCFQTMDRLYFVM
                   +

                            .                   .
421-480  EYVNGGDLMYHIQQVGRFKEPHAVFYAAEIAIGLFFLQSKGIIYRDLKLDNVMLDSEGHI
                                                     ***

                            .                   .
481-540  KIADFGMCKENIWDGVTTKTFCGTPDYIAPEIIAYQPYGKSVDWWAFGVLLYEMLAGQAP
          ***                  ***   ***           |--------P2-------|

                            .                   .
541-600  FEGEDEDELFQSIMEHNVAYPKSMSKEAVAICKGLMTKHPGKRLGCGPEGERDIKEHAFF
                                                     + +  +

                            .                   .
601-660  RYIDWEKLERKEIQPPYKPKARDKRDTSNFDKEFTRQPVELTPTDKLFIMNLDQNEFAGF
               +                  ******

661-671  SYTNPEFVINV
```

Figure 1. Deduced Amino Acid Sequence of $PKC_{\beta 1}$

The deduced amino acid sequence begins with the first in-frame methionine codon and encodes a 671 amino acid protein with a predicted molecular weight of 76.8 kd. The arrow (R1) above the amino acid sequence designates a repeating pattern of conserved cysteine residues. The spacing between conserved cysteines is indicated. The asterisks (*) designate conserved amino acid residues found in other serine/threonine protein kinases. Crosses (+) designate consensus ATP-binding site sequences. S2 and P2 denote peptides which were determined by automated Edman degradation. For additional explanations see the text.

by the calcium phosphate co-precipitation method (Graham and van der Eb, 1973; as modified by Wigler et al., 1977). After 48 hours the medium was harvested, filtered, and used to infect recipient subconfluent Rat 6 fibroblasts. The recipient cells were grown to confluence and then transferred into selective medium containing 200 μg/ml G418. After approximately ten days of growth in selective medium, ten individual G418-resistant clones were isolated and maintained independently in G418-containing medium. These lines were designated R6-$PKC_{\beta 1}$1 through R6-$PKC_{\beta 1}$10 (abbreviated as R6-PKC1 through R6-PKC10). In parallel, a set of control Rat 6 lines was generated by transfection of the plasmid pMV7 (lacking the PKC cDNA insert) onto Ψ-2 cells, infection of recipient Rat-6 cells, and selection for G418 resistance as described above for plasmid pMV7-$PKC_{\beta 1}$. Similarly, after ten days of growth in the G418-containing medium, five individual G418-resistant clones were then isolated and maintained independently. These control lines were designated R6-C1 through R6-C5.

To determine the level of PKC activity, each of the cell lines was assayed for Ca^{2+} and phosphatidylserine-dependent protein kinase activity using partially purified cell extracts of total (membrane and cytosolic) PKC activity, as previously described (Housey et al., 1988) Total PKC activity was assayed using the synthetic peptide R-K-R-T-L-R-R-L as the phosphoacceptor substrate. This peptide corresponds to amino acid residues 651 through 658 of the epidermal growth factor (EGF) receptor (Ullrich et. al. 1984), and contains a threonine residue (T_{654}) which is an in vivo substrate for PKC (Hunter et al., 1984; Davis and Czech, 1985;). In addition, we and others have found that this synthetic peptide is a highly specific substrate for PKC activity in vitro (Walton et al, 1987; Woodgett et al., 1986; O'Brian and Housey, unpublished studies). Cell extracts were prepared 48 hours after the cells had reached confluence. As shown in Table 1, four of six cell lines generated by infection with the pMV7-$PKC_{\beta 1}$ construct (lines R6-PKC1, R6-PKC3, R6-PKC5 and R6-PKC6) contained marked increases in PKC activity when compared to the control lines (R6-C1, -C2, and -C3). It is remarkable that cell line R6-PKC3 contained a 53 fold higher level of PKC activity than that present in the control cells. Two of the lines (R6-PKC2 and R6-PKC4), however, did not display a significant increase in PKC activity, yet they presumably had integrated the pMV7-PKC construct as evidenced by their continued G418 resistance. Southern blot analyses subseqequently verified that these two cell lines contained deletions in the cDNA clone encoding $PKC_{\beta 1}$.

In additional studies (not shown here) we have provided independent evidence for the presence of a high level of PKC in these cell lines by subjecting partially purified cell extracts to phosphorylation conditions which favor the autophosphorylation of PKC. We found that the extracts prepared from four cell lines that had high PKC activity (R6-PKC1, -PKC3, -PKC5, and -PKC6, see Table 1) displayed a prominent phosphorylated protein band which was about 75 kd in size. When examined in an immunoblot assay this 75 kd band also reacted with an antibody to the β1 form of PKC (Jaken and Kiley, 1987; S. Jaken, unpublished studies). The control cell lines R6-C1, -C2 and -C3, and the cell lines R6-PKC2 and R6-PKC4, however, which did not have increased levels of PKC (see Table 1)

Table 1

PKC Activity, ^{3}H-PDBU Binding, and Growth Properties of Rat 6 Cells Infected with pMV7 or pMV7-PKC$_{\beta 1}$ Constructs

	PKC Activity		^{3}H-PDBU Binding
Cell line	Specific Activity (pmol/min/mg prot)	Fold Increase Relative to Control	(pmol/10^6 cells)
R6-C1	100	-----	1.6
R6-C2	85	-----	1.3
R6-C3	150	-----	1.5
mean ± s.d.	110 ± 34	1	
R6-PKC1	2480	23	12.7
R6-PKC2	85	1	ND
R6-PKC3	5840	53	9.9
R6-PKC4	190	2	0.7
R6-PKC5	2200	20	5.8
R6-PKC6	4600	42	7.1

		Growth in Monolayer Culture		Growth In Agar	
Cell Line	TPA Addition	Doubling Time (hours)	Saturation Density (x10^6)	Effic.(%)	Colony Size (mm)
R6C1	-	26.4	3.4	0	---
	+	24.6	4.2	0	---
R6-PKC3	-	24.2	5.7	25.1	0.10 - 0.15
	+	21.5	10.0	29.7	0.15 - 0.35
R6-PKC5	-	24.9	4.8	17.3	0.05 - 0.10
	+	22.9	7.0	34.7	0.10 - 0.15

PKC Activity and ^{3}H-PDBU Binding. The control cell lines were obtained from Rat 6 fibroblasts infected with the pMV7 vector itself (lacking the PKC cDNA insert) whereas the PKC lines were obtained from Rat 6 fibroblast cells infected with the pMV7-PKC PKC$_{\beta 1}$ construct, as described in the text. Total PKC activity was partially purified from each of the cell lines and assayed in the presence of 1 mM Ca^{2+} and 80 μg/ml phosphatidylserine, using the synthetic peptide R-K-R-T-L-R-R-L as substrate. Specific activity is reported as the amount of incorporation of ^{32}P into the synthetic peptide substrate per milligram of protein per minute. All assays were done in duplicate and varied by less than 10%. A subset of these cell lines was also assayed for high-affinity ^{3}H-PDBu binding. "ND" designates not determined.

Growth Properties. The "doubling times" relate to the initial exponential phase of cell growth and the "saturation density" represents the number of cells per 6 cm plate on day 11.

did not show this 75 kd phosphorylated band nor did they contain any bands which reacted with the antibody to the β1 form of PKC (data not shown here).

Since it has been shown that PKC is a high affinity intracellular receptor for the phorbol ester tumor promoters (for review see Nishizuka, 1986), we also assayed a subset of the cell lines for ^{3}H-PDBu binding using a previously described intact cell assay (Horowitz et al., 1981). We found that cell lines R6-PKC1, R6-PKC3, R6-PKC5 and R6-PKC6, all of which had high PKC enzyme activity, also had a marked increase in ^{3}H-PDBu binding, whereas the R6-PKC4 cell line, which did not display a significant increase in PKC activity, did not show an increase in ^{3}H-PDBu binding when compared to the two control cell lines R6-C1 and R6-C2 (Table 1). Scatchard analyses of the control cell line R6-C1 and of the R6-PKC3 cell line, performed as previously described (Horowitz et al., 1981), indicated that the number of high affinity receptors in the two cell lines was 1.6 X 10^5 and 1.4 X 10^6, respectively. The affinity constants were approximately the same in both cell lines (K_d = 16 nM). Thus, under the assay conditions used, the R6-PKC3 cells contain about ten times the level of high affinity phorbol ester binding sites as the control cells. The fact that this increase is not as great as the 53-fold increase in PKC enzyme activity detected in R6-PKC3 cells (Table 1) may simply reflect the fact that in the intact cell assay ^{3}H-PDBu does not have access to all of the PKC molecules expressed in these cells. Since high affinity phorbol ester binding requires the association of PKC with lipids or membranes (Leach et al, 1983), this apparent discrepancy might also be due to the fact that a major fraction of the large number of PKC molecules present in these cells is not membrane-associated. These aspects are currently being investigated. Nevertheless, it is apparent that the cell lines that express very high levels of PKC also have a significant increase in phorbol ester binding sites. Thus, taken together, the results discussed above clearly indicate that the R6-PKC1, R6-PKC3, R6-PKC5, and R6-PKC6 cell lines contain elevated levels of PKC.

Cell Lines which Overproduce PKC Display Altered Growth Control

Growth Curves. We have previously shown that the PKC-overproducing cell lines R6-PKC3 and R6-PKC5 are highly responsive to TPA treatment, in contrast to the control R6-C1 cell line (Housey et al., 1988). To further characterize the phenotypic changes which have occurred in the R6-PKC3 and R6-PKC5 cell lines, which contain 53- and 20-fold elevations of PKC activity, respectively, we examined their growth properties in monolayer culture in detail. As shown in Figure 2 and Table 1, in the absence of TPA, the R6-C1 control cell line displayed the longest doubling time (26.4 hours) and the lowest saturation density (3.4 x 10^6 cells/plate); the R6-5 cell line had a shorter doubling time (24.9 hours) and a higher saturation density (4.8 x 10^6 cells/plate); and the R6-PKC3 cell line had the shortest doubling time (24.2 hours) and the highest saturation density (5.7 x 10^6 cells/plate). The presence of TPA decreased the doubling times, and also increased the saturation densities of all three cell lines, but the enhancement by TPA was particularly striking in the case of R6-PKC3 (Figure 2 and Table 1). When the cell lines were maintained in the presence of TPA for a longer period of time, the R6-PKC3 cells, but not the R6PKC5 or R6-C1 cells, showed a decline in cell density (Figure 2). The latter effect was due to the fact

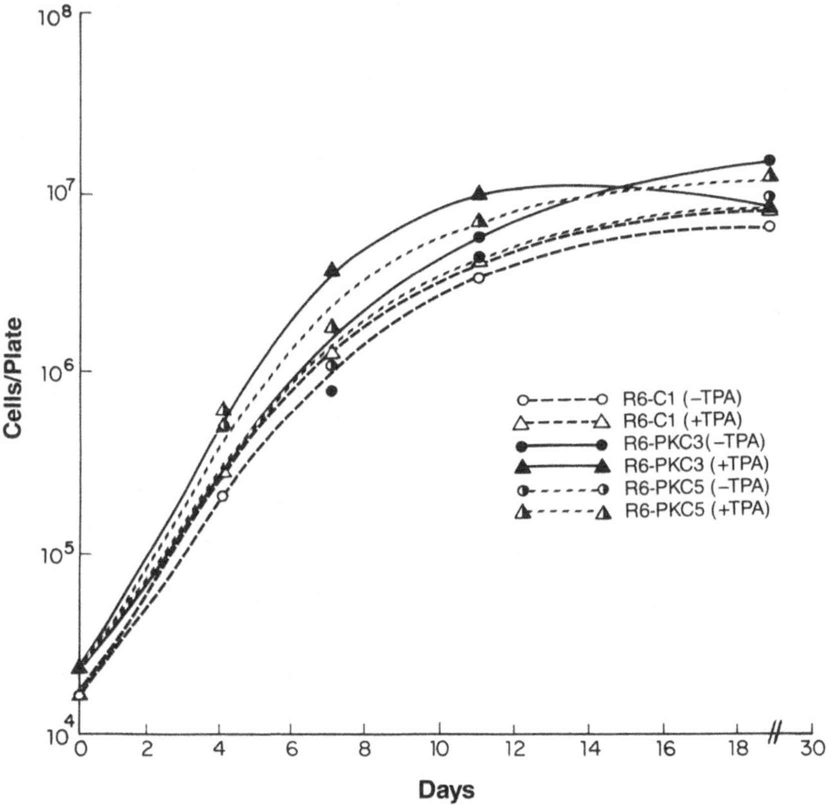

Figure 2. Growth Curves of Control and PKC-Overproducing Cell Lines. The indicated cell lines were seeded at 1 x 10^4 per 6 cm plate in DMEM plus 10% CS, in the presence ("+TPA") and absence ("-TPA") of 100 ng/ml TPA. Cell numbers were determined in replicate plates during the subsequent 21 day growth period. The values given indicate the means of triplicate determinations, which varied by less than 10%.

that when the R6-PKC3 cells were maintained at high cell density in the presence of TPA they became less adhesive and tended to detach from the plate. Thus, even in the absence of TPA the R6-PKC3 cells, (which have the highest level of PKC), and to a lesser extent the R6-PKC5 cells (which have an intermediate level of PKC), exhibit an enhancement of their growth properties which is even greater than that seen when the R6-C1 control cells are grown in the presence of TPA. Furthermore, in the presence of TPA, the differences in growth properties between the R6-C1 control cells and the PKC-overproducing R6-PKC3 cells are even more striking.

Tissue culture studies. In additional studies, monolayer cultures were maintained at post-confluence for an extended period of time (28 days), with media changes every 3 days, in the absence of TPA. Whereas the control R6-C2 cell line remained a fairly uniform monolayer, after about 21 days the R6-PKC3 cell line developed numerous dense foci. An example of one of these foci, which were approximately 0.1-0.3 mm in diameter, is shown in Figure 3. When the dense foci seen in the R6-PKC3 culture were picked and further passaged they grew like the parental R6-PKC3 cells and did not display a morphology typical of malignantly transformed cells. We suspect, therefore, that these dense foci reflect physiologic rather than genetic changes which are being induced by the high level of PKC activity, but this requires further study.

Anchorage-independent growth. We also assayed these cell lines for their ability to form colonies in soft agar, since with rodent cells the acquisition of anchorage-independent growth often correlates with tumorigenicity (Freedman and Shin, 1974). When 2×10^4 cells were plated in 0.3% soft agar, both the R6-PKC3 cells (Figure 4) and the R6-PKC5 cells (data not shown here) formed numerous small colonies, whereas the control R6-C1 cells failed to grow and persisted as single cells (Figure 4 and Table 1). In addition, when TPA (100 ng/ml) was added to the agar medium, the colony sizes and cloning efficiencies of the R6-PKC3 and R6-PKC5 cells were enhanced but the R6-C1 cells still failed to grow in agar (Table 1). Thus it is clear that the overproduction of PKC is associated with the acquisition of anchorage-independent growth in Rat 6 cells. We should emphasize, however, that the sizes of the colonies formed in agar by the PKC cell lines are smaller than those formed by Rat 6 cells transformed by an activated c-H-*ras* oncogene, which have a diameter of about 0.5 - 1.0 mm (data not shown here).

DISCUSSION

The purpose of these studies was to begin a detailed analysis of the role of PKC in cellular growth control and the process of tumor promotion. To accomplish this, we developed a series of Rat-6 cell lines which stably overexpressed a full-length $PKC_{\beta 1}$ cDNA clone. This has permitted, for the first time, a detailed analysis of the effects of the overproduction of PKC on cellular morphology and growth control. Our studies with these cell lines have indicated that the mere overproduction of PKC causes, even in the absence of TPA, multiple phenotypic changes, including: an increased growth rate and saturation density, the formation

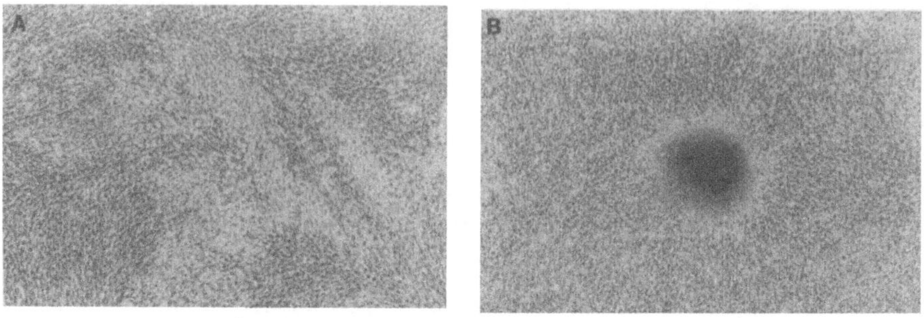

Figure 3. <u>Post-confluence Foci Formation</u>.

Control R6-C2 cells (Panel A) and R6-PKC3 cells (Panel B) were grown to confluence and then maintained for an additional 28 days in DMEM plus 10% CS (without TPA), with the addition of fresh medium every 3 days. Photographs were taken at the end of the 28 day period. Magnification : 30 X.

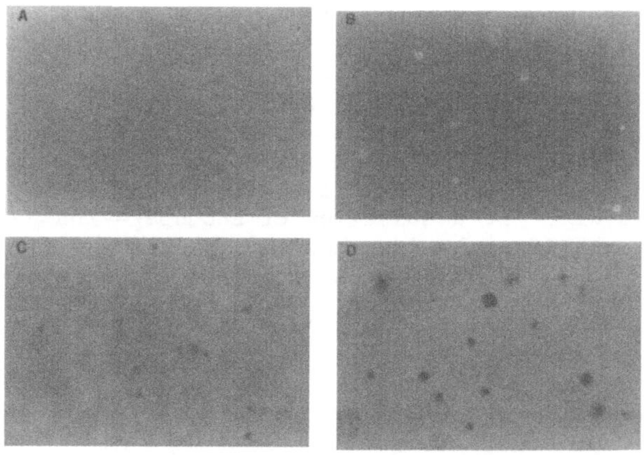

Figure 4. <u>Growth in Soft Agar</u>.

Cell lines R6-C1 and R6-PKC3 were seeded into 60 mm petri dishes in 0.3% agar containing DMEM plus 20% FBS and 50 μg/ml G418, plus or minus 100 ng/ml TPA. Photographs were taken after 21 days of growth.
A) R6-C1 + TPA (low-power field); B) R6-C1 + TPA (medium-power field); C) R6-PKC3; D) R6-PKC3 + TPA. Magnification : 80 X in panel B; 30 X in all other panels.

of dense foci in post-confluent cultures, and, perhaps most surprisingly, anchorage-independent growth. All of these properties were manifest to a greater extent in the R6-PKC3 cell line (which has the highest level of PKC) than in the R6-PKC5 cell line, and in both cell lines these properties were further enhanced by the addition of TPA. Furthermore, parallel studies with various control cell lines (R6-C1,-C2, and -C3), which had been infected with the pMV7 construct lacking the $PKC_{\beta 1}$ cDNA sequence, did not show these changes. Taken together, these results demonstrate that the above-described phenotypic changes are due to the high level of production of PKC. It is, perhaps, paradoxical that these phenotypic changes were seen in the clones that overproduce PKC even in the absence of an exogenous PKC activator like TPA. This might be explained by the presence of serum factors in the growth medium that can lead to PKC activation, presumably by increasing cellular levels of diacyglycerol (Nishizuka, 1986). Alternatively, marked increases in cellular levels of PKC might lead to accummulation of the constitutively active carboxyl terminal catalytic domain of this enzyme (Kikkawa et al., 1982; Pontremoli et al., 1987) due to proteolytic degradation (see below). Thus far, assays of R6-PKC3 cell extracts have not provided evidence for an accummulation of such fragments (unpublished studies), but this requires further study. An additional possibility is that some of the intracellular substrates of PKC involved in the control of cell growth may be phosphorylated by the enzyme without activation by diacylglycerol. The latter would be analogous to the Ca^{2+}- and phospholipid <u>in</u>dependent phosphorylation of protamine sulfate by PKC). However, regardless of the mechanism, our results indicate that disturbances in cellular levels of PKC can, in themselves, cause abnormalities in growth control, and this may be of relevance to mechanisms of cancer causation.

The previous evidence that PKC is involved in the action of the phorbol ester tumor promoters was largely based on the ability of these compounds to bind to this enzyme and activate its protein kinase activity (Nishizuka, 1984). The present results, indicating that overproduction of PKC mimics several of the effects of treating cells with TPA, and exaggerates the usual responses of cells to TPA, provide independent evidence that PKC mediates the major, and perhaps all, of the effects of the phorbol ester tumor promoters on cellular growth and morphology.

It seems likely that the altered growth properties of the cell lines that produce high levels of PKC, and their exaggerated morphologic responses to TPA, reflect increased phosphorylation of critical cellular proteins involved in the control of cellular growth and morphology. These cell lines should, therefore, be extremely valuable for highlighting and identifying these proteins, thus providing insights into mechanisms that normally control these complex phenotypic properties.

Since PKC is encoded by a multigene family, the present studies specifically address the phenotypic effects of the $\beta 1$ form of PKC. Using the strategies employed in the present studies, however, it will be of interest, therefore, to develop similar cell lines that overexpress other specific members of the PKC gene family. Indeed, this might be the most definitive method of determining whether or not these individual forms differ with respect to their function and responses to various activators and inhibitors.

Finally, the present studies also provide insights into the role of PKC in the multistage carcinogenic process. Although we found that the cell lines that overexpress PKC display significant changes in their growth properties, these cells do not seem to have the typical characteristics of malignantly transformed rodent fibroblasts. It would appear, therefore, that overproduction of an intact form of PKC per se may not be sufficient to cause cancer. Nevertheless, the changes that we have observed suggest that altered expression of PKC could enhance the carcinogenic process. It is also possible that mutational changes in the structure of the PKC genes might reveal complete transforming activity. In this sense the PKC genes might be proto-oncogenes. These considerations are consistent with the central role that PKC plays in signal transduction, the action of certain growth factors and phorbol ester-mediated tumor promotion. It will be of interest, therefore, to determine whether cell lines that overproduce PKC are more sensitive to malignant transformation by specific viral, chemical and physical agents. These studies are now in progress.

ACKNOWLEDGEMENTS

This research was supported by a National Cancer Institute Grant CA 02656 (to I.B.W.). G.M.H. is in the Medical Scientist Training Program.

REFERENCES

Ashendel, C. The Phorbol Ester Receptor: a phospholipid-regulated protein kinase. (1984) Biochim. Biophys. Acta 822, 219-242.

Berg, J. (1986) Potential Metal-Binding Domains in Nucleic Acid Binding Proteins. Science 232, 485-487

Bollag, G. E., Roth, R. A., Beaudoin, J., Mochly-Rosen, D., Koshland, D. E. Jr. (1986) Protein kinase C directly phosphorylates the insulin receptor in vitro and reduces its protein-tyrosine kinase activity. Proc. Natl. Acad Sci. USA 83, pp 5822-5824

Coussens, L., Parker, P. J., Rhee, L., Yang-Feng, T. L., Chen, E., Waterfield, M. D., Francke, U., & Ullrich, A. (1986) Multiple, distinct forms of bovine and human protein kinase C suggest diversity in cellular signalling pathways. Science 233, 859-866.

Davis, R. J., and Czech, M. P. (1985) Platelet-derived growth factor mimics phorbol diester action on epidermal growth factor receptor phosphorylation at threonine 654. Proc. Natl. Acad. Sci. 82, 4080-4084.

Freeman, A. E., Price, P. J., Igel, H. J., Young, J. C., Maryak, J. M. Huebner, R. J. (1970) Morphological transformation of rat embryo cells induced by diethylnitrosamine and murine leukemia viruses. J. Natl. Cancer Inst. 44, pp 65-78.

Freedman, V. H. and Shin, S. (1974) Cellular Tumorigenicity in nude Mice: Correlation with Cell Growth in Semi-Solid Medium. Cell 3, 355-359.

Graham, F. L., and van der Eb, A. J. (1973) A new technique for the assay of infectivity of human adenovirus DNA. Virology 52, 456-467.

Gould, K. L., Woodgett, J. R., Cooper, J. A., Buss, J. E., Shalloway, D., Hunter, T. (1985) Protein Kinase C Phosphorylates $pp60^{src}$ at a novel site. Cell 42, pp 849-857.

Horowitz, A. D., Greenebaum, E. and Weinstein, I. B. (1981) Identification of receptors for phorbol ester tumor promoters in intact mammalian cells and of an inhibitor of receptor binding in biologic fluids. Proc. Natl. Acad. Sci. USA 78, pp 2315-2319

Housey, G. M., Kirschmeier, P., Garte, S. J., Burns, F., Troll, W., & Weinstein, I. B. (1985) Expression of long terminal repeat (LTR) sequences in carcinogen-induced murine skin carcinomas. Biochem. Biophys. Res. Commun. 127, 391-398.

Housey. G. M., O'Brian, C. A., Johnson, M. D., Kirschmeier, P., and Weinstein, I. B. (1987) Isolation of cDNA clones encoding protein kinase C: Evidence for a protein kinase C-related gene family. Proc. Natl. Acad. Sci. USA 84, pp 1065-1069

Housey. G. M., O'Brian, C. A., Johnson, M. D., Kirschmeier, P., and Weinstein, I. B. (1988) Overproduction of Protein Kinase C Causes Disordered Growth Control in Rat Fibroblasts. Cell, in press

Hsiao, W.-L. W., T. Wu, Weinstein, I. B., (1986) Oncogene-Induced Transformation of a Rat Embryo Fibroblast Cell Line is Enhanced by Tumor Promoters. Mol. Cell. Biol. 6, pp 1943-1950

Hunter, T., Ling, N., Cooper, J. A. (1984) Protein kinase C phosphorylation of the EGF receptor at a threonine residue close to the cytoplasmic face of the plasma membrane. Nature 311, 480-3

Hunter, T. & Cooper, J. A. (1985) Protein Tyrosine Kinases. Ann. Rev. Biochem. 54, 897-930.

Jaken, S. and Kiley, S. (1987) Purification and characterization of three types of protein kinase C from rabbit brain cytosol. Proc. Natl. Acad. Sci. USA 84, pp 4418-4422

Jeng, A. Y., Srivastava, S. K., Lacal, J. C., Blumberg, P. M. (1987) Phosphorylation of ras oncogene product by protein kinase C. Biochem. Biophys. Res. Commun. 145, pp 782-8.

Kikkawa, U., Takai, Y., Minakuchi, R., Inohara, S., and Nishizuka, Y. (1982) Calcium-activated, phospholipid-dependent protein kinase from rat brain. Subcellular distribution, purification, and properties. J. Biol. Chem. 257, 13341-13348.

Kirschmeier, P. T., Housey, G. M., Johnson, M. D., Perkins, A. S., & Weinstein, I. B. (1988) Construction and characterization of a retroviral vector demonstrating efficient expression of cloned cDNA sequences. DNA, in press.

Knopf, J. L., Lee, M-H, Sultzman, L. A., Kriz, R. W., Loomis, C. R., Hewick, R. M. & Bell, R. (1986) Cloning and expression of multiple protein kinase C cDNAs. Cell 46, 491-502.

Laemmli, U. K. (1970) Cleavage of structural proteins during the assembly of the head of bacteriophage T4. Nature 227, 680-685.

Leach, K. L., James, M. L., & Blumberg, P. M., (1983) Characterization of a specific phorbol ester aporeceptor in mouse brain cytosol. Proc. Natl. Acad. Sci. USA 80, 4208-4212.

Makowske, M., Birnbaum, M. J., Ballester, R., Rosen, O. M. (1986) A cDNA encoding PKC identifies two species of mRNA in brain and GH3 cells. J. Biol. Chem. 261, pp 13389-13392

Mann, R., Mulligan, R. C., and Baltimore, D. (1983) Construction of a retrovirus packaging mutant and its use to produce helper-free defective retrovirus. Cell 33, 153-159

Nishizuka, Y. (1986) Studies and Perspectives of Protein Kinase C. Science 233, 305-312.

Nishizuka, Y. (1984) The Role of Protein Kinase C in Cell Surface Transduction and Tumour Promotion. Nature (London) 308, 693-698.

O'Brian, C. A., Lawrence, D. S., Kaiser, E. T., & Weinstein, I. B. (1984) Protein kinase C phosphorylates the synthetic peptide Arg-Arg-Lys-Ala-Ser-Gly-Pro-Pro-Val in the presence of phospholipid plus either Ca^{2+} or a phorbol ester tumor promoter. Biochem. Biophys. Res. Commun. 124, 296-302.

Ono, Y., Kurokawa, T., Fujii, T., Kawahara, K., Igarashi, K., Kikkawa, U., Ogita, K., Nishizuka, Y. (1986) Two types of complementary DNAs of rat brain protein kinase C. FEBS 206, 347-52

Ono, Y., Kikkawa, U., Ogita, K., Tomoko, F., Kurokawa, T., Asaoka, Y, Sekiguchi, K., Ase, K., Igarashi, K., Nishizuka, Y. (1987) Expression and Properties of Two Types of Protein Kinase C: Alternative Splicing from a Single Gene. Science 236, pp 1116-1120.

Ohno, S., Kawasaki, H., Imajoh, S., Suzuki, K., Inagaki, M., Yokohura, H., Sakoh, T., Hidaka, H. (1987) Tissue-specific expression of three distinct types of rabbit protein kinase C. Nature (London) 325, pp 161-6.

Parker, P. J., Coussens, L., Totty, N., Rhee, L., Young, S., Chen, E., Stabel, S., Waterfield, M. D., & Ullrich, A. (1986) The complete primary structure of protein kinase C--the major phorbol ester receptor. Science 233, 853-858.

Pontremoli, S., Melloni, E., Michetti, M., Sparatore, B., Salamino, F., Sacco, O., and Horecker, B. L. (1987) Phosphorylation and proteolytic modification of specific cytoskeletal proteins in human neutrophils stimulated by phorbol-12-myristate 13-acetate. Proc. Natl. Acad. Sci. 84, 3604-3608.

Ullrich, A., Coussens, L., Hayflick, J. S., Dull, T.J., Gray, A., Tam, A. W., Lee, J., Yarden, Y., Libermann, T. A., Schlessinger, J., Downward, J., Mayes, E. L. V., Whittle, N., Waterfield, M. D., and Seeburg, P. H. Human epidermal growth factor receptor cDNA sequence and abberrant expression of amplified gene in A431 epidermoid carcinoma cells. Nature 309, 418-425.

Walton, G. M., Bertics, P. J., Hudson, L. G., Vedvick, T. S., Gill, G. N. (1987) A Three-Step Purification Procedure for Protein Kinase C: Characterization of the Purified Enzyme. Anal. Biochem. 161, 425-437.

Weinstein, I. B., (1987) Growth Factors, Oncogenes, and Multistage Carcinogenesis. J. Cell. Biochem. 33, pp 213-224.

Wigler, M., Silverstien, S., Lee, L.-S., Pellicer, A., Cheng, Y.-C, and Axel, R. (1977). Transfer of purified herpes virus thymidine kinase gene to cultured mouse cells. Cell 11, 223-232.

THE *ras* PATHWAY:

A MODEL FOR THE CONTROL OF PROLIFERATION IN ANIMAL CELLS

Dennis W. Stacey

The Roche Institute of Molecular Biology
Roche Research Center
Nutley, N.J. 07110

INTRODUCTION

Scrupulous control over cellular proliferation is essential for the survival of a multicellular organism, yet little is known about the mechanism of this control. A variety of factors are known to influence cell division in both positive and negative ways. These include diffusible factors and the results of cell-cell contact. While the ultimate decision concerning cell division is most likely made at the molecular level within a specific cell, this decision must be based upon the signals received from outside the cell. In this study data and hypotheses are presented primarily regarding the means by which extracellular proliferative signals are transferred into the cell. This discussion, therefore, represents an important but limited aspect of the control of proliferation within the organism. While the discussion will center primarily upon a presentation of data, hypotheses relating to these data will also be included. In many cases, the hypotheses are only speculative at present.

Both physiological and molecular aspects of proliferation have been carefully studied for some time. It is clear that upon mitogenic induction a number of biochemical changes take place within a cell. These include alterations in phospholipid metabolism, morphology, membrane structure, sugar and ion transport, intracellular pH, and energy metabolism. While each of these and numerous other effects is involved in cell division, it is not clear which of them is the primary effect of the mitogenic signal from outside the cell. Coupled with biochemical studies, molecular analyses indicate some cellular genes which might participate in mitogenesis. The cellular genes transduced by RNA tumor viruses are known to be involved in tumor formation, and are therefore believed to be involved in control of proliferation. While the structure of these genes is well known in some cases, their relationships to cellular metabolism is unclear. The challenge in understanding proliferation would then seem to center upon identifying which genes are involved, how they interact and how they relate to cellular metabolism.

Classification of oncogenes

Numerous oncogenes have been identified both in tumor viruses and in tumor cells. These generally share the ability to alter the proliferative properties of cultured cells. The most well characterized of these tend to fall into several general categories. The largest such group includes the membrane-associated tyrosine kinase molecules. These are related to or are directly derived from the receptors for growth factors (Doolittle *et al.*, 1983; Downward *et al.*, 1984; and Scherr *et al.*, 1985). Considerable effort has been expended in identification of the biologically active substrate(s) of these molecules. At present the phospholipase inhibitory protein, lipocortin, appears to be the best candidate, although it is unclear what significance this substrate might have (Brugge, 1986). Related to this group is the *sis* oncogene which itself is derived from the gene for platelet-derived growth factor. A second, smaller group of kinase oncogenes is located within the cytoplasm and recognize serine and threonine residues. Only two examples of this group have been positively identified: *mos* which is expressed in a restricted set of developing and reproductive tissues, and *raf* which appears to be expressed in all proliferating tissues.

The third group of oncogenes include *fos* and *myc* and make up a well-characterized nuclear oncogene category. These are not efficient in transforming fibroblasts, as are the others described, and appear to be expressed primarily at different stages of mitogenic stimulation or differentiation. The final well-characterized oncogene class includes the three members of the homologous *ras* gene family. These membrane proteins bind and hydrolize GTP as do the G-regulatory proteins involved in controlling adenylate cyclase (Scolnick *et al.*, 1979; Sweet *et al.*, 1984). The experiments to be described hereafter were initially designed to study the biological function of *ras* proteins (termed p21*ras*). Because of the central role played by these cellular proteins, we have also been able to gain some indication of the activities of other cellular protooncogene proteins.

IDENTIFICATION OF A NEUTRALIZING ANTI-*ras* ANTIBODY

In order to analyze the biological function of *ras* proteins, it was our goal to identify a neutralizing antibody which would eliminate *ras* activity from a microinjected cell. The search for this antibody was aided by the availability of purified, bacterially synthesized p21*ras*. When this protein was injected into NIH3T3 cells, they became morphologically transformed (Plate 1) and were induced to initiate an unscheduled cycle of DNA synthesis (Stacey and Kung, 1984). A number of monoclonal antibodies previously prepared by others (Furth *et al.*, 1982) were combined with the purified p21*ras* prior to microinjection. When several of these antibody mixtures were injected, all but one retained its ability to transform the recipient cells. That one antibody, Y13-259, had apparently blocked the transforming activity of the co-injected p21*ras*. Other antibodies tested, including Y13-238, were unable to inhibit the *ras*-induced transformation (Kung *et al.*, 1986).

Monoclonal antibody 259 (Y13-259) is a particularly valuable reagent since it recognized an epitope common to all known, naturally occurring *ras* proteins. To further analyze its activity, antibody 259 was injected into NIH3T3 cells transformed by high level expression of cellular *ras* protein. Not only was the transformed morphology of the recipient cells abolished, they were unable to continue proliferation after the antibody injection. Non-neutralizing antibody 238 (Y13-238) had no detectable effect upon microinjected cells. It is therefore clear that antibody 259

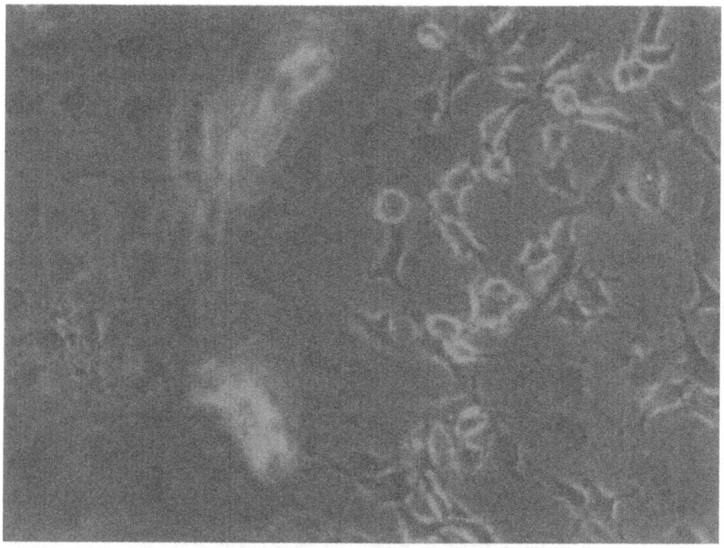

Plate 1. Transforming potential of p21*ras*. Purified bacterial *ras* protein was microinjected into the Balb 3T3 cells at the right of the area photographed above. This phase-contrast photograph was taken 15 hrs after injection. The rounded, refractile appearance of the injected cells can be compared to the flattened, non-transformed appearance of the uninjected cells (at the left).

is able to neutralize not only purified p21*ras* but also the *ras* protein made within cells (Kung *et al.*, 1986). That an antibody can perform such a function was suggested by early injection experiments which indicated that anti-hemoglobin antibody was able to remove hemoglobin from the cytoplasm of microinjected cells (Stacey and Allfrey, 1976).

Even though it is clear that antibody 259 can inhibit the proliferation of *ras*-transformed cells, it is not clear that this is due to its direct interaction with *ras* proteins, rather than with some unknown target which is also necessary for cellular transformation. To rule out this possibility, we were fortunate to have available to us a mutant *ras* gene specifically prepared to lack the epitope recognized by antibody 259. The mutant protein retained the ability to transform cells. When cells transformed by this mutant gene received injections of antibody 259, they continued to proliferate almost without inhibition (Papageorge *et al.*, 1986). This result indicated not only that the antibody was inhibitory due to its interaction with *ras* protein, but also that the mutant viral *ras* could substitute during proliferation for the cellular *ras* proteins that would have been neutralized by the microinjected antibody (see below).

BIOLOGICAL ANALYSES OF CELLULAR *ras* FUNCTION

We therefore conclude that anti-*ras* monoclonal antibody 259 specifically interacts with *ras* proteins and is able to efficiently neutralize their biological activity. This was observed in cells where *ras* protein was expressed as part of a retrovirus and would therefore be

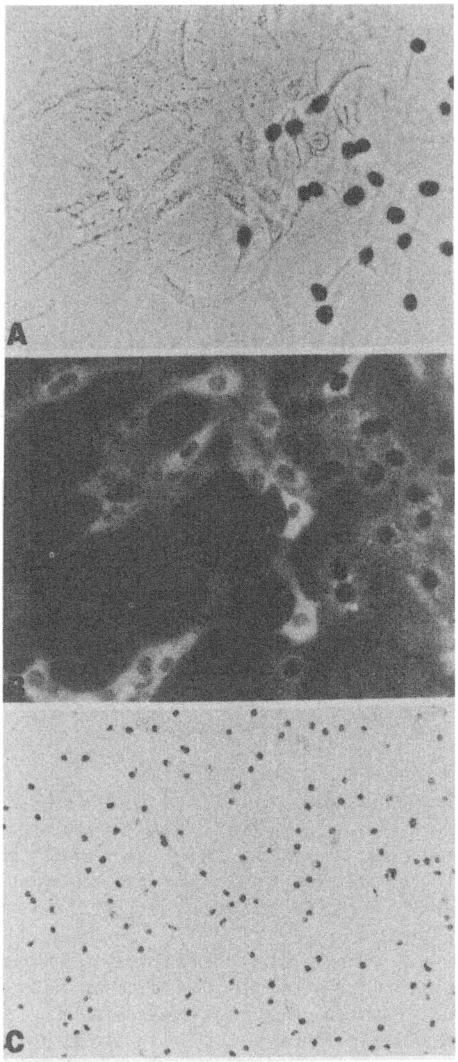

Plate 2. Inhibition of thymidine labelling by microinjected anti-*ras*. These NIH3T3 cells received injections of antibody 259 (A and B) or a non-neutralizing antibody (C) 15 hrs prior to thymidine labelling, staining with a fluorescent antibody to recognize the injected immunoglobulin, and autoradiography. Cells at the left of each photograph were injected, as indicated by their positive staining in the fluorescent photograph (B). Note from photographs of the autoradiographs (A,C) that the cells injected with antibody 259 (A and B, 400X) are efficiently inhibited in thymidine labelling. Non-neutralizing antibody-injected cells, pictured under low power (100X) at the left of the area photographed (C), are not inhibited in thymidine incorporation.

present at extremely high levels. It is assumed, therefore, that sufficient antibody is microinjected to neutralize all the *ras* protein normally produced by cultured cells. With the neutralizing anti-*ras* antibody it was possible to continue with the biological analysis proposed above by functionally removing *ras* activity from injected cells in order to study its biological function.

The requirement for *ras* proteins during cellular proliferation

We first microinjected antibody 259 into NIH3T3 cells that had been rendered quiescent by serum deprivation. After the injection, serum was added to the recipient culture which was then pulsed with labelled thymidine for the next 18 hrs. The injected cells were efficiently inhibited from incorporating thymidine (Plate 2). Cells injected in parallel with the non-neutralizing antibody 238, on the other hand, were not inhibited to any detectable extent in thymidine labelling (Mulcahy *et al.*, 1985). It appeared, therefore, that *ras* activity was required for mitogenic stimulation of NIH3T3 cells by serum growth factors. This inhibition was apparent even when the antibody was injected up to 6 hrs after the addition of serum to quiescent cells. No inhibition was observed if antibody was injected into cells already in S-phase. *ras* activity must, therefore, be present for the cellular transition from the G-1 phase of the cell cycle to S-phase, but is not apparently required at the time of DNA synthesis.

Previous studies by others with an antibody against the P53 protein indicated that partial inhibition of proliferation was observed in quiescent cells stimulated by the addition of serum, but not in rapidly growing cells. This indicated that P53 might function only to remove cells from the quiescent, G-0 phase of the cell cycle (Mercer *et al.*, 1984). In contrast, *ras* appears to be required later in the cell cycle and would therefore be needed for both quiescent and rapidly proliferating cells. To test this possibility, anti-*ras* antibody was injected into untreated NIH3T3 cultures which were then pulsed with thymidine at different times thereafter. Little inhibition of labelling was observed in cultures labelled soon after injection as expected, since cells able to incorporate thymidine at this time would most probably have been in S-phase at the time of injection. After 12 to 15 hrs, however, injected cells were efficiently inhibited from incorporating labelled thymidine. Cells labelled at this later time would have to initiate a new cycle of DNA synthesis after the injection. This reinitiation of DNA synthesis was efficiently blocked by the injected antibody as expected (Mulcahy *et al.*, 1985).

Injected cells were identified by staining with a fluorescent antibody able to recognize the injected monoclonal antibody within the recipient cell. With this procedure it was clear that antibody was stable within the cells for at least 24 hrs, but was generally not apparent after 40 hrs. Cells pulsed with thymidine after the antibody had disappeared from the cells were no longer inhibited in thymidine labelling.

These observations demonstrate that *ras* activity plays a critical role in controlling the proliferation of NIH3T3 cells. Purified p21*ras* is able to induce DNA synthesis in the absence of other factors. Even in the presence of all the growth factors present in calf serum, however, proliferation does not occur if *ras* proteins are inactivated by injected neutralizing antibody. In NIH3T3 cells, therefore, *ras* activity is both sufficient and necessary for proliferation. Furthermore, the *ras* activity is required during proliferation just prior to initiation of S-phase as might be expected for a critical control factor. We therefore propose that *ras* proteins function after the action of serum growth factors to perform an essential step in initiation of DNA synthesis (Diagram 1).

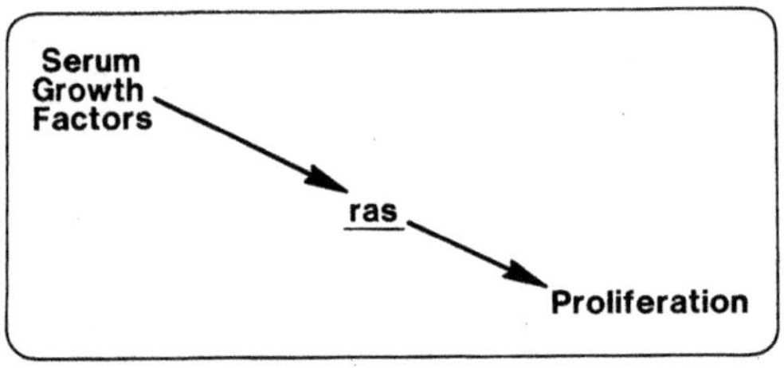

Diagram 1. Requirement for p21*ras* during serum-stimulated growth. This is the first of five diagrams which sequentially illustrate the data and models presented. In each case, information added from the previous diagram is darkened. This illustration emphasizes the fact that non-tumorous cells cultured in the presence of calf serum are blocked from proliferating by anti-*ras* antibody.

The requirement for *ras* during transformation by other oncogenes

From the foregoing discussion, it is clear that cellular *ras* proteins are required during mitogenic stimulation. It is also clear that a number of oncogenes are able to confer new proliferative properties upon cells. These oncogenes are able to promote proliferation in the absence of growth factors or cellular attachment. We wondered if these oncogene-induced proliferative properties were dominant or recessive to the requirement for *ras* during cellular mitogenesis. To test this we obtained NIH3T3 cells which had been transformed by a variety of oncogenes. These cells were then injected with anti-*ras* antibody 259 as previously described. As before, the injected cells were actively proliferating at the time of injection. At different times after injection (between 12 and 24 hrs) the injected cultures were pulsed with thymidine for 3 hrs, stained with a fluorescent antibody against the injected immunoglobulin, and finally autoradiographed.

When neutralizing antibody 259 was injected into NIH3T3 cells transformed by the oncogene *src*, it was apparent that the morphologic characteristics of transformation present in the culture prior to injection had been abolished. This reversion from the transformed to the normal phenotype was apparent within approximately 10 hrs after injection and persisted for up to 30 to 40 hrs before returning to the previous transformed phenotype. In addition, the ability of the injected cells to incorporate thymidine was dramatically inhibited (Smith *et al.*, 1986).

In order to determine the extent of proliferative inhibition resulting from antibody injection, the percentage of injected cells able to incorporate thymidine was determined. Injected cells were identified with the fluorescent antibody stain against the injected immunoglobulin. The percentage of fluorescent, thymidine-labelled cells was then determined. For comparison, the percentage of uninjected cells that were able to incorporate thymidine was also determined by observing cells near the injected area. The percentage of inhibition was then determined by dividing the percentage of labelled-injected cells by the percentage of labelled-uninjected cells, and then subtracting the result from 100%. This number, therefore, indicates the proportion of injected cells able to incorporate label compared to the proportion expected if no injection had occured. In the *src*-transformed cells injected with anti-*ras* antibody the inhibition of thymidine incorporation averaged near 80% in several different determinations. When non-neutralizing antibody 238 was injected into a number of cell types described below, there was never any indication that thymidine labelling had been inhibited (Smith *et al.*, 1986).

The oncogene *src* is a membrane-associated tyrosine kinase which is similar in structure to a number of other oncogenes including those derived from the receptors for growth factors. Further studies involved anti-*ras* antibody microinjection into NIH3T3 cells transformed by two other tyrosine kinases. Cells transformed by both of these oncogenes were inhibited as were the *src*-transformed cells described above. The morphologic characteristics of cellular transformation were abolished in cells transformed by the *fes* oncogene and the *fms* oncogene (which has been shown to be derived from the receptor for colony-stimulating factor-1). In addition, thymidine labelling was inhibited in these cells by approximately 80% as with the *src*-transformed cell. The *sis* oncogene is not a tyrosine kinase but is derived from the gene for platelet-derived growth factor and should therefore be functionally related to the tyrosine kinase oncogenes. When NIH3T3 cells transformed by *sis* were tested, they too were dependent upon cellular *ras* activity for maintenance of the transformed state as were the other cells described (Smith *et al.*, 1986).

These results emphasize the central role played by cellular *ras* proteins in regulating proliferation. *ras* activity is required even for the action of a group of unrelated oncogenes. The group of tyrosine kinase oncogenes is large, and we have tested only three of these. These oncogenes are known to be related to growth factor receptors, however, and from the initial studies it is clear that the action of the complex mixture of growth factors present in serum also depends upon *ras* activity for mitogenic stimulation. We therefore assume that growth factors, their receptors, and related oncogenes depend upon cellular *ras* proteins for biological activity. This result is considered particularly important since it establishes for the first time a functional connection between different oncogene classes and suggests that cellular oncogenes might function together in controlling cellular proliferation. A possible biochemical mechanism for the biological association between tyrosine kinases and *ras* proteins will be described below.

While the membrane-associated tyrosine kinases described above represent the largest class of oncogenes, they are not the only oncogene class with kinase activity. The oncogenes *mos* and *raf* constitute a class of cytoplasmic serine/threonine kinase molecules. When NIH3T3 cells transformed with either of these oncogenes were injected with anti-*ras* antibody 259, there was no apparent inhibition of thymidine labelling and no alteration of the transformed morphology. The distinction between these two classes of oncogenes was dramatic. Less than 10% inhibition was observed in thymidine labelling of injected *mos*- or *raf*-transformed cells, while no determination resulted in less than a 70% inhibition with the tyrosine kinases. It is therefore clear that *ras* protein is required for the activity of some retroviral oncogenes but not for others (Smith *et al.*, 1986).

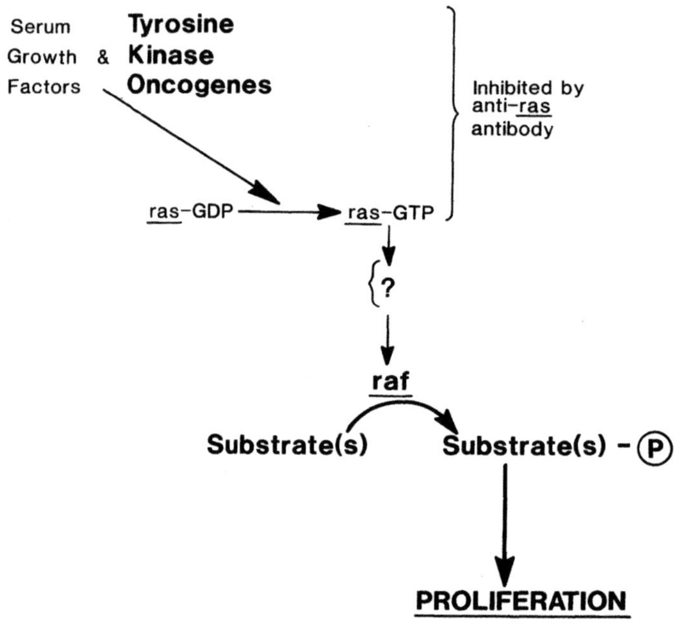

Diagram 2. The interrelationship of proto-oncogenes. The proliferative potential of serum growth factors, like the action of tyrosine kinase oncogenes, depends upon cellular *ras* activity. Because anti-*ras* inhibits the transforming potential of tyrosine kinase oncogenes but not the *raf* oncogene, it is proposed that p21*ras* functions to transfer a proliferative signal between these two types of proto-oncogene proteins during proliferative stimulation. It is believed that p21*ras* is activated biologically only when binding a GTP molecule.

Model of proliferative signal transduction. The observations described above directly analyze the functional interrelationships between cellular oncogenes. The results form the basis for a model proposed to explain how proliferation might be controlled by such interactions between different cellular oncogene classes. The model is based upon an apparent analogy with the mechanism of signal transduction involving cyclic AMP (Gilman, 1984). It is known that transmembrane binding proteins associated with the appropriate ligand are able to induce alterations in the GTP binding properties of G-regulatory proteins. The alterations result in a greater proportion of the α-subunits of these proteins becoming associated with GTP. When bound to GTP, the α-subunit is able to induce the activity of adenylate cyclase. The cyclic AMP produced is then able to stimulate a cytoplasmic cyclic AMP-dependent protein kinase. The activity of this kinase is believed to be the primary mediator of the biological response induced by the binding of the original ligand to its receptor.

The results of anti-*ras* injection into cells transformed by different oncogene classes suggests that the cellular oncogenes might be involved in a similar signal transduction mechanism. It is known that *ras* proteins bind and hydrolyze GTP as do the G-proteins described above (although *ras* does not apparently bind to other subunits as do the α-subunits of G-proteins). It has been proposed that growth factor receptors and related oncogenes initiate a proliferative signal which is transferred into the cell by *ras* proteins just as G-proteins mediate the signal from a different set of binding proteins. This would explain the requirement for *ras* activity during oncogene transformation and serum-stimulated growth (Diagram 2; Smith *et al.*, 1986, Stacey *et al.*, 1987).

Furthermore, we propose that activated *ras* proteins affect the activity of an unknown factor able to generate a signal which is received by cytoplasmic kinase molecules such as the cellular *raf* or *mos* oncogenes. (Just as G-proteins induce the production of cyclic AMP which stimulates other cytoplasmic kinases). Since these molecules are proposed to function downstream of *ras* proteins in the hypothetical signal transduction system, they would not depend upon *ras*, if mutated to become constitutively active oncogenes (Diagram 2). It is significant that there are only two known oncogenes which function independently of *ras*. The *mos* oncogene is expressed only during development and in reproductive organs, while cellular *raf* appears to be expressed in all proliferating tissues. If the above scheme is correct, it is clear that cellular *raf* proteins function downstream of *ras* in proliferative signal transduction.

It is important to point out that other explanations of these data are possible. It is possible, for example, that both *ras* and *raf* function in separate, parallel biochemical pathways which are necessary for proliferation. It would be difficult, however, to explain the fact that purified *ras* protein alone is able to induce proliferation if such a scheme were correct. While it is known that separate classes of oncogenes do cooperate during the transformation of some cell types, *ras* and *raf* apparently belong to the same class (Ruley, 1983). They do not complement each other. We are anxious to prepare a neutralizing anti-*raf* antibody that would allow us to obtain direct evidence for the action of this cellular oncogene downstream of *ras*. Until such an antibody is found, however, the signal transduction scheme presented will remain only speculative.

The role of nuclear oncogenes

The discussion above relates to only three of the four general classifications of retroviral oncogenes previously outlined. It is not possible to include an analysis of the nuclear oncogenes in the discussion above because these are not able to efficiently transform the NIH3T3 cells employed. There are cases where nuclear oncogenes do apparently transform fibroblast cells in culture. A few of these transformed cell lines have been tested. In each case it appears that the anti-*ras* antibody does block proliferation of the transformed cells (M.R. Smith and D.W. Stacey, unpublished data). These results are not considered as definitive, however, as the results with tyrosine kinases. It is not clear, therefore, where the action of nuclear oncogenes might fit into the model presented above.

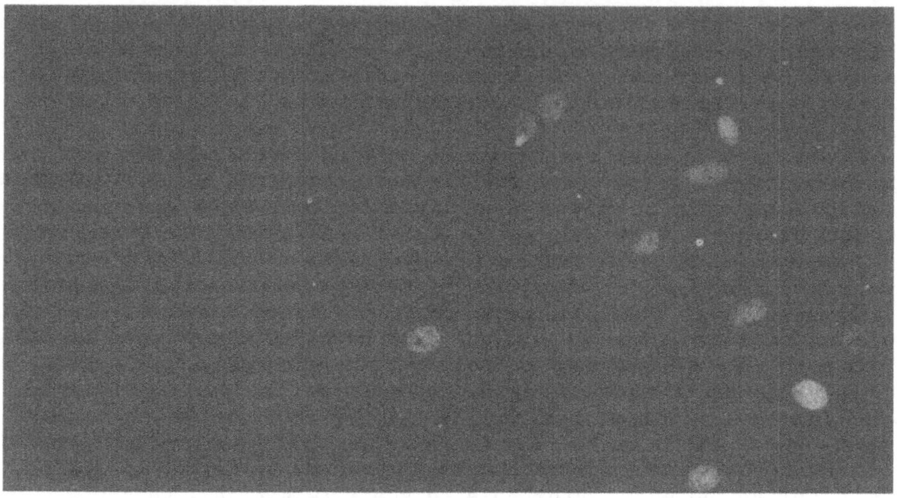

Plate 3. Induction of *fos* expression by p21*ras* injection. Cells at the right of the area photographed were injected with p21*ras* 10 hrs prior to fixation and staining with a fluorescent antibody to cellular *fos* protein. In this fluorescence photograph only the nuclei of injected cells are strongly stained.

It would be tempting to speculate that the signal initiated by tyrosine kinase molecules which passes through *ras* might proceed to the nucleus and influence nuclear oncogenes. This does not fit with the biological data, however. Nuclear proto-oncogenes are known to be expressed early in the cell cycle. In the case of cellular *fos*, protein expression is observed between one and three hrs after addition of serum to quiescent cultures (Muller *et al.*, 1984). The cellular *myc* proteins are also expressed early in the G-0 to G-1 transition. Since *ras* activity is required later in the cell cycle than expression of these nuclear oncogenes, it would seem unlikely that the expression of these nuclear oncogenes would be controlled by *ras* activity.

Nevertheless, in collaboration with H.-F. Kung (who participated in the identification of the neutralizing anti-*ras* antibody) and T. Curran, we tested the possibility that *ras* activity might influence the expression of the nuclear *fos* protein. In this context, it is important to note that since p21*ras* is stable within the cell and not expressed with major differences at different stages of the cell cycle, it is likely to be present within the cell at all times. Its activity must therefore be regulated by altering the properties of pre-existing protein. The situation with the nuclear oncogenes is different. Their expression is highly regulated and the proteins are labile. It is, therefore, likely that these nuclear proteins are active whenever they are present within the cell. Their activity would then be controlled at the level of protein production.

In order to test the possible involvement of *ras* activity in *fos* expression, purified p21*ras* protein prepared from a bacterial expression vector was microinjected into NIH3T3 cells which had been deprived of serum. At various times thereafter the cells were fixed and stained with a fluorescent antibody able to recognize cellular *fos* proteins. We found that the injected *ras* protein was able to strongly induce *fos* protein synthesis (Plate 3). *fos* protein was observed within 90 min of the injection and persisted for up to 20 hrs (Stacey *et al.*, 1987). The brief delay in *fos* expression following microinjection of p21*ras* compared to that observed following serum addition might be due to the time required for the cell to modify the injected p21*ras*. The fact that *fos* expression persisted for so long is surprising since this is not normally observed following serum addition.

It is significant that the purified protein was so active in *fos* induction. There are many treatments which stimulate the synthesis of this protein, however. It is possible that the stimulation observed had nothing to do with the biological function of p21*ras*. To test this possibility injections were made of both the normal and mutant, transforming *ras* proteins. We found that the transforming protein was much more stimulatory than the protein with the non-mutant sequence. This indicates that *fos* induction is related to the biological function of *ras*. In fact, the concentrations of injected protein required for the stimulation of *fos* expression was almost identical to that required for mitogenic stimulation with both the normal and transforming protein (Stacey *et al.*, 1987). This observation draws a strong correlation between mitogenic activity and *fos* protein expression induced by microinjected *ras* protein.

We next addressed the question of the normal involvement of *ras* activity in the induction of *fos* expression. For this purpose we injected neutralizing anti-*ras* antibody 259 into quiescent cells and then added serum. This tests the ability of the added serum to stimulate *fos* expression in cells from which *ras* activity had been neutralized. We consistently found that the antibody-injected cells had a reduced level of *fos* expression. In few cases, however, was *fos* expression totally blocked by the injected anti-*ras* (Stacey *et al.*, 1987). Since other experiments to be described suggest that the antibody is extremely efficient in inhibiting *ras* activity, this result suggests that while *ras* might be involved in the stimulation of *fos* synthesis, factors present in serum are able to stimulate its synthesis in the absence of *ras* activity. The fact that a number of different mechanisms might stimulate *fos* is consistent with the observations of others. It is clear, however, that *ras* protein does play a role in controlling the expression of nuclear oncogenes.

Injected p21*ras* was able to induce a prolonged period of *fos* expression suggesting that in *ras*-transformed cells the level of *fos* protein might be constantly elevated. This, however, was not the case. We found that in cells transformed by a *ras*-containing virus the level of *fos* synthesis was as low as in resting, nontransformed cells. Addition of serum to the *ras*-transformed cells, however, was able to induce *fos* synthesis as was the microinjection of purified p21*ras* protein. Furthermore, the induction of *fos* expression in *ras*-transformed cells was inhibited by injected anti-*ras* antibody (Stacey *et al.*, 1987). It therefore appears that the control over *fos* expression is highly complex and results from positive and negative control elements. While *ras* can induce a strong and prolonged period of expression, that level apparently must eventually revert to non-induced levels without destroying the ability of the cells to respond to a subsequent *ras* stimulus.

On the basis of these results, it is clear that there is a connection between *ras* activity and the expression of at least one nuclear oncogene. It is also clear that this relationship is highly complex and controlled at a number of levels. The rigorous nature of this control indicates the importance of *fos* expression. At present, however, it is not clear what function nuclear oncogenes play. They are known to be involved in differentiation particularly of lymphoid cells, to cooperate with *ras* in

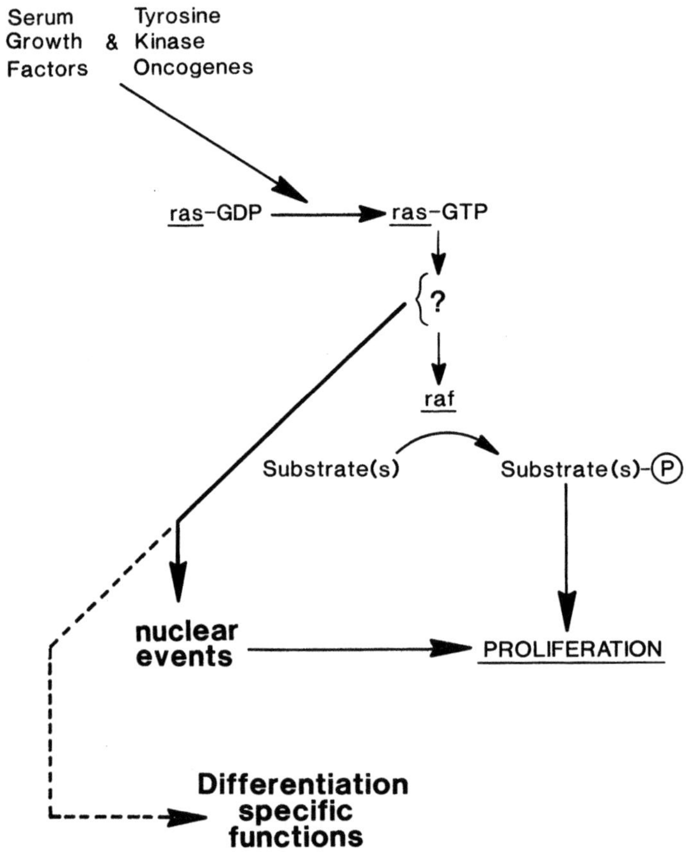

Diagram 3. Role of nuclear oncogenes. p21*ras* is able to induce *fos* production and anti-*ras* inhibits serum-stimulated *fos* protein synthesis. It is therefore proposed that *ras* activity might play a role in controlling nuclear proto-oncogene expression. If so, p21*ras* might function twice in mitogenic stimulation. *ras* activity would induce nuclear proto-oncogene expression early in mitogenic stimulation to prepare the cell for a subsequent *ras*-stimulus to initiate S-phase. Nuclear proto-oncogenes might also be involved in cellular differentiation.

the transformation of non-established cell types, and to be connected with the action of competence growth factors. It is proposed on the basis of all these factors that nuclear oncogenes are involved in controlling the physiological state of cells. This might involve the transition from a resting to a proliferative state or from one differentiation state to another. If the induction of this transition involves the action of growth factors, then the induction also involves the action of *ras* proteins since, as previously proposed, *ras* proteins are believed to mediate the intracellular signals initiated by growth factors (Diagram 3).

This model fits well with what is known about the action of different growth factor types. Two classes of such factors have been identified. Competence factors are not mitogenic but are able to prepare a cell to respond to the action of the second type of factor, the progression growth factor. This would indicate that two signals might be needed to induce proliferation and would indicate that *ras* activity might therefore be required twice in mitogenic stimulation. The signal initiated by competence growth factors (or perhaps other types of stimuli) occurs early in mitogenic stimulation. The second growth factor-initiated signal comes when the cell is prepared to divide, just prior to the beginning of S-phase (as observed earlier in the antibody injection studies).

These observations also suggest a possible explanation for the observed requirement for two different oncogenes in the transformation of non-established cell types (Ruley, 1983). It is possible that these cells must be rendered competent for proliferation by the expression of nuclear oncogenes. *ras* protein alone can induce these proteins and presumably lead to proliferation. After a period of prolonged, high level *ras* expression, however, the negative control factors within a cell inhibit nuclear proto-oncogene expression. The cell would then stop proliferating due to the lack of these nuclear factors even in a cell containing high levels of transforming *ras* protein. If the activity of *ras* were regulated in the proper sequence, however, it might become sufficient in itself to transform even established cell types.

ras ACTIVITY AND PHOSPHOLIPID METABOLISM

The studies discussed above indicate some of the biological functions performed by cellular *ras* proteins. First, it is clear that *ras* activity plays a critical role in controlling cell proliferation. This activity is necessary for cell division in a number of normal cell types. In NIH3T3 cells the activity of transforming *ras* protein is sufficient to induce mitogenesis in the absence of other growth factors. Second, it is also apparent that cellular *ras* protein is essential for the activity of tyrosine kinase, growth factor receptor-like oncogenes. On the basis of these observations we conclude that *ras* proteins form an essential part of a proliferative signal transduction pathway initiated by growth factor receptor molecules. There is no indication from these biological studies, however, of the biochemical mechanism for this signal transduction.

The nature of the proliferative pathway in which *ras* proteins might function has been suggested by a number of biochemical studies. These indicate that the addition of growth factors to cells results in a significant increase in the rate of phosphatidyl inositol metabolism. It is known, furthermore, that the action of phospholipase C, which hydrolyzes phosphatidyl inositol, is controlled by a GTP binding protein such as p21*ras* (Cockcroft and Gomperts, 1985). The hydrolysis of phosphatidyl inositol results in the generation of two important second messengers, diacylglycerol and inositol triphosphate. These molecules activate protein kinase C and intracellular calcium mobilization, respectively (Berridge, 1984). If, as postulated, p21*ras* functions by controlling phospholipase C, its action would result in the activation of protein kinase C and an increase in intracellular calcium concentrations.

Phospholipase C and *ras* activity

We tested the possibility that *ras* proteins control phospholipase C activity by attempting to artifically duplicate the effects of this enzyme within cells. We then determined if this treatment was able to overcome the requirement for *ras* activity during cellular proliferation. We duplicated the action of phospholipase C by treating the cells with the phorbol ester TPA and in some cases with calcium ionophore, ionomycin. These treatments are known to activate protein kinase C and to raise intracellular calcium concentrations just as the two second messengers produced by phosphatidyl inositol turnover. Swiss 3T3 cells were induced to proliferate by this treatment following microinjection of the neutralizing anti-*ras* antibody #259. We found that the injected cells were efficiently inhibited in proliferation by the injected antibody (Yu *et al.*, in review). This result indicates that treatments designed to mimic phospholipase C activity were unable to overcome the proliferative need for cellular p21*ras*.

This observation suggests that *ras* activity does not function to control the action of phospholipase C. It is of course possible that the added TPA and ionomycin did not faithfully duplicate the action of this phospholipase. If, however, the action of these materials upon cells does mimic phospholipase C activity to any extent, and if p21*ras* does function to control phospholipase C, then the resulting proliferation would be independent of cellular *ras* to the extent by which the treatment mimics the phospholipase activity. The fact that the proliferation induced is almost completely inhibited by the injected antibody suggests that *ras* functions by another mechanism.

We next tested the possibility that p21*ras* functions to control phospholipase A_2 (Bar-Sagi and Feramisco, 1986). This enzyme cleaves the fatty acid from the #2 position of phospholipids generally to release arachadonic acid which is then a precursor in the formation of leukotrienes and prostaglandin synthesis. Again we attempted to duplicate the action of this enzyme by adding arachadonic acid or other metabolites to cells. We (Chun-Li Yu and Men-Hwei Tsai) were successful in efficiently inducing proliferation in the NIH3T3 cells only with prostaglandin F2α. As before, NIH3T3 cells were induced to proliferate with this prostaglandin following microinjection of anti-*ras* antibody 259. The prostaglandin was relatively efficient in its mitogenic induction but was unable to induce proliferation in antibody-injected cells. In fact, the antibody inhibited proliferation with unexpected efficiency following prostaglandin treatment. When NIH3T3 cells are induced to proliferate with serum, the anti-*ras* antibody was able to inhibit proliferation by 85-95%. The prostaglandin-induced proliferation, however, was inhibited by 99% in the same cell cultures (Yu *et al.*, in review). Again, this result indicated that p21*ras* does not function to control phospholipase A_2.

The mitogenic potential of phosphatidic acid

The above experiments fail to provide support for the popular notion that p21*ras* functions to control the activity of either phospholipase C or A_2. It is possible that another phospholipase is the target of *ras* activity. It is known that phosphatidic acid (PA) can function with the properties of a growth factor (Moolenaar *et al.*, 1986). We tested this with our NIH3T3 cells and found that its mitogenic potential was great. When vesicles of pure PA were added to the cells, they were induced to undergo a profound morphologic alteration from their normal flattened appearance to the rounded, refractile appearance of transformed cells. In addition, the added PA was able to induce thymidine incorporation in quiescent cells with similar kinetics and efficiency as the addition of 10% calf serum (Yu *et al.*, in review). These observations indicated that the action of PA is closely related to the action of growth factors and therefore might be closely connected to *ras* activity.

We tested the possibility that p21*ras* functions to contol PA production within cells as before by microinjecting anti-*ras* antibody prior to treatment of quiescent cells with PA. We were surprised to observe again that the antibody was highly efficient in blocking PA-induced mitogenesis. The injected cells were inhibited in thymidine incorporation by 99% as was observed with prostaglandin-treated cells. In addition, the profound morphologic alteration produced by PA treatment was abolished by the injected antibody (Yu *et al.*, in review). It was clear that *ras* does not function to control the production of PA. In fact, it was clear that the phospholipid metabolites we tested were even more dependent upon *ras* activity than were the growth factors present in serum. This observation is totally inconsistent with the notion that *ras* proteins control phospholipid metabolism.

p21*ras* appears to be controlled by a phospholipid metabolite

The clear connection between cellular phospholipid metabolism and mitogenic induction indicates that activity of p21*ras* is related to that of a phospholipase. The the fact that the phospholipid metabolites tested were remarkably dependent upon *ras* activity for their mitogenic potential indicates that p21*ras* does not function to control a phospholipase. We were therefore forced to consider the possibility that rather than controlling phospholipid metabolism, p21*ras* activity is itself controlled by phospholipid metabolism.

This novel suggestion might be supported by work other than that described above. It is known that the tyrosine kinase oncogenes phosphorylate the phospholipase inhibitor lipocortin. This phosphorylation is believed to release its inhibitory potential (Hirata, 1981). It is, therefore, reasonable to predict that the activity of these tyrosine kinases might lead to increased phospholipase activity. Biologically it is known that the activity of these tyrosine kinase, receptor-like molecules depends upon *ras* activity. It is now predicted that *ras* activity is controlled by a phospholipase metabolite. These considerations form the basis for a model to explain the biochemical mechanism of proliferative signal transduction (Diagram 4).

The "*ras* PATHWAY", a model of proliferative signal transduction

It is proposed, therefore, that cellular proliferation is controlled by the interaction of a number of cellular oncogenes. The proliferative signal is initiated by an occupied growth factor receptor or a related oncogene. The resulting tyrosine kinase activity leads to the production of a phospholipid metabolite which is able to stimulate the activity of cellular p21*ras*. The active p21*ras* then transfers the proliferative signal into the cell. It is not known what biochemical function activated p21*ras* performs. It has been postulated that p21*ras* controls an enzyme able to produce a proliferative second messenger. This second messenger is then postulated to control the activity of the cellular oncogene *raf*.

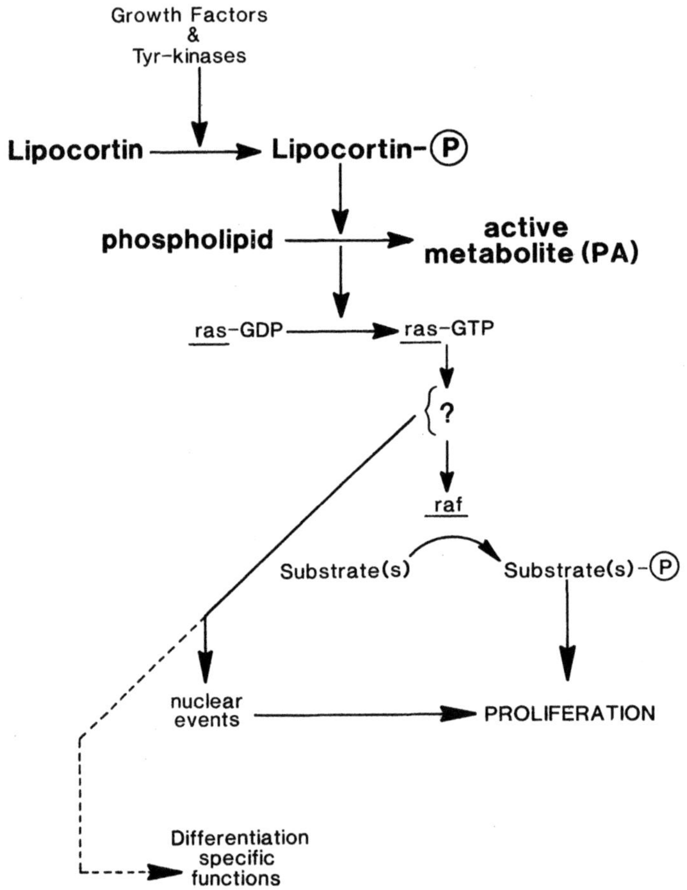

Diagram 4. Phospholipase activity in mitogenic stimulation. The mitogenic action of lipid metabolites depends totally upon *ras* activity. It is, therefore, proposed that phospholipid metabolism forms a signal transduction mechansim between tyrosine kinase, receptor-like molecules and p21*ras*. The phospholipase inhibitor lipocortin, a substrate of tyrosine kinases, might play a role in this postulated signal transduction mechanism.

No direct evidence supports these assumptions as yet. The most important part of the proposed model relates to the postulated connection between tyrosine kinases and *ras*. It is proposed that phospholipid metabolism forms a signal transduction mechanism between these two cellular oncogene classes (Diagram 4).

The verification of the model proposed above (Diagram 4) will require a great deal of work. Because of its apparent ability to biologically activate p21*ras*, it would be important to understand the metabolism of phosphatidic acid (PA). Unfortunately, the means by which PA is normally produced in the plasma membrane is not known. It is known that PA is an intermediate in the biosynthesis of phospholipids, which normally occurs in cytoplasmic membranes. PA is present at low concentrations in the plasma membranes of most animal cells where its concentration is carefully controlled. There is recent evidence that in response to insulin treatment PA production is stimulated rapidly and remains elevated (Farese *et al.*, 1987). The primary source of PA appears to be *de novo* synthesis. There is also a diacylglycerol kinase activity present in the plasma membranes of a wide variety of animal cells. This would convert the products of phospholipase C into PA.

While there is much to learn about the production of PA, there is also much to be learned about the interrelationship between different intracellular signal transduction mechanisms. We propose that p21*ras* functions with tyrosine kinases in a signal transduction mechanism which specifically controls proliferation. This signal transduction mechanism, we have shown, is distinct from that involving phosphatidyl inositol turnover. There is support from others for the existence of these two signalling systems within cells. Bombesin has growth factor-like potential in some cells and is known to act through phosphatidyl inositol metabolism. It has been shown that the mitogenesis resulting from bombesin treatment involves a distinct signalling pathway from that involving growth factors such as insulin (Chambard *et al.*, 1987).

Further support for the idea that *ras* functions to transduce the signals initiated by growth factors has been found in amphibian oocytes. In these cells both insulin and progesterone lead to maturation, as does the action of microinjected *ras* proteins. Anti-*ras* antibody is able to block the effect of insulin upon the oocytes, but has no effect upon the action of progesterone (Deshpande and Kung, 1987). *ras* activity within the oocytes is therefore not required for the process of maturation but is necessary for the action of insulin.

These observations support our proposal that growth factors like insulin must function through the "*ras* Pathway" primarily to control proliferation in a manner independent of phosphatidyl inositol turnover. It is clear, however, that treatments such as TPA or bombesin which activate protein kinase C are mitogenic. This might be due to the ability of protein kinase C, like the receptor kinases, to phosphorylate lipocortin leading to *ras* activation. It should also be noted, however, that the action of lipocorins (or calpactins) is controversial.

Lipocortins

There is some evidence that lipocortins do not affect phospholipases directly but bind their substrates (Davidson *et al.*, 1987). It has also been proposed that lipocortins might actually function in attaching structural proteins to the membrane. Neither of these possibilities argues against the model above (Diagram 4). If lipocortins are the biologically important substrates of tyrosine kinases (and this is not clear at present) our model would propose that phosphorylation would lead to the production of an important phospholipid metabolite. When one considers the nature of membranes and the action of phospholipases, any alteration in the structure of intimately associated proteins might have profound consequences upon the activity of a critical phospholipase.

In relationship to the possibility that lipocortins do function in linking structural molecules to membranes, there is one important consideration. If lipocortins function in the maintenance of membrane structure, and if tyrosine kinases do recognize these molecules, it might be attractive to postulate that the morphologic alterations seen in transformed cells result from lipocortin phosphorylation. Our data do not support this conclusion. We have shown that the morphologic alterations resulting from tyrosine kinase transformation do not occur following anti-*ras* antibody injections. The morphologic alterations, therefore, depend upon *ras* activity rather than upon the fundamental biochemical properties of tyrosine kinases. Furthermore, PA was able to induce morphologic alterations only in a totally *ras*-dependent manner. It is difficult to form a model in which the phosphorylation of lipocortin by tyrosine kinases can be directly responsible for morphological alterations under these circumstances.

All the foregoing analyses assume that p21*ras* functions normally to control proliferation. We have tested a number of normal cell types both of epithelial and connective tissue origin. In all cases the anti-*ras* antibody efficiently inhibited proliferation. In support of this conclusion, it has been found that *ras* protein is expressed in all proliferating tissues. It has also been found that *ras* is expressed in the brain, a nonproliferating tissue (Furth *et al.*, 1987). Furthermore, anti-*ras* antibody 259 has been shown to block the morphological alterations caused by nerve growth factor in PC12 cells (Hagag *et al.*, 1986). It is clear, therefore, that *ras* activity might have non-proliferative functions. Our primary thesis suggests that growth factor receptor-like molecules initiate an intracellular signal that passes through *ras* protein. In proliferating tissues this signal ultimately controls mitogenesis. In non-proliferating tissues the same signal transduction pathway might affect other important cellular functions as long as these are controlled by the interaction between growth factor-like molecules and their binding proteins.

STUDIES OF *ras* ACTIVITY AND TUMOR CELL PROLIFERATION

Any discussion of the control of proliferation is incomplete without consideration of the possible alterations of this control in tumor cells. The data presented indicate that *ras* activity is required for the proliferation of several cell types. Furthermore, *ras* activity is needed during the transformation of NIH3T3 cells by a variety of oncogenes related to growth factor receptors or growth factor molecules. These observations suggest that p21*ras* functions to transduce the proliferative signal generated by growth factor receptors. It is also clear that *ras* activity is not necessary for the transforming potential of cytoplasmic, serine kinase oncogenes. This not only suggests that the proliferative signal which passes through p21*ras* might then proceed to a cellular oncogene related to the *raf* oncogene, but also indicates that anti-*ras* antibody injections are able to distinguish the activities of two separate classes of oncogenes. It is not necessary to understand the molecular basis of this distinction. It is simply clear that anti-*ras* antibody injection allows classification of oncogenes without any other knowledge of their structure.

Dependence of DNA tumor viral transformation upon p21*ras*

Retroviral oncogenes (which were derived from cellular genes) were found either to be almost totally dependent upon cellular *ras* activity for expression of their transforming potential, or to function almost totally independently of *ras* activity. Another class of viral transforming genes are associated with a variety of DNA tumor viruses. These oncogenes do not appear to be directly related to any particular cellular genes and are often present within the virus in combination with another oncogene. It was of interest for us to determine the *ras*-dependence of transformation by these oncogenes. Accordingly, NIH3T3 cells transformed by the polyoma, SV40, and bovine papilloma virus (BPV) were obtained and received anti-*ras* antibody injections.

It was clear that both the BPV-transformed cells and the polyoma-transformed cells were morphologically reverted from the transformed to the non-transformed phenotype following the injection of anti-*ras*. In addition, the ability of these cells to incorporate thymidine was reduced by the injected antibody. No such inhibitions were apparent when non-neutralizing antibody #238 was injected. It is important to note that while the anti-*ras* inhibitions were distinct, the thymidine incorporation was not reduced to the extent seen with tyrosine kinase oncogenes. With polyoma, for example, the inhibition was from 50-75%, while *src*-transformed cells averaged 80% inhibitions. Similar observations were made with BPV. No alteration of the transformed morphology was observed in injected SV40-transformed cells, and thymidine incorporation was inhibited less than 10% in these cells (Smith *et al.*, 1986).

The observations above lead to several conclusions. First, it is clear that like the retroviral oncogenes, DNA tumor viral oncogenes may function in either a *ras*-dependent or in a *ras*-independent manner. Second, it suggests that while the polyoma and SV40 viruses are related molecularly, their transforming genes function differently. Finally, these data raise a question about the mechanism of inhibition by anti-*ras* antibody. Why were inhibitions of thymidine incorporation by polyoma and BPV only partial? It is possible that insufficient antibody had been injected to neutralize all the cellular p21*ras* within the cells. Yet in cells where the expression of this protein was driven by a highly efficient retroviral promoter, the inhibition was virtually complete (Kung *et al.*, 1986). It does not appear, therefore, that insufficient antibody had been injected.

It might also be argued that for some reason the localization or conformation of *ras* protein within these cells had been altered so that the antibody became inefficient in neutralization. This does not appear possible since, as described above, in those NIH3T3 cells which had been passaged long enough to develop a *ras*-independent proliferative phenotype, the PA and prostaglandin-induced proliferation remained 99% inhibited by anti-*ras*. The antibody had been able to completely inhibit the efficient proliferation induced by phospholipid metabolites but remained unable to totally inhibit serum-induced mitogenesis. The *ras*-independent proliferative phenotype must, therefore, be due to a fundamental property of the oncogenes within these cells rather than to the inability of injected antibody to efficiently neutralize *ras* protein activity.

Based upon the above considerations, it was assumed that the partial *ras* dependence of polyoma and BPV transformation upon *ras* activity represented an important characteristic of transformation within these cells. Since there are two transforming genes in each virus, we postulated that the partial requirement for *ras* activity during transformation is a composite phenotype resulting from the interaction of a *ras*-dependent and a *ras*-independent oncogene. As a preliminary test of this hypothesis one of the polyoma tranforming genes was shown to be almost completely dependent upon *ras* activity for expression of the transforming phenotype. NIH3T3 cells transformed by only the polyoma middle-T antigen (obtained from T. Benjamin) were inhibited by 95% in thymidine incorporation (unpublished data). This observation indicates that one of the polyoma oncogenes is as *ras*-dependent as are the retroviral tyrosine kinase oncogenes. The hypothesis would, therefore, suggest that the other transforming activity of polyoma must function more *ras*-independently. This hypothesis has not yet been tested. These data, however, provide reason to consider the possibility that if two genes from different classes were functioning together to promote proliferation, the resulting *ras*-dependence might be of an intermediate phenotype.

Antibody injection into carcinoma cell lines

Anti-*ras* antibody injections have been employed above to classify retroviral oncogenes that are derived from distinct cellular genes and to classify DNA tumor viral oncogenes whose structures are known but cannot be related directly to cellular genes. With these data in mind, an analysis is made of tumor cell lines. While some of these are known to contain mutations in the cellular *ras* genes, the identity of the transforming genes of most tumor cell lines are not known. These have no genes able to promote the transformation of NIH3T3 cells. The first set of experiments involved 14 carcinoma cell lines. Of these, five cell lines have been shown to contain mutations in a *ras* gene. No mutations were identified in the other nine.

Anti-*ras* antibody #259 was injected into the cells of each carcinoma line. After 15-21 hrs the recipient cultures were labelled with thymidine, stained to detect the injected immunoglobulin, and autoradiographed as described previously. It is known that the antibody is able to recognize and efficiently neutralize the mutant *ras* protein since NIH3T3 cells transformed by human transforming *ras* genes are efficiently blocked in proliferation. If the mutant *ras* gene were totally responsible for the altered, mutant proliferative phenotype of the tumor cells, it would be expected that the microinjected antibody would totally inhibit proliferation. Instead, we (S. DeGudicibus and M.R. Smith) found that each of the mutant *ras*-containing cell lines was only partially inhibited in thymidine incorporation. Inhibition ranged between 50% and 85% as observed with polyoma and BPV-transformed cells. In no case was thymidine incorporation observed to be less than 50% (Stacey *et al.*, 1987).

It is therefore apparent that while the *ras* gene mutation might ensure that *ras* retains a role in controlling the proliferation of the tumor, other genes must also be involved. These genes are able to induce proliferation in a *ras*-independent manner as do the *mos* and *raf* oncogenes. The partial dependence upon *ras* activity might, therefore, result from the interaction of *ras*-dependent and *ras*-independent genes, as previously observed in DNA tumor viral-transformed cells.

Of the nine carcinoma cell lines which did not contain identifiable mutant genes, two were partially inhibited in proliferation as were those with *ras* mutations. In the seven other cell lines without mutations there was no detectable inhibition of thymidine incorporation by the neutralizing anti-*ras* antibody. These cell lines incorporated thymidine equally as well following injection with neutralizing antibody #259 as with the non-neutralizing antibody #238 (Stacey *et al.*, 1987). We have tested the possibility that tumor cell lines lose *ras*-dependence as a result of placement in culture. For this purpose several epithelial cell lines that have been extensively cultured received injections of anti-*ras* antibody. Of six lines tested, only one was able to proliferate efficiently following the microinjection (Plate 4; Stacey *et al.*, 1987). It is therefore believed that the process of placement in culture could not be responsible for the totally *ras*-independent proliferative phenotype of seven of the nine tumor cell lines tested (without identifiable mutations).

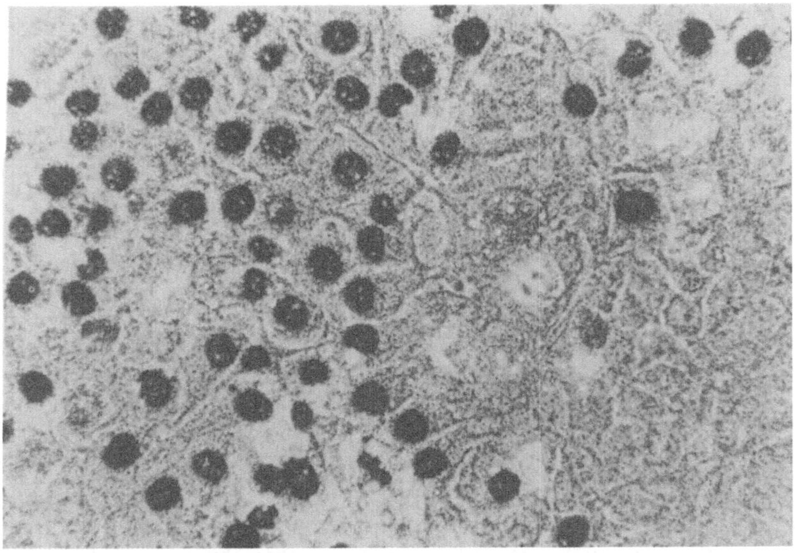

Plate 4. Inhibition of thymidine incorporation in epithelial cell culture. The BS-C-1 cells at the right of the area photographed were microinjected with anti-*ras* antibody 259 and pulsed with labelled thymidine after 18 hrs. It is apparent in this autoradiograph that uninjected cells are much more likely to be labelled.

Of the 14 carcinoma cell lines tested none were as dependent upon *ras* activity as were the NIH3T3 cells previously tested. This suggested to us that the process of tumor development had in some way overcome the normal requirement for *ras* activity during proliferation. This conclusion is weakened by the fact that it is not possible to test the cells which gave rise to the tumor lines tested. It is possible that the normal progenitor cell was also able to proliferate in a *ras*-independent manner. To address this possibility normal epithelial cells were tested to determine if they, like the NIH3T3 cells, depend upon *ras* activity during proliferation. Normal keratinocytes were prepared with the help of S. Yuspa and normal human mammary cells were provided by M. Stampfer. Both cell lines were highly dependent upon *ras* activity. Injected cells were inhibited by nearly 99% in thymidine incorporation. The observations with mammary cells are particularly important since the mammary tumors we tested were never totally inhibited by the microinjected antibody and in some cases were not inhibited to any detectable extent.

Analysis of sarcoma cell lines

To extend the comparison between normal and tumor cells, we analyzed six sarcoma lines. Of these, two were known to contain mutant *ras* genes. As with the carcinoma cells, the two *ras* mutant-containing sarcoma cell lines were inhibited approximately 75% in thymidine incorporation. This is a significant inhibition, but is not as great as is normally seen with NIH3T3 cells. Injected anti-*ras* inhibited thymidine incorporation in the four other cell lines by only an average of aproximately 20-25%. When several diploid, non-established fibroblast cell lines were tested, each was inhibited approximately 90% by the antibody (Stacey *et al.*, 1987).

The results with cells of connective tissue origin, therefore were similar to the results obtained with cells of epithelial origin. In both cases normal cells were dependent upon *ras* activity in proliferation. The epithelial cells were particularly *ras*-dependent. In tumor cells derived from either tissue type, those with *ras* mutations were partially sensitive to the antibody. Inhibitions were always more than 50% but normally less than 80%. Of the cells without identifiable mutant genes, some were also partially inhibited by the antibody but most proliferated almost without inhibition following anti-*ras* microinjection. We therefore conclude that for some unknown reason, the dependence of proliferation upon *ras* activity is the characteristic of normal cells that is lost during tumor formation.

The data above suggest a possible fundamental distinction between the proliferative activity of normal and tumor cells. This distinction is of potential significance since it involved a gene known to play a critical role in proliferation. In NIH3T3 cells this gene is both sufficient and necessary for proliferation. The practical value of this observation will have to await demonstration that it represents a general distinction between normal and transformed cells. If so, and if *ras* activity can be blocked pharmacologically, it might be possible to make use of this observation in cancer treatment. If *ras* activity were inhibited, normal but not tumor cells would stop proliferating. The tumor cells would then be comparatively more susceptible to antiproliferative treatments. In other words, the anti-*ras* treatment might be able to specifically spare the normal cells of antiproliferative drugs.

Tumor formation and *ras* activity

A simple explanation for the loss of *ras* dependence during tumor formation relates to previous studies of the biological function of p21*ras*. It has been shown that *ras* is necessary for the action of growth factors. Since tumor cells do not rely upon growth factors for proliferation, they do not need *ras* to transmit the growth factor-generated proliferative signal.

Because our present understanding of the function of *ras* is incomplete, it might not be possible to fully understand why tumor cells lose their *ras*-dependence. In fact, it is possible that the tumor cell studies reveal that the control over proliferation is highly complex. To explain the tumor cell studies it is important to consider several factors. First, it is known that a single mutation is never sufficient to lead to tumor formation. Otherwise the number of tumors would be overwhelming. Second, it is clear that some genes are involved not so much in promoting tumor formation as in keeping tumors from developing. These recessive, anti-tumor genes have not yet been characterized but will undoubtedly hold important keys in understanding tumor formation. Any model for proliferative control and related tumor formation must involve negative control mechanisms or multiple positive signals.

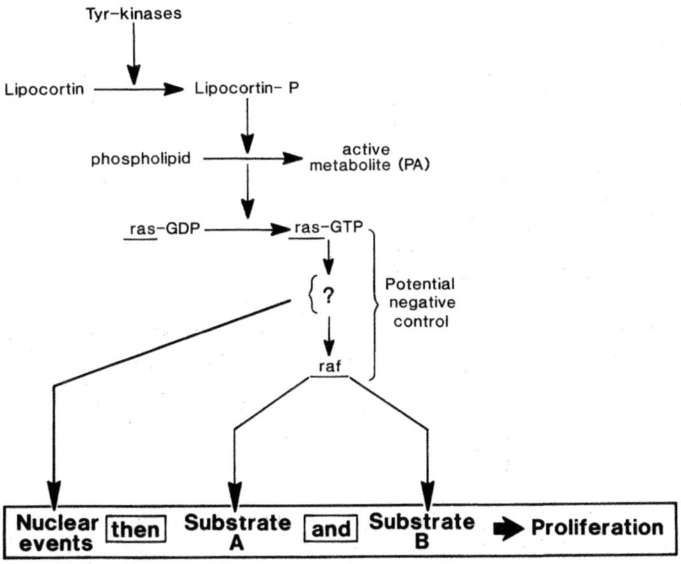

Diagram 5. Proliferative events subsequent to *ras* action. It is proposed that negative control mechanisms regulate *ras* activity and, subsequent to the action of *ras*, the prolifertive control mechanism must involve multiple, perhaps parallel, pathways. This hypothesis is based not only on the *ras*-independence of tumor cell proliferation but also upon the fact that oncogenes of most tumors are not identifiable in an assay which requires a single, dominant transforming gene.

A model concerning tumor formation To explain our data and those considerations described above, the following model is proposed (Diagram 5). The initial studies indicate that retroviral oncogenes might participate in a linear signal transduction pathway. These oncogenes are unique in that they are able to induce the transformation of NIH3T3 cells alone. Since these genes function in a dominant positive manner, there most be some cellular mechanism to control their function. Since *ras* is intimately involved in proliferation and is required for the transforming potential of many retroviral oncogenes, it might be the focus of negative control mechansims. If negative proliferative mechanisms focus upon p21*ras*, then a proliferative phenotype which does not rely upon *ras* activity would no longer be under this control. This fact might be responsible for the *ras*-independence of tumor proliferation. If a cell can develop a proliferative phenotype which does not depend upon *ras*, it might be difficult for the organism to exercise normal restraints.

The nature of a *ras*-independent proliferative phenotype is obviously highly speculative at this time. It clearly does develop in some cells since it has been observed in NIH3T3 cells after extensive culture. That this *ras*-independent proliferation does not involve phospholipid metabolites has been established earlier. It is also apparent that the *ras*-independent proliferative phenotype often does not involve the generation of oncogenes detectable in the NIH3T3 transfection assay since most tumor cells have no identifiable oncogenes. Therefore, while the genes involved must be *ras*-independent as are *mos* and *raf*, their transforming potiential is not as great.

To explain the data, a preliminary working model is described in Diagram 5. As before it is postulated that the positive stimulus for proliferation originates with an occupied growth factor receptor. This receptor perturbs phospholipid metabolism to generate a metabolite such as phosphatidic acid which activates p21*ras*. The active *ras* protein sends a signal to *raf* proteins. The activity of these two cellular proteins must, however, be the focus of strong negative control mechanisms. Otherwise, activated oncogenes in this pathway could lead to tumor formation directly. After the action of the "*ras* pathway," which is strictly controlled, proliferation depends upon the action of multiple pathways. This could be accomplished by assuming that multiple substrates of cellular *raf* are required to induce proliferation. In this way, no single mutation can be sufficient to induce proliferation. Those genes with the ability to function in a dominant proliferative manner are under strict control. Those genes which function downstream of these control mechanisms to induce proliferation must not be able to function independently.

SUMMARY

The studies described above have led to some conclusions as well as some speculations regarding the participation of oncogenes in proliferation. It is important to make a clear distinction between the data described along with the resulting conclusions, and the highly speculative models which have been proposed here to describe these results. On the basis of the results described, several conclusions appear to be indicated. First, it is clear that in normal cells of many types *ras* proteins are required for proliferation. Second, these proteins are needed for the maintenance of the transformed phenotype induced by some, but not all, oncogenes. Third, the activity of *ras* proteins is not apparently involved in controlling the activity of phospholipase C or any other identifiable phospholipase. Finally, tumor formation appears to involve the development of a proliferative phenotype which functions independently of *ras* activity.

To explain these data, a model of proliferative control is presented. This model is highly speculative at this time and is based upon the assumption that the "*ras* pathway" involves the sequential action of cellular genes related to the retroviral oncogenes. The function of this pathway is to pass the proliferative signal generated by an occupied growth factor receptor into the cell. It is clearly needed late in the G-1 phase of the cell cycle, but the "*ras* pathway" might also be involved in the early events associated with mitogenic stimulation. Due to its central role in the control of proliferation and the fact that tumor cells circumvent its action, *ras* proteins are postulated to a site of negative proliferative control. After the action of *ras* and related cellular oncogenes, the proliferative control signal no longer involves the action of a simple linear sequence of protein activities, but might involve multiple, interdependent pathways. This model is primarily of value as a working hypothesis and does not account for many observations central to proliferative control such as the involvement of cell-cell contact, differentiation, and the action of factors which inhibit rather than promote proliferation, such as interferons. The model does summarize the data described and even in the simplest form represents a novel approach to explain proliferative control in terms of the activities of known genes.

REFERENCES

Barbacid, M., 1987, *ras* genes, Annual. Rev. Biochem., (in press).

Bar-Sagi, D. and Feramisco, J.R., 1986, Induction of membrane ruffling and fluid-phase pinocytosis in quiescent fibroblasts by *ras* proteins. Science, 233:1061-1068.

Berridge, M.J., 1984, Inositol triphosphate and diacylglycerol as second messengers, Biochem. J., 220:345-360.

Brugge, J.S., 1986, The p35/p36 substrates of protein-tyrosine kinases as inhibitors of phospholipase A_2, Cell, 46:149-150.

Chambard, J.C., Paris, S., L'Allemain, G., and Pouyssegur, J., 1987, Two growth factor signalling pathways in fibroblasts distinguished by pertusis toxin, Nature, 326:800-803.

Cockcroft, S. and Gomperts, B.D., 1985, Role of guanine nucleotide binding protein in the activation of polyphosphoinositide phosphodiesterase, Nature, 314:534-536.

Davidson, F.F., Dennis, E.A., Powell, M., and Glenney, J.R., 1987, Inhibition of phospholipase A_2 by "lipocortins" and calpactins,an effect of binding to substrate phospholipids, J.Biol. Chem., 262:1698-1705.

Deshpande, A.K., and Kung, H.-F., 1987, Insulin induction of *Xenopus laevis* oocyte maturation is inhibited by monoclonal antibody against p21*ras* proteins, Mol. Cell. Biol., 3:1285-1288.

Doolittle, R.F., Hunkapiller, M.W., Hood, L.E., Devare, S.G., Robbins, K.C., Aaronson, S.A. and Antoniades, H.N., 1983, Simian sarcoma virus *onc* gene, v-*sis*, is derived from the gene (or genes) encoding a platelet-derived growth factor, Science, 221:275-277.

Downward, J., Yarden, Y. Mayes, E. Scrace, G. Totty, N., Stockwell, P., Ullrich, A., Schlessinger, J. and Waterfield, M.D., 1984, Close similarity of epidermal growth factor receptor and v-*erb* B oncogene protein sequences, Nature, 307:521-527.

Farese, R.V., Konda, T.S., Davis, J.S., Standaert, M.L., Pollet, R.J., and Cooper, D.R., 1987, Insulin rapidly increases diacylglycerol by activating de novo phosphatidic acid synthesis, Science, 236:586-589.

Furth, M.E., Davis, L.J., Fleurdelys, B. and Scolnick, E.M., 1982, Monoclonal antibodies to the p21 products of the transforming gene of Harvey murine sarcoma virus and the cellular *ras* gene family, J. Virol., 43:294-304.

Furth, M.E., Aldrich, T.H. and Cordon-Cordo, C., 1987, Expression of *ras* proto-oncogene in normal human tissues, Oncogene, 1:47-58.

Gilman, A.G., 1984, G-proteins and dual control of adenylate cyclase, Cell, 36:577-579.

Hirata, F., 1981, The regulation of lipomodulin, a phospholipase inhibitory protein, in rabbit neurophils by phosphorylation, J. Biol. Chem, 256:7730-7733.

Kung, H.F., Smith, M.R., Bekesi, E., Manne, V. and Stacey, D.W., 1986, Reversal of transformed phenotype by monoclonal antibodies against Ha-*ras* p21 proteins, Exp. Cell Res., 162:363-371.

Lacal, J.C., Santos, E., Notario, V., Barbacid, M., Yamazaki, S., Kung, H.-F., Seamans, C., McAndrew, S., and Crowl,R., 1984, Expression of normal and transforming H-*ras* genes in *Escherichia Coli* and purification of their encoded p21 proteins, Proc. Natl. Acad. Sci. USA, 81:5305-5309.

Mercer, W.E., Avignolo, C., and Baserga, R., 1984, Role of the p53 protein in cell proliferation as studied by microinjection of monoclonal antibodies, Mol. Cell. Biol., 4:276-281.

Moolenaar, W.H., Kruijer, W. Tilly, B.C., Verlaan, I. Bierman, A.J. and deLaat, S.W. (1986). Growth factor-like action of phosphatidic acid. Nature 323:171-173.

Mulcahy, L.S., Smith, M.R. and Stacey, D.W., 1985, Requirement for *ras* proto-oncogene function during serum-stimulated growth of NIH3T3 cells, Nature, 313:241-243.

Muller, R., Bravo, R, Burckhardt, J., and Curran, T., 1984, Induction of c-*fos* gene and protein by growth factors precedes activation of c-*myc*, Nature, 312: 716-720.

Papageorge, A.G., Willumsen, B.M., Johnsen, M., Kung, H.-F., Stacey, D.W., Vass, W.C. and Lowy, D.R., 1986, A transforming *ras* gene can provide an essential function ordinarily supplied by an endogenous *ras* gene, Mol. Cell.Biol., 6:1843-1846.

Ruley, H.E., 1983, Adenovirus early region 1A enables viral and cellular transforming genes to transform primary cells in culture, Nature, 304:602-606.

Scolnick, E.M., Papageorge, A.G. and Shih, T.Y., 1979, Guanine nucleotide binding activity as an assay for *src* protein of rat-derived murine sarcoma viruses,Proc. Natl. Acad. Sci. USA, 76:5355-5359.

Sherr, C.J., Rettenmier, C.W., Sacca, R., Roussel, M.F., Look, A.T. and Stanley, E.R., 1985, The c-*fms* proto-oncogene product is related to the receptor for the mononuclear phagocyte growth factor, CSF-1, Cell, 41:665-676.

Smith, M.R., DeGudicibus, S.J. and Stacey, D.W., 1986, Requirement for c-*ras* proteins during viral oncogene transformation, Nature, 320:540-543.

Stacey, D.W., and Allfrey, V.G., 1977, Evidence for the autophagy of microinjected proteins in HeLa cells, J. Cell Biol., 75:807-817.

Stacey, D.W., and Kung, H.-F., 1984, Transformation of NIH 3T3 cells by microinjection of Ha-*ras* p21 protein, Nature, 310:508-511.

Stacey, D.W., Watson, T., Kung, H.-F., and Curran, T., 1987, Microinjection of transforming *ras* protein induces c-fos expression, Mol. Cell. Biol., 7:523-527.

Stacey, D.W., DeGudicibus, S.R., and Smith, M.R., 1987, Cellular *ras* activity and tumor cell proliferation, Exptl. Cell Res., (in press).

Sweet, R.W., Yokoyama, S., Kamata, T., Feramisco, J.R., Rosenberg, M. and Gross, M. (1984). The product of *ras* is a GTPase and the T24 oncogenic mutant is deficient in this activity. Nature 311:273-275.

Yu, C.-L., Tsai, M.-W., and Stacey, D.W., Cellular *ras* activity and phospholipid metabolism, (in review).

THE Ca^{2+}/PHOSPHOLIPID-BINDING PROTEINS OF THE SUBMEMBRANEOUS SKELETON

Tony Hunter

Molecular Biology and Virology Laboratory
The Salk Institute, P.O Box 85800
San Diego, California 92138

Among the best characterized substrates for the protein-tyrosine kinases are two proteins with molecular weights between 34 and 39 kDa and pI's about 7. These proteins have been given a variety of names (Table 1), but I will refer to them here as p36 and p35. It has recently emerged that p36 and p35 are structurally and functionally related to each other and to at least two additional proteins, p67 and p33. All four proteins contain multiple copies of a related 75 amino acid repeat, and all of them bind to phospholipid bilayers in the presence of μM concentrations of Ca^{2+} and are localized to the inner face of cytoplasmic membranes. The true functions of these proteins are not known, but there are two major schools of thought. The first proposes that these proteins have structural roles in the cortical submembrane skeleton of the cell. The second holds that these proteins act as inhibitors of phospholipase A_2, and have called them lipocortins. Here I will review the properties of these proteins, and discuss the evidence for their possible functions.

Table 1. **Nomenclature of Ca^{2+}/phospholipid binding proteins**

p67	p36	p35	p33
73K	Calpactin I Lipocortin II	Calpactin II Lipocortin I	
Protein III	Protein I		Protein II Endonexin
Calelectrin 67	Calelectrin 36		Calelectrin 32.5
Lymphocyte 68	Lymphocyte 33		
Calcimedin 67			Calcimedin 35
Chromobindin 20	Chromobindin 8	Chromobindin 9	

The relevant literature for this classification is given in the text. Each of the four columns represents proteins which, despite their different names, are likely to be products of a single gene. Individual calmotilins (Rocha et al., 1986) probably also correspond to p67, p36, p35, and p33. The numbers following calelectrin, calcimedin and lymphocyte are estimated molecular weights in kDa. Chromobindins are numbered according to the scheme of Creutz et al. (1983,1987).

p36

A basic protein of ~36 kda (p36) was the first substrate to be identified for pp60$^{v\text{-}src}$ (Radke and Martin, 1979). p36 was subsequently shown to be among the most prominent substrates for many of the retroviral protein-tyrosine kinases (Radke et al., 1980; Erikson and Erikson, 1980; Kobayashi and Kaji, 1980; Cooper and Hunter, 1981a,b; Cheng and Chen, 1981; Erikson et al., 1981a; Decker, 1982; Nakamura and Weber, 1982; Greenberg and Edelman, 1983a), as well as for the EGF and PDGF receptor protein-tyrosine kinases (Hunter and Cooper, 1981; Erikson et al., 1981b; Cooper et al., 1982)(for review see Cooper and Hunter, 1983a).

Subsequently several proteins apparently identical to p36 have been described independently in other systems (Table 1). The following proteins are synonymous with p36: protein II (Gerke and Weber, 1984); the 36 kDa mammalian calelectrin (Geisow et al., 1984; Südhof et al., 1984), one of a family of proteins which are defined by their antigenic relatedness to a membrane-associated protein first characterized in the electric organ of the electric eel (Walker, 1982; Südhof, 1984a,b); chromobindin 8 (Creutz et al., 1987), one of a set of proteins associated with chromaffin granule membranes in a Ca^{2+}-dependent fashion (Creutz et al., 1983; Geisow and Burgoyne, 1982); lipocortin II, a protein identified based on its ability to inhibit phospholipase A_2 activity in vitro, which was shown to be identical to p36 through molecular cloning (Huang et al., 1986); and a 33 kDa lymphocyte protein bound to plasma membranes in a Ca^{2+}-dependent fashion (Davies and Crumpton, 1985). The implications of these identities for the function of p36 will be discussed later. In the following discussion of p36, I will make use of the reported properties of these proteins assuming that they are p36.

In fibroblasts p36 is an abundant protein, accounting for 0.1-0.4% of total cell protein (Radke et al., 1980, Cooper and Hunter, 1983b; Isacke et al., 1986). Using immunochemical and cell fractionation techniques p36 was localized in cultured cells to the cortical skeleton, a protein layer which underlies the plasma membrane (Amini and Kaji, 1983; Cooper and Hunter, 1983b; Courtneidge et al., 1983; Greenberg and Edelman, 1983b; Nigg et al., 1983; Lehto et al., 1983; Radke et al., 1983). A more recent study using monoclonal antibodies shows that there are two populations of p36 (Zokas and Glenney, 1987). One of these populations is bound tightly to the detergent insoluble cytoskeleton, whereas the other is detergent soluble. The cytoskeletal population of p36 is associated with a small 11 kDa subunit, p11, in the form of a $(p36)_2(p11)2$ tetramer (see below). The nature of the cytoskeletal interaction is unknown, but does not appear to involve actin, tubulin, or intermediate filament proteins, since the pattern of p11 localization is unlike that of any of these filamentous structures (Zokas and Glenney, 1987), and since drugs which affect cytokeratin, microtubule and actin filament structures do not alter the distribution of p36 in the cell (Alitalo et al., 1984). Whether p11 itself binds to whatever cytoskeletal protein is involved, or whether p11 modifies the activity of the p36 subunits to cause binding is not known. The "soluble" pool of p36 is monomeric and diffusely localized to membranes (Zokas and Glenney, 1987). The independence of the two pools of p36 is emphasized by the fact that the cytoskeletal p36 is aggregated into patches following microinjection of an anti-p11 monoclonal antibody, without affecting the soluble pool of p36 (Zokas and Glenney, 1987). p36 has also been shown to colocalize with the spectrin-like molecule in the surface lamina of fibroblasts (Lehto et al., 1983). However, there does not appear to be a strong interaction between p36 and spectrin, since the microinjection of either anti-spectrin or anti-p36 antibodies fails to result in the co-aggregation of spectrin and p36 (Mangeat and Burridge, 1984; Zokas and Glenney, 1987).

```
                                            *  *
p36         (1)  STVHEILCKLSLEGDHSTPPSAYG.SVKPYT...........
p35         (1)  AMVSEFL.KQAWFIENEEQ..EYVQTVKSSKGGPGSAVSPYP
p33                                (1) AAKGGTVKAAS...........
p67                                (1) AKPAQGAKYR....GS.IHDFP

p67 (Linker)     DAAGQFFPEAAQVAYQMWELSAVARVELKGTVRPAN

p36 (Repeat 1)         NFDAERDALNIETAV.KTKGVDEVTIVNILTNRSNVQRQDIAFAYQRRTKKE.LPSALKSALSGHLETVILGLL
p35 (Repeat 1)         TFNPSSDVAALHKAIM.VKGVDEATIIDILTKRNNAQRQQIKAAYLQETGKP.LDETLKKALTGHLEEVVLALL
p33 (Repeat 1)         GFNAAEDAQTLRKAM.KGLGTDEDAIISVLAYRSTAQRQEIRTAYKSTIGRD.LLDDLKSELSGNFEQVILGMM
p67 (Repeat 1)         GFDPNQDAEALYTAM.KGFGSDKEAILDIITSRSNRQRQEVCQSYKSLYGKD.LIADLKYELTGKFERLIVGLM
p67 (Repeat 5)         DFNPDADAKALRKAM.KGLGTDEDTIIDIITHRSNVQRQQIRQTFKSHFGRD.LMTDLKSEISGDLARLILGLM

p36 (Repeat 2)        KTPA.QYDASEL.KASMKGLGTDEDSLIEIICSRTNQELQEINRVYKE.MYKTDLEKDIISDTSGDFRKLMVALAKGRR
p35 (Repeat 2)        KTPA.QFDADEL.RAAMKGLGTDEDTLIEILASRTNKEIRDINRVYRE.ELKRDLAKDITSDTSGDFRNALLSLAKGDR
p33 (Repeat 2)        .TPTVLYDVQEL.RRAMKGAGTDEGCLIEILASRTPEEIRRINQTYQL.QYGRSLEDDIRSDTSFMFQRVLVSLSAGGR
p67 (Repeat 2)        RPPA.YCDAKEI.KDAISGIGTDEKCLIEILASRTNEQMHQLVAAYKD.AYERDLEADIIGDTSGHFQKMLVVLLQGTR
p67 (Repeat 6)        MPPA.HYDAKQL.KKAMEGAGTDEKALIEILATRTNAEIRAINEAYKE.DYHKSLEDALSSDTSGHFRRILISLATGHR

p36 (Repeat 3)   AEDGSVIDYELIDQDARELYDAGVKRKGTDVPKWISIMTERSVCHLQKVFERYKSYSPYDMLE.SIKKEVKGDLENAFLNLVQCIQN
p35 (Repeat 3)   SEDFGVNE.DLADSDARALYEAGERRKGTDVNVFNTILTTRSYPQLRRVFQKYTKYSKHDMNK.VLDLELKGDIEKCLTAIVLCATS
p33 (Repeat 3)   .DEGNYLDDALVRQDAQDLYEAGEKKWGTDEVKFLTVLCSRNRNHLLHVFDEYKRISQKDIEQ.SIKSETSGSFEDALLAIVKCMRN
p67 (Repeat 3)   .EEDDVVSEDLVQQDVQDLYEAGELKWGTDEAQFIYILGNRSKQHLRLVFDEYLKTTGKPMKA.SIRGELSGDFEKLMLAVVKCIRS
p67 (Repeat 7)   .EEGGENLDQARE.DAQVAAEILEIADTPSGDKFMTILCTRTYPHLRRVFQEFIKMTNYDVEH.TIKKEMSGDVRDAFVAIVQSVKN
                                                ↑
                                              TSLETR

p36 (Repeat 4)         KPLYFAD.R.LYDS.MKGKGTRDKVLIRIMVSRSEVDMLKIRSEFKR.KYGKSLYYYIQQDTKGDYQKALLYLCGGDD (338)
p35 (Repeat 4)         KPAFFAE.K.LHQA.MKGVGTRHKALIRIMVSRSEIDMNDIKAFYQK.MYGISLCQAILDETLGDYEKILVALCGGN  (345)
p33 (Repeat 4)         KSAYFAE.R.LYKS.MKGLGTDDNTLIRVMVSRAEIDMMDIRANFKR.LYGKSLYSFIKGDTSGDYRKVLLILCGGDD (318)
p67 (Repeat 4)         TPEYFAE.R.LFKA.MKGLGTRDNTLIRIMVSRSELDMLDIREIFRT.KYEKSLYSMIKNDTSGEYKKTLLKLSGGDD
p67 (Repeat 8)         KPLFFAD.K.LYKS.MKGAGTDEKTLTRIMVSRSEIDLLNIRREFIE.KYDKSLHQAIEGDTSGDFLKALLALCGGED (673)
                          ↑   ↑        ↑        ↑            ↑   ↑          ↑     ↑
Conserved residues       d/e 1/i       G        i/v  R       i/l f/y        i/l    G
```

The sequences of p36, p35, p33 and p67 are aligned to demonstrate the 75 amino acid repeats. Gaps (.) are introduced to maximize the similarity between repeats. A 6 amino acid insertion below repeat 7 of p67 is also omitted for this reason. Repeats 4-8 of p67 have been aligned as if the p67 gene was derived by duplication, i.e. repeat 1 being homologous to repeat 5, etc. The Tyr 23 and Ser 25 phosphorylation sites in p36 are depicted by asterisks. (1) is the N-terminus of each protein and (xxx) the C-terminus. Residues conserved in each repeat are given at the bottom (upper case residues are conserved in >19/20 repeats), and the highly conserved region is underlined (Geisow et al., 1986; Geisow and Walker, 1986). The sequences are: murine p36 (Saris et al., 1986); human p35 (Huang et al., 1986); porcine p33 (Weber et al., 1987) and human p67 (Südhof et al., 1988).

Figure 1 **SEQUENCE ALIGNMENT OF p36, p35, p33 and p67**

p36 is associated peripherally with the membrane rather than integrally. It is synthesized on soluble ribosomes and the mature protein is the same size as the primary translation product (Courtneidge et al., 1983). Although there is one report that p36 is modified by the covalent attachment of myristate (Soric and Gordon, 1985), this has not been confirmed by other investigators (K. L. Gould and J. E. Buss, personal communication; Magee and Courtneidge, 1985). Indeed myristylation is believed to occur uniquely at the N-termini of proteins, and p36 has been shown to be N-terminally acetylated (Glenney et al., 1986). Much of p36 can be eluted from membranes and the cytoskeleton with chelators of divalent cations in the absence of detergent (Cooper and Hunter, 1983; Courtneidge et al., 1983; Greenberg and Edelman, 1983b; Radke et al., 1983). This behaviour would be atypical of an integral membrane protein, and strongly supports the idea that p36 is peripherally associated. There is one report that a fraction p36 is present in ribonucleoprotein particles associated with membranes and can be cross-linked to RNA in the cell (Arrigo et al., 1983). The significance of this observation is unclear, but may reflect the basic nature of p36.

In the animal p36 is present at high levels in certain epithelial cells, all endothelial cells, and fibroblastic cells (Gould et al., 1984; Greenberg et al., 1984). Although a fuller description of the tissues in which p36 is expressed is given below, it is worth noting here that in the columnar epithelial cells of the intestine, which are particularly rich in p36, p36 is concentrated in the brush border cytoskeletal region beneath the microvillar array (Gould et al., 1984; Greenberg et al., 1984). This situation is analogous to the cortical location of p36 in cultured cells.

Use has been made of the ability of chelating agents to release p36 from particulate fractions to purify milligram amounts of homogeneous p36 from mammalian intestine and lung (Gerke and Weber, 1984; Glenney and Tack, 1985; Glenney et al., 1987). p36 has also been purified from placenta (Huang et al., 1986). Smaller quantities have been obtained from fibroblasts (Erikson and Erikson, 1980; Cooper and Hunter, 1983; Courtneidge et al., 1983; Greenberg and Edelman, 1983a; Soric and Gordon, 1986). The observed properties of purified p36 go some way to explaining its subcellular location and give clues to its function.

The protein is isolated from intestine in the form of a heterotetramer consisting of two p36 subunits and two 11 kDa subunits (p11) (Gerke and Weber, 1984; Glenney and Tack; 1985; Glenney et al., 1987). In fibroblasts only about half of p36 is found as the tetramer, with the rest being monomeric (Erikson et al., 1984). The formation of a p11 dimer is apparently the critical force stabilizing the tetramer (Gerke and Weber, 1985; Glenney et al., 1986; Johnsson et al., 1986b). A 33 kDa C-terminal core fragment, which no longer binds p11, can be generated by proteolysis from p36, localizing the major p11 binding site to the N-terminal 3 kDa of p36 (Glenney and Tack, 1985; Glenney 1986b; Johnsson et al., 1986b). A synthetic peptide corresponding to the N-terminal 29 residues of p36 abrogates binding of purified p11 to p36, confirming the assignment of this binding site (Glenney et al., 1986). The sensitivity of p36 to a variety of proteases between residues 25 and 30 implies that this region forms an accessible linker joining two domains, comprising a C-terminal 33 kDa resistant core, and a small 3 kDa N-terminal domain.

The tetramer has been shown to bind to F-actin and non-erythroid spectrin in a Ca^{2+}-dependent fashion, albeit only at mM Ca^{2+} concentrations, which are clearly nonphysiological (Gerke and Weber, 1984). In vitro monomeric p36 can bundle actin filaments in the presence of Ca^{2+} suggesting that there is more than one actin binding site per monomer (Gerke and Weber, 1985a; Glenney et al., 1987). The number of binding sites for spectrin is unknown, but as indicated above there does not appear to be a strong interaction between these two molecules in vivo. Likewise the fact that p36 does not codistribute with actin filaments in the cell implies that the affinity of p36 for actin in vivo is low.

p11 is of interest because it has strong homology with S-100 (Glenney and Tack, 1985; Gerke and Weber, 1985b; Hexham et al., 1986; Saris et al., 1987), a prominent brain Ca^{2+}-binding protein with two Ca^{2+}-binding sites of the E-F hand type (Tufty and Kretsinger, 1975), characteristic of calmodulin and troponins. p11 proves to be a member of a burgeoning family of proteins, which includes S100-α and S100-β, 2A9 (Calabretta et al., 1986), and the CF antigen (Dorin et al., 1987). Despite this homology to known Ca^{2+}-binding proteins, however, p11 cannot be demonstrated to bind Ca^{2+} directly (Gerke and Weber, 1985a,b; Glenney 1986b). Analysis of the sequence of p11 shows that both the putative E-F hand structures are degenerate and would not be expected to bind Ca^{2+} (Saris et al., 1987). Despite the loss of the Ca^{2+}-binding activity, p11 has retained the ability to dimerize, which was originally noted for S100. The regions through which p11 dimerizes and binds to p36 have not yet been defined. In many ways p11 has attributes which suggest that it is a regulator of p36 function (Gerke and Weber, 1985b; Johnsson et al., 1986a,b; Glenney et al., 1986,1987).

The $(p36)_2(p11)_2$ complex, monomeric p36, and the 33 kDa core fragment of p36, bind to phospholipid vesicles in the presence of μM concentrations of Ca^{2+}, showing a preference for phosphatidylserine and phosphatidylinositol (Glenney, 1985; Glenney, 1986a,b; Glenney et al., 1987; V. C. Carter and G. S. Martin, personal communication). Based on its ability to bind <u>cal</u>cium, <u>p</u>hospholipid and <u>actin</u> the $(p36)_2(p11)_2$ complex has been termed calpactin I (Glenney, 1986a), in which p36 corresponds to the heavy chain, and p11 the light chain. This complex has also been called protein I (Gerke and Weber, 1984). The binding of Ca^{2+} has been shown to cause a conformational change in the p36 subunit, with the K_d for Ca^{2+} being 100 μM (Gerke and Weber, 1985a). Since the Ca^{2+} concentration in the cell never reaches 100 μM, this might imply that p36 does not bind Ca^{2+} in vivo. In the presence of phospholipid, however, both p36 and its 33 kDa core bind greater than 2 moles of Ca^{2+}/mole with a K_d of 5 μM, a value much nearer physiological Ca^{2+} concentrations (Glenney, 1986b). More recent experiments suggest that the maximum number of Ca^{2+}-binding sites may be four (Glenney and Zokas, 1988), but there does not appear to be marked cooperativity in the binding of Ca^{2+} to p36. In the presence of Ca^{2+}, p36 and the 33 kDa core can aggregate phospholipid vesicles, implying the presence of multiple phospholipid binding sites (Glenney et al., 1987). The affinity of the p36 complex for phospholipid liposomes is found to be higher than that of the monomer (Powell and Glenney, 1987). In addition it has recently been shown that the $(p36)_2(p11)_2$ complex, but not p36 monomer, will aggregate chromaffin granule membranes in the presence of μM Ca^{2+} (Drust and Creutz, 1988). In the same assay the 33 kDa core of p36 causes aggregation, although it requires a higher Ca^{2+} concentration than the p36 complex. All these findings strengthens the idea that p11 regulates p36 function.

Based on these assembled properties it has been proposed that p36, either as the monomer or the tetramer, might function in a structural manner in the submembraneous or cortical skeletion, either linking the cytoskeleton to the plamsa membrane or else cross-linking vesicles to the membrane. Obviously a critical question is whether p36 is directly bound to phospholipids and/or cytoskeletal proteins in the cell, and if so under what conditions. In early experiments with purified p36 the Ca^{2+} ED_{50} for p36 binding to phospholipids or membranes was estimated to be 2-5 μM (Glenney, 1986b; Drust and Creutz, 1988), which is high compared to normal resting intracellular Ca^{2+} levels (0.1 μM). This suggested that the interaction of p36 with the membrane might only occur following treatment by agents, such as growth factors, which cause an increase in intracellular Ca^{2+} above 1 μM. However, recent experiments carried out at higher phospholipid concentrations, which probably more nearly reflect the situation in vivo, indicate that p36 can bind to phospholipids in the absence of detectable Ca^{2+} (Powell and Glenney, 1987). The fact that much of p36 is associated with particulate fractions of untreated cells and can be released by EGTA certainly reinforces the notion that p36 is normally bound to phospholipids in a Ca^{2+}-dependent fashion in vivo. It is currently unclear whether the binding to phospholipids or the cytoskeleton is modulated by Ca^{2+} in vivo. Agents which affect intracellular Ca^{2+} levels do not cause a dramatic redistribution of p36, as determined by immunofluorescence studies (Alitalo, et al., 1984). Nevertheless it still seems probable that the function of p36 is modulated by Ca^{2+}. Regulation could also be exerted by phosphorylation (see below), by p11, or by interaction of p36 with other membrane components.

The structure of p36 has been investigated through the isolation of cDNA clones from murine, human, bovine and avian sources (Saris et al., 1986; Huang et al., 1986; Kristensen et al., 1986; S. Jones and R. L. Erikson, personal communication). All of them predict proteins of 338 amino acids following removal of the N-terminal methionine. Comparison of the predicted sequences for p36 shows that it is very highly conserved among avian and mammalian species, with 90% homology between the chicken and human proteins (Saris et al., 1986; S. Jones and R. L. Erikson, personal communication). The most striking feature of the predicted sequence is the presence of a 4-fold 75 amino acid repeat, occupying the C-terminal 300 amino acids, with identity between the repeats ranging from 28-32% (Saris et al., 1986)(Figure 1). The N-terminal 30 amino acids do not form part of this repeating unit. This is entirely consistent with the two-domain model for p36 (Glenney and Tack, 1985; Johnsson et al., 1986b), and with the finding that this sequence is responsible for p11 binding (Glenney et al., 1986; Johnsson et al., 1986b). The 4-fold repeat structure suggests an evolutionary mechanism for the p36 gene in which an archetypal repeat was duplicated, and then the dimer was reduplicated to give the present core structure with four repeats. This idea is supported by the presence of a short 10 residue extension at the N-terminal end of repeat 3, which shows a limited similarity to the 10 residues at the end of the N-terminal domain.

Although no formal proof is yet available, for many reasons it seems likely that each repeat represents a Ca^{2+}/phospholipid binding domain. First, there is a rough correspondence between the number of Ca^{2+}-binding sites and the number of repeats. Second, similar repeat structures have been identified in a series of other Ca^{2+}/phospholipid binding proteins (see below). Third, site-directed fragmentation studies on the p35 (lipocortin I) family member suggest that the minimal fragment which can bind Ca^{2+} and phospholipid comprises a single repeat (Huang et al., 1987; R. B. Pepinsky, personal communication). No canonical E-F hand structure(s) are present in the repeat, however, implying that we are dealing with a novel type of Ca^{2+}-binding site (Saris et al., 1986). Analysis of the repeats for conserved secondary structure by prediction paradigms reveals a potential α-helix/loop/β-sheet structure in the N-terminal 30 residues (Saris et al., 1986), a structure which has similarities to that of the E-F hand Ca^{2+}-binding site. A conserved Gly is present in the predicted loop structure, which could be important for Ca^{2+} binding. This region also coincides with the most highly conserved part of the repeat that was first identified as an antigenic epitope recognized by antisera raised against the 34 kDa *Torpedo* calelectrin, a protein with similar Ca^{2+}/phospholipid-binding properties (Walker, 1982; Südhof et al., 1984; Geisow et al., 1984).

Overall p36 is a rather hydrophilic and highly charged protein, but a hydropathy plot indicates that there is a relatively hydrophobic region at the N- and C-terminal ends of each repeat (Saris et al., 1986), which could form the basis for a hydrophobic interaction with the lipid bilayer. Hydrophobic residues also tend to occur at a spacing of 4-5 residues over each repeat (Südhof et al., 1988). Since p36 binds to phenothiazine in the presence of Ca^{2+} (Carraway et al., 1987), Ca^{2+} binding by p36 apparently results in the exposure of a hydrophobic surface, as is the case for other Ca^{2+}-binding proteins such as calmodulin. It is currently unclear whether there are separate Ca^{2+}- and phospholipid-binding sites, or whether there is a combined Ca^{2+}/phospholipid binding site. p36 binds to phospholipids in the absence of Ca^{2+} providing high concentrations of phospholipid are used (Powell and Glenney, 1987), suggesting there is a separate phospholipid binding site. Conversely, p36 binds Ca^{2+} weakly in the absence of phospholipid (Gerke and Weber, 1985a). The simplest model would be that each repeat contains a combined Ca^{2+}/ phospholipid binding site. These four sites must be arrayed on the p36 molecule in such a fashion that at least two phospholipid vesicles can be bound.

The exact nature of the actin- and spectrin-binding sites are unknown, although a very weak similarity to the actin-binding site in gelsolin has been noted near the C-terminus of the consensus repeat (Burgoyne, 1987)). The N-terminal p11-binding domain must be joined to the p36 core by an exposed linker, since many proteases cleave between residues 20 and 30 (Glenney and Tack, 1985; Johnsson et al., 1986b), and this region also contains two identified phosphorylation sites (see below). Clearly many of the uncertainties about the localization of binding sites will be resolved by analysis of site-directed mutants of p36. The highly conserved residues within all four repeats of p36, and in those of other family members are obvious targets for such analysis (Figure 1). Final proof of the location of the binding sites, however, will have to come from a crystal structure.

The regulation of p36 and p11 expression have been examined through surveys of tissues and cell types that express these genes. This has been accomplished both at the mRNA and at the protein level. As described above p36 is expressed in certain epithelial and endothelial cells, as well as in fibroblastic cells, cardiac muscle cells and macrophages (Gould et al., 1984; Huang et al., 1986). When total tissues are assayed, p36 is present at the highest levels in lungs and intestine, at intermediate levels in spleen, lymph nodes, thymus and testes, and at low levels in heart and kidney (Gould et al., 1984; Greenberg et al., 1984; Shadle et al., 1985; Glenney et al., 1987). p36 is not detected in brain, skeletal muscle or red blood cells. p36 is present in certain but not all lymphoid cell lines (Cooper et al., 1983; Gould et al., 1984). The distribution of p36 mRNA (size 1.6 kb) parallels that of the protein rather closely (Gould et al., 1984; Saris et al., 1986).

The expression of p36 is probably regulated at the transcriptional level, although this has not been formally shown. The p36 gene, which maps to mouse chromosome 9, has been estimated to be over 30 kb (B. F. Tack and T. Hunter, unpublished results), but the promoter region has not been cloned and characterized. Based on what is known about the cell type-specific regulation of other genes, it seems likely that there will be binding sites for tissue-specific transcription factors upstream of the p36 gene. There is one interesting case where the expression of p36 has been shown to be inducible (Carter et al., 1986). Myoblasts in day 5 chick embryo limb buds express undetectable levels of p36 protein. Following dispersion of limb bud myoblasts and subsequent attachment to tissue culture plastic there is the rapid appearance of p36 protein within 1-2 hr. Presumably this reflects induction of p36 mRNA synthesis, although this remains to be shown. If the myoblasts are dispersed and suspended in methocel rather than being allowed to attach to plastic, p36 synthesis is still induced but no protein accumulates. Thus p36 is apparently unstable in unattached cells. Since it is known that p36 dissociates from the cytoskeleton when cells are detached from their substratum (Cooper and Hunter, 1982), this may mean that p36 is unstable unless it is associated with some cytoskeletal structure. This notion is reinforced by the finding that the noncytoskeletal monomeric pool of p36 has a much shorter half life than the cytoskeletal complexed pool of p36 (Zokas and Glenney, 1987)

p11 appears to be expressed only in the tissues and cell types where p36 is present (Zokas and Glenney, 1987), and the level of p11 relative to p36 appears to be fairly constant. In fibroblasts and some other cell lines, however, there is clearly an excess of p36 (Erikson et al., 1984; Gerke and Weber, 1985a). Unlike p36 there are some discrepancies between the level of p11 mRNA and p11 protein. For instance brain has a low level of p11 mRNA but no detectable protein (Zokas and Glenney, 1987; Saris et al., 1987). Possibly p11 is very unstable if it is not complexed with p36. This idea is borne out by the observation that p11 is never present in excess over p36 (Zokas and Glenney, 1987).

The phosphorylation of p36 has been extensively studied in a number of systems. Phosphorylated p36 (pp36) is detected both in cells transformed by viruses, which encode protein-tyrosine kinases (Radke and Martin, 1979; Erikson and Erikson, 1980; Kobayashi and Kaji, 1980; Radke et al., 1980; Cooper and Hunter, 1981a,b; Erikson et al., 1981a; Cooper et al., 1983) and in certain cells treated with growth factors, such as epidermal growth factor (EGF) (Hunter and Cooper, 1981; Erikson et al., 1981b; Isacke et al., 1986) and platelet-derived growth factor (PDGF) (Cooper et al., 1982; Isacke et al., 1986), whose receptors are protein-tyrosine kinases. p36 is found phosphorylated both on serine and tyrosine (Radke et al., 1980; Erikson and Erikson, 1980; Hunter and Cooper, 1981a), although any one molecule has just a single phosphate moiety, and only between 1 and 10% of all p36 molecules contain phosphate at steady state (Radke et al., 1980, Cooper and Hunter, 1983b). There is evidence for multiple serine and tyrosine phosphorylation sites in p36 (Isacke et al., 1986). In vitro pp60$^{v\text{-}src}$ phosphorylates bovine intestinal p36 predominantly at Tyr 23 (Glenney and Tack, 1985). Ca^{2+} at high concentrations by itself, or at low concentrations in the presence of phospholipids, stimulates the phosphorylation of p36 at Tyr 23 by pp60$^{v\text{-}src}$ (Glenney, 1986). This same tyrosine is the major tyrosine phosphorylation site in pp36 isolated from RSV-transformed cells (Erikson and Erikson, 1980; Isacke et al., 1986). In PDGF-treated human fibroblasts, and in EGF-treated A431 cells, both Tyr 23 and a second unidentified tyrosine phosphorylation site become phosphorylated to an equal extent (Isacke et al., 1986). p36 is an excellent substrate for the Ca^{2+}/phospholipid-dependent protein kinase, protein kinase C (Gould et al., 1986; Khanna et al., 1986b) and a major serine phosphorylation site for protein kinase C has been identified as Ser 25 (Gould et al., 1986). The phosphorylation of Ser 25 is also increased in TPA-treated human fibroblasts showing that p36 is phosphorylated by protein kinase C in vivo. There appears to be a second site for protein kinase C phosphorylation in monomeric p36 within the N-terminal 20 residues (Johnsson et al., 1986a). This site can apparently also be phosphorylated by the multifunctional calmodulin-dependent protein kinase and by cAMP-dependent protein kinase. There is evidence that phosphorylation at this site blocks association with p11 (Johnsson et al., 1986a). Phosphorylation at a serine other than Ser 25 is induced in PDGF-treated human fibroblasts (Isacke et al., 1986). In adrenal chromaffin cells phosphorylation of p36 on serine can be induced by nicotine, a drug which causes exocytosis (Creutz et al., 1987). Given the localization of Tyr 23 and Ser 25 at the end of the N-terminal domain, the other unmapped in vivo phosphorylation sites could lie in the N-terminal 20 residues or in the N-terminal extension of repeat 3, which also contains a Ser and a Tyr.

The effects of these phosphorylations on the functions of p36 are as yet poorly defined. When RSV-transformed cells are fractionated, phosphorylated p36 (pp36) is predominantly membrane bound, and shows no difference in distribution to the unphosphorylated form of p36 (Cooper and Hunter, 1982; Amini and Kaji, 1983). p36 phosphorylated at Tyr 23 binds to phospholipid vesicles with decreased affinity in the absence of Ca^{2+} (Powell and Glenney, 1987), but it is unclear whether this phosphorylation alters the affinity for Ca^{2+}. It is worth noting that the effects of phosphorylation in general on Ca^{2+} binding could be very hard to measure if each phosphate moiety affected exclusively the neighbouring binding site, since this would only reduce the stoichiometry of Ca^{2+} binding by 25%. Phosphorylation of Tyr 23 does not affect binding of p11, but, as indicated above, phosphorylation at an unidentified residue in the N-terminal 20 residues does block p11 binding (Johnsson et al., 1986a). If a similar phosphorylation occurs in vivo, then the ratio of monomeric to complexed p36 might change. This in turn could alter the amount of p36 bound to membranes, since the complex has a higher affinity for phospholipid than the monomer. Identification of the additional phosphorylation sites in p36, and site-directed mutation of these sites individually and in combination may help to unravel the effects of phosphorylation on p36 function, when it is elucidated.

p35

A 35 kDa basic protein, initially identified as a substrate for the EGF receptor protein-tyrosine kinase (Fava and Cohen, 1984; Sawyer and Cohen, 1985; Giugni et al., 1985), has recently been shown to be closely related to p36 both functionally (Glenney, 1986a) and structurally (Wallner et al., 1986; Huang et al., 1986). For instance p35 binds to F-actin and phospholipid in a Ca^{2+}-dependent manner, and consequently has been called calpactin II (Glenney, 1986a). Like p36, proteins that are synonymous with p35 have been uncovered in other guises (Table 1). These are: chromobindin 9 (and its proteolytic product chromobindin 6), both of which have been shown to be related to p35 (Creutz et al., 1987); and lipocortin I, which was shown directly to be identical to p35 (Wallner et al., 1986; Sinclair and Pepinsky, 1986).

The structure of p35 predicted from the cDNA clone for lipocortin I (mRNA size 1.6 kb) (Wallner et al., 1986) has four 75 amino acid repeats, each of which is clearly homologous to the repeats in p36 (Saris et al., 1986)(Figure 1). Historically, it should be noted that p35 was called 35 kDa because it migrated slightly faster than p34 (Fava and Cohen, 1984), the size estimated by Erikson's group for p36 (Erikson and Erikson, 1980). In fact p36 has 338 amino acids, while p35 has 345 amino acids, consistent with their SDS gel electrophoretic mobilities. However, both should have molecular masses near 38,500, which is actually near our original estimate for p36 (Cooper and Hunter, 1981a). Many of the residues conserved between the p36 repeats are also conserved in the four p35 repeats (Figure 1). Rather than discuss this issue here, I will consider it later in light of all known examples of this repeat structure, of which there are now over 20 available. The overall identity between p36 and p35 is 50% (Huang et al., 1986; Saris et al., 1986), but there is little homology between the N-terminal 30 residues of the two proteins. Since this region of p36 is required for p11 binding, one would predict that p35 would not bind p11, and this is indeed the case (Glenney, 1986a)). Upon purification of p35 several groups identified a spontaneously occuring proteolytic product which lacks the N-terminal 12 residues of the intact protein (De et al., 1986; Glenney et al., 1987; Haigler et al., 1987; Creutz et al., 1987).

p35 can also be cleaved by trypsin at Lys 26 (Haigler et al., 1987). Thus p35 has a two domain structure very like that of p36, with a short accessible N-terminal domain, and a protease resistant C-terminal core.

p35 was first detected based on its phosphorylation in membrane preparations of EGF-treated A431 cells (Fava and Cohen, 1984). p35 was found to be associated with membranes in a Ca^{2+}-dependent fashion, and was purified from membranes following elution with EGTA. Larger quantities of p35 (calpactin II) have been purified from bovine (De et al., 1986) and porcine lung (Glenney et al., 1987), as well as from placenta (Huang et al., 1986; Haigler et al., 1987; Valentine-Brown, et al., 1986). The isolated protein shows Ca^{2+}-dependent phospholipid binding activity, with a preference like that of p36 for anionic phospholipids (Glenney et al., 1987; Schlaepfer and Haigler, 1987). Although there is not total conformity among the data, it appears that p35 binds a maximum of 4 molecules of Ca^{2+} in the presence of phosphatidylserine (Glenney et al., 1987; Schlaepfer et al., 1987) with a K_d of 10-75 μM. Glenney et al. (1987), who originally found evidence for cooperativity between 2 Ca^{2+} binding sites (Glenney et al., 1987), have now reinvestigated this question and come to the conclusion that there are 4 independent Ca^{2+}-binding sites (Glenney and Zokas, 1988), in agreement with Schlaepfer et al. (1987). Like p36, p35 can aggregate phospholipid vesicles (Glenney et al., 1987).

Cell fractionation studies and immunofluorescence staining show p35 to be present in the submembraneous cortical skeleton. However, double immunofluorescence staining indicates that p35 and p36 are not precisely colocalized, although their distributions clearly overlap (Glenney et al., 1987; C. M. Isacke, personal communication). In fibroblasts the pattern for p35 is more uniform over the cell and covers the nucleus, whereas staining for p36 excludes the nucleus and is less intense towards the edge of the cell. Since p35 does not appear to be tightly associated with the cytoskeleton, the difference in localization may in part be explained by the cytoskeletal population of complexed p36. In human cells p35 is an abundant protein, although somewhat scarcer than p36 (Glenney, 1986a). p35 is expressed in many cell types and tissues (De et al., 1986; Huang et al., 1986; Pepinsky et al., 1986; Wallner et al., 1986; Glenney et al., 1987). The highest levels are found in spleen, thymus, lung and placenta, and lower levels are present in kidney, smooth muscle and heart. The single report that p35 is present in platelets (Touqui et al., 1986) has not been confirmed by others, and should probably be discounted. p35 is not detected in intestine, brain, liver, and red blood cells. Many cell types express p35, including fibroblasts, macrophages and their precursors, and epithelial cells. Unlike p36, p35 does not appear to be expressed in lymphoid cells. p35 and p36 are coexpressed in many cell types and tissues (Huang et al., 1986; Glenney et al., 1987). There are cell types that express p36 but not p35, but with the possible exception of the HL-60 myeloid cell line p35 is not found in cells which lack p36.

p35 is phosphorylated on tyrosine in EGF-treated A431 cells (Sawyer and Cohen, 1985), and in EGF-treated permeabilized human fibroblasts (Giugni et al., 1985). Phosphoserine can also be detected in p35, following EGF treatment (Sawyer and Cohen, 1985). The major tyrosine phosphorylation site for the EGF receptor in p35 has been identified as Tyr 20 (De et al., 1986; Haigler et al., 1987), which lies in a region equivalent to that in which the major phosphorylation site of p36 is located. However, there is little homology between the primary amino acid sequences in this region (Figure 1), implying that recognition by protein-tyrosine kinases is more likely to involve secondary and tertiary structure rather than primary sequence. In this regard it is interesting to note that while p35 is an excellent substrate for the EGF receptor (Fava and Cohen, 1984; Pepinsky and Sinclair, 1986; Valentine-Brown et al., 1986; Sheets et al., 1987;

J. R. Glenney and G. N. Gill, personal communication), it is a poor substrate for pp60$^{v\text{-}src}$ (Fava and Cohen, 1984) and the insulin receptor (J. R. Glenney and R. A. Roth, personal communication). Interestingly, p36 shows the converse specificity, being a poor substrate for the EGF receptor, but a good substrate for pp60$^{v\text{-}src}$ and the insulin receptor. These specificity differences may be explained by the fact that Tyr 20 is preceded by several glutamic acids, which generally favour phosphorylation, while there are no glutamic acids in the vicinity of Tyr 23 in p36 (Figure 1). As was the case for the phosphorylation of p36, Ca^{2+} stimulates the phosphorylation of p35 by the EGF receptor particularly in the presence of phospholipids (Fava and Cohen, 1984; Valentine-Brown et al., 1986; Sheets et al., 1987; Schlaepfer et al., 1987; J. R. Glenney and G. N. Gill, personal communication). This may well reflect the ability of Ca^{2+} to cause the association of p35 with phospholipids. p35 (chromobindin 9) is a substrate for protein kinase C in vitro (Summers and Creutz, 1985; Khanna et al., 1986a), and undergoes serine phosphorylation when chromaffin cells are stimulated with TPA or nicotine (Michener et al., 1986). Two sites for protein kinase C phosphorylation of p35 in vitro have been mapped, Thr 23 and Ser 26/27 (L. C. Cantley, personal communication). It is not yet known whether these sites correspond to sites phosphorylated in vivo.

There is evidence that phosphorylation of p35 affects its binding functions. Phosphorylation of p35 at Tyr 20 reduces by 5-fold the amount of Ca^{2+}required for half-maximal association with phospholipid (Schlaepfer and Haigler, 1987), which is opposite to the effects of Tyr 23 phosphorylation on the binding of p36 to phospholipids (Powell and Glenney, 1987). p35 can bind to an endosome vesicle fraction containing internalized EGF receptor, and is phosphorylated in this situation (Cohen and Fava, 1985). Whether this occurs in cells treated with EGF is unclear.

p33

Several closely related if not identical acidic proteins will be discussed under the rubric p33 (Table 1). These are: endonexin, a protein isolated from bovine liver or adrenal medulla (Geisow et al., 1984; Südhof et al., 1984; Geisow et al., 1986); protein II, a protein initially purified from porcine intestine (Gerke and Weber, 1984; Shadle et al., 1985; Shadle and Weber, 1987; Weber et al., 1987); the 32 kDa form of calelectrin (Geisow et al., 1984; Südhof et al., 1984); Peak G from placenta (Haigler et al., 1987); the 35 kDa calcimedin (Moore and Dedman, 1982; Smith and Dedman, 1986; the nature of the 33 kDa calcimedin is unclear), one of a group of proteins which bind to phenothiazine in the presence of Ca^{2+}; and a 34 kDa placental "growth factor" (Sen-Majumdar et al., 1986; R. J. Robb, personal communication). Comparison of partial protein sequence data for endonexin (Geisow and Walker, 1986) and the 34 kDa placental "growth factor" (R. J. Robb, personal communication) with the complete sequence of protein II (Weber et al., 1987) supports the idea that these proteins are identical.

In the presence of micromolar Ca^{2+} p33 binds to phospholipids, showing the highest affinity for phosphatidylethanolamine and phosphatidylinositol (Geisow et al., 1986; Shadle and Weber, 1987). There is some disagreement whether p33 binds to phosphatidylserine alone; porcine intestinal p33 binds to phosphatidylserine vesicles (Shadle and Weber, 1987), whereas bovine liver p33 does not (Geisow et al., 1986). One high affinity binding site for Ca^{2+} (K_d ~25 μM) and one low affinity Ca^{2+} binding site in p33 have been detected in the absence of phospholipid (Shadle et al., 1985). The affinity for Ca^{2+} is apparently not changed by phospholipid, but p33 only binds to phosphatidylserine in the presence of Ca^{2+} (Shadle and Weber, 1987). The number of binding sites in the presence of phospholipid has not been determined (Shadle and Weber, 1987), and it is unclear whether p33 aggregates phospholipid vesicles. However, p33 will aggregate chromaffin granule membranes in the presence of Ca^{2+} (Südhof et al., 1984). Unlike p36 and p35, p33 does not bind to F-actin (Gerke and Weber, 1985b; Shadle et al., 1985). p33 is present more diffusely in the cell than p36, but is not retained in the detergent insoluble cytoskeleton (Geisow and Walker, 1986). p33 appears to be associated with the endoplasmic reticulum (Geisow et al., 1984, 1986), but some p33 can be recovered from the plasma membrane fraction of the cell.

The complete sequence of p33 (protein II) has been determined by protein sequencing (Weber et al., 1987). No p33 cDNA clone has yet been reported. The p33 sequence has four 75 amino acid repeats with striking homology to the repeats in p36 and p35 (Figure 1). The overall homology with either protein is about 50%. p33 has a very short N-terminal extension of 11 residues, compared with 30 for p36 and 39 for p35. There is no evidence that p33 associates with p11, or any other protein. However, the N-terminal sequence of p33 is susceptible to trypsin digestion, which removes the N-terminal 8 residues (Weber et al., 1987); the rest of the protein forms a protease resistant core.

p33 is expressed in many different tissues (Geisow et al., 1984; Shadle et al., 1985; Silva et al., 1986), appearing to be more widely distributed than p36, but by no means ubquitous. p33 is present at high levels in intestine, liver, spleen, and lower levels in kidney (Shadle et al., 1985). Most other tissues appeared to have very low levels of p33 (Geisow et al., 1984). p33 has been purified from porcine intestine (Gerke and Weber, 1984), beef adrenals and liver (Südhof et al., 1984), and human placenta (Haigler et al., 1987). p33 is present in fibroblast cell lines, as well as in other cells of epidermal and neural origin (Geisow et al., 1984). Like p36 and p35, p33 appears to be absent from mature erythroid cells (Geisow et al., 1984; Shadle et al., 1985).

Very little is known about the phosphorylation of p33. There are no reports of phosphorylation of p33 in vivo, but it is a good substrate for protein kinase C in vitro, being phosphorylated on threonine (Weber et al., 1987). This site has been identified as Thr 6, which is homologous to Ser 25 in p36 and Thr 23 in p35 (Figure 1). It will be interesting to see whether p33 is phosphorylated in TPA-treated cells. p33 is not a substrate for the EGF receptor, in keeping with the fact that there is no tyrosine in the N-terminal 8 amino acids.

p67

Like the other family members p67 has a number of aliases (Table 1). These are: the 67 kDa calelectrin, identified based on its cross-reactivity to the 34 kDa *Torpedo* calelectrin (Südhof, 1984; Südhof et al., 1984); p68, a 68 kDa Ca^{2+}-binding protein from lymphocytes (Owens and Crumpton, 1984; Owens et al., 1984); protein III (Shadle et al., 1985); the 67 kDa calcimedin (Moore and Dedman, 1982; Moore, 1986); chromobindin 20 (Creutz et al., 1987); synhibin (Pollard and Scott, 1982); and the intestinal 73K protein (Glenney, 1986c).

In the presence of Ca^{2+} p67 binds to phospholipid, with a higher affinity than either p35 or p36 (Davidson et al., 1987). This may reflect a greater number of binding sites (see below). The phospholipid binding specificity of p67 appears to be broad like that of p33, and p67 can bind to phosphatidylinositol (Geisow and Walker, 1986). The number of Ca^{2+} binding sites has not been accurately determined. Although there appears to be only one high affinity Ca^{2+}-binding site (K_d 1 μM) in the absence of added phospholipid (Owens and Crumpton, 1984), there could well be more Ca^{2+}-binding sites in the presence of phospholipid. Indeed by analogy with p35 one might predict eight Ca^{2+}-binding sites, based on the structure of p67 (see below). Although it binds to phospholipid vesicles in the presence of Ca^{2+}, p67 does not aggregate phospholipid vesicles (J. R. Glenney, personal communication). Indeed, although there is one report that p67 by itself aggregates chromaffin granule membranes (Südhof et al., 1984), p67 inhibits aggregation of chromaffin granule membranes by other family members (Creutz et al., 1987), and hence was termed synhibin (Pollard and Scott, 1982). Probably this inhibitory activity is a result of the ability of p67 to mask phospholipid-binding sites for the other proteins.

Two groups have recently isolated cDNA clones for p67 (mRNA size 2.5 kb)(Südhof et al., 1988; M. Crompton and M. J. Crumpton, personal communication), and partial protein sequences for p67 have been reported (Martin et al., 1987). There appear to be two forms of p67 differing by the presence or absence of a 5 amino acid exon, which are detected as a protein doublet (M. Crompton and M. J. Crumpton, personal communication). The predicted structure contains eight 75 amino acid repeats, with a short 17 amino acid N-terminal extension, and a 36 residue linker between repeats 4 and 5 (Figure 1). Inspection of the sequence shows that p67 arose by duplication of a four repeat gene (Südhof et al., 1988). Further aspects of the structure will be discussed below.

p67 is expressed widely, and is detected in most cells and tissues, possibly having the broadest distribution of all the family members (Geisow et al., 1984; Owens et al., 1984; Silva et al., 1986). p67 is present at high levels in liver and spleen, and at lower levels in kidney intestine, brain, skeletal and cardiac muscle, and many other tissues (Shadle et al., 1985, Glenney, 1986c). p67 is particularly prominent in lymphocytes (Owens et al., 1984; Shadle et al., 1985), but it is also present in most cell lines (Owens et al., 1984; Geisow et al., 1984). However, like the other family members, p67 is absent from late stage erythroid cells (Shadle et al., 1987; Geisow et al., 1984). p67 is retained in the detergent insoluble matrix and is therefore presumably associated with the cytoskeleton (Owens et al., 1984). Immunofluorescence staining shows p67 to be localized to the underside of the plasma membrane (Owens et al., 1984), which is in keeping with cell fractionation studies showing that p67 can be isolated from purified plasma membranes.

Phosphorylation of p67 in vivo has not been reported. p67 lacks an N-terminal protein kinase C site homologous to the other family members (Figure 1). It has a N-terminal tyrosine at position 9, but the basic residues on both sides may preclude its phosphorylation, since acidic residues are usually required for recognition by protein-tyrosine kinases.

<u>Are there other family members?</u>

At least one other protein has been found which is built up of a similar 75 amino acid repeat. The sequence of cDNA clones for anchorin CII, a 31 kDa glycoprotein first characterized as a membrane-associated type II collagen-binding protein from chondrocytes (Mollenhauer and von der Mark, 1983, 1986; Mollenhauer, et al., 1984), reveal a protein built up of four 75 amino acid repeats with clear homology to those in the other family members (J. Mollenhauer, personal communication). The N-terminus of anchorin CII has not been identified in the available cDNA clones, but the 16 amino acids to the N-terminal side of repeat 1 are distinct from the N-terminal domains of the other family members. Since the collagen-binding function and immunolocalization studies indicate that anchorin CII is an extracellular protein, it will obviously be important to determine whether or not the N-terminal sequence of anchorin CII has a signal peptide, which would allow its secretion. Anchorin CII has Ca^{2+}-dependent phospholipid binding activity (J. Mollenhauer, personal communication). Given the high extracellular Ca^{2+} concentration, one might expect anchorin CII to be attached to the plasma membrane through binding to the outer phospholipid leaflet, but this may depend on the phospholipid specificity of anchorin CII. There is no convincing evidence that p36, p35, p33 and p67 are actively secreted, and their sequences lack signal peptides. Anchorin CII, however, could be the first of a series of proteins belonging to this family of proteins that normally function outside the cell.

General structure and evolution

The key feature of these proteins is the 75 amino acid repeat, of which there are now 24 examples. Within this domain the N-terminal half is most highly conserved and, with the possible exception of repeat 7 in p67, there is an invariant Gly, present in all of these repeats, about 17 residues from the N-terminus of the repeat in the most highly conserved part of the repeat (Geisow, 1986; Geisow and Walker, 1986; Kretsinger and Creutz, 1986; Kristensen et al., 1986; Saris et al., 1986; Weber and Johnsson, 1986)(Figure 1). This Gly may be at the head of a loop that forms a region of conserved secondary structure, which has also been called the endonexin fold (Geisow and Walker, 1986). The only other invariant residue is an Arg 13 residues downstream of the conserved Gly, which could be involved in phospholipid binding (Sühof et al., 1988). Unlike the classic E-F hand structure in Ca^{2+}-binding proteins such as calmodulin, there are no totally conserved acidic residues that might be involved in the binding of Ca^{2+}. This may imply that the Ca^{2+} is bound via the phospholipid head groups. There is a second Gly 16 residues from the C-terminus of the repeat, which is found in every repeat except in repeat 2 in p33 and anchorin CII (Figure 1). As has already been noted there tends to be a hydrophobic amino acid every fifth residue (Südhof et al., 1988), and a more pronounced hydrophobic region just to the C-terminal side of each of the two conserved glycines (Saris et al., 1986), which could be involved in membrane binding.

Recently a model for the three-dimensional structure of the 75 amino acid Ca^{2+}/phospholipid binding reapeat has been proposed (Taylor and Geisow, 1987), which was generated through energy minimization starting with the known three-dimensional structure of the bovine intestinal vitamin D-dependent Ca^{2+}-binding protein (ICaBP). In this model the repeat is formed into a four-helical bundle with pseudo 2-fold symmetry with the two conserved glycines in the loops between each pair of helices. Although ICaBP binds two Ca^{2+} ions, which are well buried in E-F hand structures, the model for this repeat can only bind one Ca^{2+}, which is partly exposed in a deep cleft and is coordinated to 4 peptide bond carbonyls and to 2 acidic residues. The cleft could accomodate a phospholipid head group, which would interact directly with the Ca^{2+} forming a combined Ca^{2+}/phospholipid binding site. The solution of a crystal structure will be needed to determine whether this model is correct.

It is obvious that this gene family arose by duplication at several levels. To start with there are hints that the 75 amino acid domain might have arisen itself by a duplication. Inspection of the sequence for p36 shows that repeats 1 and 3 are more closely related to each other than to repeats 2 and 4, which are themselves more similar to each other. Thus p36 probably arose by a duplication of a two repeat protein, which in turn was generated by the duplication of a single repeat. Examination of the p67 sequence, which has 8 repeats, indicates that it was generated by a duplication of a four-fold repeat structure, i.e. repeat 1 is homologous to repeat 5, etc. When a comparison is made between proteins it appears as though repeats 2 and 4 are most highly conserved, while repeats 1 and 3 are less well conserved (Weber et al., 1987). The signficance of this is unclear. Possibly the four repeats are not all functionally equivalent, but at least for p35 there is experimental evidence that they can all bind Ca^{2+} in the presence of phospholipid (Schlaepfer and Haigler, 1987). As discussed above the phospholipid requirement for Ca^{2+} binding coupled with the ability to aggregate phospholipid vesicles places topological constraints on the organization of phospholipid binding sites. For the four repeat proteins, one possibility is that they occupy the faces of a tetrahedron. Finally, it is interesting to note that the synexins and protein kinase C, which have similar Ca^{2+}-dependent phospholipid binding properties, do not contain sequences with obvious similarity to the 75 amino acid repeat.

The N-terminal extensions of each protein, as well as the linker between repeats 1 and 5 in p67 are unique. p36, p35, p33, and p67 all lack signal sequences, and for p36, p35, and p33 it has been shown that the N-terminal methionine is removed and the exposed N-terminus is acetylated (Glenney et al., 1986; R. B. Pepinsky, personal communication; Weber et al., 1987). These regions may dictate the individual specificities of these proteins. The N-terminal domains of p36 and p35 are targets for protein kinases, and may well be sites for regulation.

General functions

Given the similarities of the structures of these proteins it appears extremely likely that they will all have analogous functions. The one striking common property is the ability of these proteins to bind to phospholipid bilayers in the presence of micromolar Ca^{2+}. This binding activity is a property of the basic 75 amino acid repeat from which all of these proteins are built. Since all of the known proteins with this repeat have a $(2n)_x$ structure, in principle they can all bind multiple phospholipid molecules. There is direct evidence for this in the case of p36 and p35, both of which aggregate phospholipid liposomes in the presence of Ca^{2+} (Glenney et al., 1987). p33 also appears to have this property (Südhof et al., 1984). There is some disagreement whether p67 will aggregate synthetic liposomes, even though it clearly binds to them. However, p67 does cause Ca^{2+}-dependent aggregation of chromaffin granule and synaptic vesicle membranes (Südhof et al., 1984), which may imply that the phospholipid composition is critical for aggregation or else that p67 interacts with some other component of these natural membranes. In some ways it would be paradoxical if p67, which has 8 potential Ca^{2+}/phospholipid-binding sites could not aggregate phospholipid vesicles. How phospholipid binding is induced by Ca^{2+} is not known. These proteins will bind to hydrophobic compounds such as phenothiazine or phenyl-Sepharose following the binding of Ca^{2+} (Moore and Dedman, 1982; Südhof et al., 1984; Shadle et al., 1985; Carraway et al., 1986; Moore, 1986), indicating that these proteins expose a hydrophobic surface following Ca^{2+} binding, which might be important for interaction with the acyl chains of phospholipids.

This common Ca^{2+}-dependent phospholipid-binding property implies that these proteins are involved in membrane-membrane interactions, which are likely to be regulated by intracellular Ca^{2+} concentration. There do not appear to be other common properties that are consistently observed. Although p36 and p35 bind to F-actin at high Ca^{2+} concentrations, p33 and p67 do not. Furthermore there is no strong evidence that phospholipid lowers the Ca^{2+} requirement for the binding of p36 and p35 to F-actin, nor is there any indication that p36 and p35 are associated with F-actin in the cell. However, p36 clearly interacts with some as yet undefined component of the cytoskeleton, although this may occur through p11. p67 is also associated with the cytoskeleton, and whereas the other family members are not strongly bound to the detergent insoluble matrix, they form part of a submembranous protein lamina that is in contact with the cytoskeleton.

Although it seems likely that these proteins have analogous functions, they must also have distinct properties, since the genes for at least four such proteins have been highly conserved among vertebrates over a long evolutionary span. The existence of multiple genes could be explained if individual family members play analogous roles in different cell types. This may well be the case, but there are many cells in which more than one of these proteins is present, implying separate if related functions. Another possibility, which is supported by their different subcellular localizations, is that their phospholipid-binding preferences dictate the targetting of these proteins to specific membranes, based on the known differential distribution of phospholipids among different cellular membranes. Individual proteins may have different binding valencies, which could be important for their function. Potentially p36 monomer, p35 and p33 could have four Ca^{2+}/phospholipid-binding sites, while p67 and $(p36)_2(p11)_2$ complex could have eight sites. These sites may be regulated independently allowing different topologies of interaction between membranes.

What membrane functions might these proteins be required for? The enrichment of p36 and its homologues in the membranes of the electric organ and chromaffin granules has prompted the suggestion that these proteins have a role in secretion, possibly mediating vesicle-membrane fusion. Indeed the $(p36)_2(p11)_2$ complex or the 33 kDa core fragment of p36, like synexin, can promote the fusion of chromaffin granule membranes in the presence of arachidonic acid (Drust and Creutz, 1988). Possibly a pair of apposing membranes bound to a single p36 molecule are brought close enough for fusion to be mediated by arachidonic acid. A group of Ca^{2+}-binding proteins, called calmotilins, have been identified in mammary epithelial cells, and have been implicated in cell motility (Rocha et al., 1986). The presence of high concentrations of p36 and p35 in the brush border of intestinal epithelial cells, lung epithelial cells, and the placental microvillar region (Edwards et al., 1987), all of which are absorptive cells, suggests a structural function or a role in vesicle-mediated uptake.

A quite different membrane function is implied by the observation that all of these proteins will inhibit phospholipase A_2 in vitro (Pepinsky et al., 1986; Wallner et al., 1986; Haigler et al., 1987; Davidson et al., 1987). There is a large body of literature showing that the anti-inflammatory effects of steroids are mediated through the secretion of a proteinaceous inhibitor of phospholipase A_2 (Flower and Blackwell, 1979; Blackwell, 1980; Flower et al., 1981; Hirata et al., 1980, Hirata, 1984). Such inhibitors are now called lipocortins, and by inhibiting phospholipase A_2 they block the formation of leukotrienes and prostaglandins, which are responsible for inflammation. A lipocortin was isolated from rat peritoneum based solely on its ability to inhibit phospholipase A_2 activity on ^{14}C-ethanolamine labeled *E. coli* (Pepinsky et al., 1986; Wallner et al., 1986). Following the isolation of a cDNA clone it quickly became clear that this lipocortin was identical to p35 (Pepinsky and Sinclair, 1986). Shortly thereafter a second lipocortin isolated from placenta was shown to be identical to p36 (Huang et al., 1986), raising the possibility that the true function of all these proteins might be to act as phospholipase A_2 inhibitors.

What are the arguments for and against this idea? First, one needs to know whether these proteins really have anti-inflammatory effects in the animal. There are conflicting results here. Bacterially-expressed p35 has anti-inflammatory effects in a lung effusion assay (Cirino et al., 1987), but purified placental p35 lacks such activity in a paw edema assay (Hollenberg et al., 1987). Second, it has become clear that p35 and p36 inhibit phospholipase A_2 in an indirect fashion. Phospholipase A_2 requires Ca^{2+} for its activity, and in the presence of Ca^{2+} p35 and p36 bind to the phospholipids thus masking the substrate from the enzyme (Davidson et al.,

1987; Haigler et al., 1987; M. J. Finnen, R. D.Taylor and D. Moon, personal communication). At high phospholipid concentrations approaching those present in the cell, no inhibitory effects of p35 and p36 are observed (Davidson et al., 1987). Third, all the proteins in this family lack an N-terminal signal peptide and are rather highly charged. Indeed there is no good experimental evidence that these proteins are normally secreted, even following treatment with steroids. In our own experiments with mouse and human fibroblasts, and with the human myeloid cell lines, HL-60 and U937, we have never detected secretion of either p35 or p36 following dexamethasone treatment (C. M. Isacke and R. A. Lindberg, unpublished observations). Neither have we been able to detect a steroid-induced increase in the mRNA level for either protein, nor an increase in the rate of synthesis or accumulation of these proteins. This is in contrast to the initial report of Wallner et al. (1986) who found that dexamethasone and TPA did increase the level of p35 mRNA in peritoneal macrophages and U937 cells, although the latter data were not shown. Strikingly, however, these workers did not observe any increase in the level of p35 as determined by immunoblotting.

Of course p35 and p36 may be secreted by some novel mechanism, but even if this were the case, it is not obvious how these proteins can inhibit cellular phospholipase A_2 activity from the outside of the cell, which is the classical mode of action of lipocortins. Some of the agonist-inducible phospholipase A_2 activity may reside on the extracellular surface of the plasma membrane, but according to the competition hypothesis of inhibition, the added p35 or p36 would have to coat all the accessible phospholipids to achieve effective inhibition. This does not rule out that the p35 and p36 could act to inhibit intracellular phospholipase activity. Certainly these proteins are abundant enough to occupy a significant fraction of the plasma membrane phospholipid (p35 and p36 are from 0.1-0.5% of total cell protein -- roughly 10^7 molecules per cell). However, since p36 and p35 are present constitutively their activity as inhibitors would have to be modulated. The inhibitory activity of lipomodulin (another name for lipocortin) has been reported to be decreased by both serine and tyrosine phosphorylation (Hirata, 1981; Hirata et al., 1984). The fact that both p36 and p35 can be phosphorylated under circumstances where phospholipase A_2 activity is known to be increased (e.g. following growth factor treatment of quiescent cells) would be consistent with their role as lipcortins. However, there is no evidence that phosphorylation alters their activity as inhibitors. Indeed, for p35 phosphorylation has been reported to increase its affinity for phospholipids, rather than causing the decrease one would expect. Moreover, growth factor treatment only induces the phosphorylation of a rather small fraction of p35 or p36 molecules, which would leave the majority of these molecules in the active inhibitory state.

Thus the question of whether these proteins are truly lipocortins is still unanswered. In my opinion these proteins are probably not lipocortins, but rather have roles in membrane functions. This idea is supported by the high abundance of these proteins. Whatever function these proteins have, it is likely to be regulated by Ca^{2+} concentration and by phosphorylation. It is interesting to note that none of these proteins appears to be present in erythroid cells. Possibly a specialized protein, such as ankyrin, replaces this type of protein in the erythroid submembraneous skeleton.

Insight into the true function of these proteins may come from the use of microinjected high affinity monoclonal antibodies specific for different family members, either singly or in combination. The hope would be that these antibodies would block a defined cellular function by sequestering the target protein. Some insights may also be gained from expression of mutant forms of these proteins created by site-directed mutagenesis.

However, depending on the cell type the presence of the endogenous protein may thwart ready interpretation of such experiments.

None of the current names for this family of proteins seems satisfactory, yet based on their structural relatedness there is clear need for a common name. Lipocortin does not seem appropriate, unless better evidence can be provided to show that these proteins are secreted physiological inhibitors of phospholipase A_2. Calelectrin is not suitable because these proteins are not exclusive to electrogenic cells. Calpactin falls down as a name because there is no evidence that these proteins bind to actin in the cell. Calmotilin is a poor choice because these proteins are certainly not restricted to motile cells. Endonexin, is really only pertinent to p33, which is bound to the endoplasmic reticulum. Calcimedin is the most reasonable name, since it does not imply anything about function except Ca^{2+}-binding. Annexin has been suggested as a common term (Geisow, 1986; Taylor and Geisow, 1987), but this may be premature unless it can be shown that these proteins are involved in membrane/membrane interactions in the cell. I would propose calnexin as a compromise name, which would imply a Ca^{2+}- dependent binding protein.

REFERENCES

Alitalo, K., Ralston, R. R., and Keski-Oja, J. (1984). Distribution of the 36,000 dalton tyrosine protein kinase substrate in drug and epidermal growth factor-treated epithelial cells. Exp. Cell Res. **150**:177-185.

Amini, S., and Kaji, A. (1983). Association of pp36, a phosphorylated form of the presumed target protein for the *src* protein of Rous sarcoma virus, with the membrane of chicken cells transformed by Rous sarcoma virus. Proc. Natl. Acad. Sci. USA **80**:960-964.

Arrigo, A.-P., Darlix, J.-L., and Spahr, P.-F. (1983). A cellular protein phosphorylated by the avian sarcoma virus transforming gene product is associated with ribonucleoprotein particles. EMBO J. **2**:309-315.

Blackwell, G. J., Carnuccio, R., Di Rosa, M., Flower, R. J., Parente, L., and Persico, P. (1980) Macrocortin: a polypeptide causing the anti-phospholipase effect of glucocorticoids. Nature **287**:147-149.

Burgoyne, R. D. (1987). Gelsolin and p36 share a similar domain. TIBS **12**:85-86.

Calabretta, B., Battini, R., Kaczmarek, L., de Riel, J. K., and Baserga, R. (1986) Molecular cloning of the cDNA for a growth factor-inducible gene with strong homology to S-100, a calcium-binding protein. J. Biol. Chem. **261**:12628-12632.

Carraway, K. L., Liu, Y., Puett, D., Carraway, K. L., and Carraway, C. A. (1987). Phenothiazine binding by a homolog of calpactin, the pp60*src* tyrosine kinase substrate. FASEB J. **1**:46-50.

Carter, V. C., Howlett, A. R., Martin, G. S., and Bissell, M. J. (1986) The tyrosine phosphorylation substrate p36 is developmentally regulated in embryonic avian limb and is induced in cell culture. J. Cell Biol. **103**:2017-2024.

Cheng, Y. S.-E., and Chen, L. B. (1981). Detection of phosphotyrosine-containing 36,000 dalton protein in the framework of cells transformed with Rous sarcoma virus. Proc. Natl. Acad. Sci. USA **78**:2388-2392.

Cirino, G., Flower, R. J., Browning, J. L., Sinclair, L. K., and Pepinsky, R. B. (1987). Recombinant human lipocortin I inhibits thromboxane release from guinea-pig isolated perfused lung. Nature **328**:270-272.

Cohen, S., and Fava, R. A. (1985). Internalization of functional epidermal growth factor:receptor/kinase complexes in A-431 cells. J. Biol. Chem. **260**:12351-12358.

Cooper, J. A., and Hunter, T. (1981a). Changes in protein phosphorylation in Rous sarcoma virus transformed chicken embryo cells. Mol. Cell. Biol. **1**:165-178.

Cooper, J. A., and Hunter, T. (1981b). Four different classes of retroviruses induce phosphorylation of tyrosines present in similar cellular proteins. Mol. Cell Biol. **1**:394-407.
Cooper, J. A., and Hunter, T. (1982). Discrete primary locations of a tyrosine protein kinase and of three proteins that contain phosphotyrosine in virally transformed chick fibroblasts. J. Cell Biol. **94**:287-296.
Cooper, J. A., and Hunter, T. (1983a). Regulation of cell growth and transformation by tyrosine-specific protein kinases: the search for important cellular substrate proteins. Curr. Topics Microbiol. Immunol. **107**:125-161.
Cooper, J. A., and Hunter, T. (1983b). Identification and characterization of cellular targets for tyrosine protein kinases. J. Biol. Chem. **258**:1108-1115.
Cooper, J. A., Bowen-Pope, D. F., Raines, E., Ross, R., and Hunter, T. (1982). Similar effects of platelet-derived growth factor and epidermal growth factor on the phosphorylation of tyrosine in cellular proteins. Cell **31**:263-273.
Courtneidge, S., Ralston, R., Alitalo, K., and Bishop, J. M. (1983). Subcellular location of an abundant substrate (p36) for tyrosine-specific protein kinases. Mol. Cell. Biol. **3**:340-350.
Creutz, C. E., Dowling, L. G., Sando, J. J., Villar-Palasi, C., Whipple, J. H., and Zaks, W. J. (1983). Characterization of the chromobindins: soluble proteins that bind to the chromaffin granule membrane in the presence of Ca^{2+}. J. Biol. Chem. **258**:14664-14674.
Creutz, C. E., Zaks, W. J., Hamman, H. C., Crane, S., Martin, W. H., Gould, K. L., Oddie, K. M., and Parsons, S. J. (1987). Identification of chromaffin granule-binding proteins. J. Biol. Chem. **262**:1860-1868.
Davidson, F. F., Dennis, E. A., Powell, M., and Glenney, J. R. (1987). Inhibition of phospholipase A_2 by "lipocortins" and calpactins. J. Biol. Chem. **262**:1698-1705.
Davies, A. A., and Crumpton, M. J. (1985). Identification of calcium-binding proteins associated with the lymphocyte plasma membrane. Biochem. Biophys. Res. Commun. **128**:571-577.
De, B. K., Misono, K. S., Lukas, T. J., Mroczkowski, B., and Cohen, S. (1986) A calcium-dependent 35-kilodalton substrate for epidermal growth factor receptor/kinase isolated from normal tissue. J. Biol. Chem. **261**:13784-13792.
Decker, S. J. (1982). Phosphorylation of the M_r = 34,000 protein in normal and Rous sarcoma virus transformed rat fibroblasts. Biochem. Biophys. Res. Commun. **109**:434-441.
Dorin, J. R., Novak, M., Hill, R. E., Brock, D. H. J., Secher, D. S., and van Heyningen, V. (1986) A clue to the basic defect in cystic fibrosis from cloning the CF antigen. Nature **326**:614-617.
Drust, D. S., and Creutz, C. E. (1988). Aggregation of chromaffin granules by calpactin at micromolar calcium. Nature, in press.
Edwards, H. C., and Booth, A. G. (1987). Calcium-sensitive, lipid-binding cytoskeletal proteins of the human placental microvillar region. J. Cell Biol. **105**:303-311.
Erikson, E., and Erikson, R. L. (1980). Identification of a cellular protein substrate phosphorylated by the avian sarcoma virus-transforming gene product. Cell **21**:829-836.
Erikson, E., Cook, R. L., Miller, G. J., and Erikson, R. L. (1981a). The same normal cell protein is phosphorylated after transformation by avian sarcoma viruses with unrelated transforming genes. Mol. Cell. Biol. **1**:43-50.
Erikson, E., Shealy, D. J., and Erikson, R. L. (1981b). Evidence that viral transforming gene products and epidermal growth factor stimulate phosphorylation of the same cellular protein with similar specificity. J. Biol. Chem. **256**:11381-11384.

Erikson, E., Tomasiewicz, H. G., and Erikson, R. L. (1984). Biochemical characterization of a 34-kilodalton normal cellular substrate of $pp60^{src}$ and an associated 6-kilodalton protein. Mol. Cell. Biol. **4**:77-85.

Fava, R. A., and Cohen, S. (1984). Isolation of a calcium-dependent 35-kilodalton substrate for the epidermal growth factor receptor/kinase from A-431 cells. J. Biol. Chem. **259**:2636-2645.

Flower, R. J., and Blackwell, G. J. (1979). Anti-inflammatory steroids induce biosynthesis of a phospholipase A_2 inhibitor which prevents prostaglandin generation. Nature **278**:456-459.

Flower, R. J., Wood J. N., and Parente, L. (1981). Macrocortin and the mechanism of action of the glucocorticoids. Adv. Inflammation Res. **7**:61-69.

Geisow, M. J. (1986). Common domain structure of Ca^{2+} and lipid-binding proteins. FEBS Letters **203**:99-103.

Geisow, M. J., and Burgoyne, R. D. (1982). Calcium-dependent binding of cytosolic proteins by chromaffin granules from adrenal medulla. J. Neurochem. **38**:1735-1741.

Geisow, M. J., and Walker, J. H. (1986) New proteins involved in cell regulation by Ca^{2+} and phospholipids. TIBS **11**:420-423.

Geisow, M., Childs, J., Dash, H., Harris, A., Panayotou, G., Südhof, T., and Walker, J. H. (1984). Cellular distribution of three mammalian Ca^{2+}-binding proteins related to *Torpedo* calelectrin. EMBO J. **3**:2969-2974.

Geisow, M. J., Fritsche, U., Hexham, J. M., Dash, B., and Johnson, T. (1986). A consensus amino acid sequence repeat in *Torpedo* and mammalian Ca^{2+}-dependent membrane binding proteins. Nature **320**:636-638.

Gerke, V., and Weber, K. (1984). Identity of p36K phosphorylated upon Rous sarcoma virus transformation with a protein purified from brush borders; calcium-dependent binding to non-erythroid spectrin and F-actin. EMBO J. **3**:227-233.

Gerke, V., and Weber, K. (1985a). Calcium-dependent conformational changes in the 36-kDa subunit of intestinal protein 1 related to the cellular 36-kDa target of Rous sarcoma virus tyrosine kinase. J. Biol. Chem. **260**:1688-1695.

Gerke, V., and Weber, K. (1985b). The regulatory chain in the p36-kd substrate complex of viral tyrosine-specific protein kinases is related in sequence to the S-100 protein of glial cells. EMBO J. **4**:2917-2920.

Giugni, T. D., James, L. C., and Haigler, H. T. (1985). Epidermal growth factor stimulates tyrosine phosphorylation of specific proteins in permeabilized human fibroblasts. J. Biol. Chem. **260**:15081-15090.

Glenney, J. R. (1985). Phosphorylation of p36 in vitro with $pp60^{src}$: Regulation by Ca^{2+} and phospholipid. FEBS Letters **192**:79-82.

Glenney, J. R. (1986a). Two related but different forms of the 36,000 Mr tyrosine kinase substrate (calpactins) which interact with phospholipid and actin in a Ca^{2+}-dependent manner. Proc. Natl. Acad. Sci USA **83**:4258-4262.

Glenney, J. R. (1986b). Phospholipid dependent Ca^{2+}-binding by the 36Kd tyrosine kinase substrate (calpactin) and its 33Kd core. J. Biol. Chem. **261**:7247-7252.

Glenney, J. R. (1986c). Co-precipitation of intestinal p36 with a 73K protein and a high molecular weight factor. Exp. Cell Res. **162**:183-190.

Glenney, J. R., and Tack, B. F. (1985). Amino-terminal sequence of p36 and associated p10: Identification of the site of tyrosine phosphorylation and homology with S-100. Proc. Natl. Acad. Sci. USA **82**:7884-7888.

Glenney, J. R., and Zokas, L. (1988). Antibodies to the N-terminus of calpactin II (p35) affect Ca^{++}-binding and phosphorylation by the EGF receptor in vitro. Biochemistry, in press.

Glenney, J. R., Boudreau, M., Galyean, R., Hunter, T., and Tack, B. F. (1986). Association of the S-100-related calpactin I light chain with

the amino-terminal tail of the 36 kDa heavy chain. J. Biol. Chem. **261**:10485-10488.
Glenney, J. R., Jr., Tack, B., and Powell, M. A. (1987) Calpactins: two distinct Ca^{2+}-regulated phospholipid- and actin-binding proteins isolated from lung and placenta. J. Cell Biol. **104**:503-511.
Gould, K. L., Cooper, J. A., and Hunter, T. (1984). The 46,000-dalton tyrosine protein kinase substrate is widespread, whereas the 36,000-dalton substrate is only expressed at high levels in certain rodent tissues. J. Cell Biol. **98**:487-497.
Gould, K. L., Woodgett, J. R., Isacke, C. M., and Hunter, T. (1986). The protein-tyrosine kinase substrate, p36, is also a substrate for protein kinase C in vitro and in vivo. Mol. Cell. Biol. **6**:2738-2744.
Greenberg, M. E., and Edelman, G. M. (1983a). Comparison of the 34,000-Da $pp60^{src}$ substrate and a 38,000-Da phosphoprotein identified by monoclonal antibodies. J. Biol. Chem. **258**:8497-8502.
Greenberg, M. E., and Edelman, G. M. (1983b). The 34kd $pp60^{src}$ substrate is located at the inner face of the plasma membrane. Cell **33**:767-779.
Greenberg, M. E., Brackenbury, R., and Edelman, G. M. (1984). Changes in the distribution of the 34-kilodalton tyrosine kinase substrate during differentiation and maturation of chicken cells. J. Cell Biol. **98**:473-486.
Haigler, H. T., Schlaepfer, D. D., and Burgess, W. H. (1987). Characterization of lipocortin I and an immunologically unrelated 33-kDa protein as epidermal growth factor receptor/kinase substrates and phospholipase A_2 inhibitors. J. Biol. Chem. **262**:6921-6930.
Hexham, J. M., Totty, N. F., Waterfield, M. D., and Crumpton, M. J. (1986). Homology between the subunits of S100 and a 10 kDa polypeptide associated with p36 of pig lymphocytes. Biochem. Biophys. Res. Commun. **134**:248-254.
Hirata, F. (1981). The regulation of lipomodulin, a phospholipase inhibitory protein, in rabbit neutrophils by phosphorylation. J. Biol. Chem. **256**:7990-7997.
Hirata, F. (1984). Roles of lipomodulin: a phospholipase inhibitory protein in immunoregulation. Adv. Inflammation Res. **7**:71-78.
Hirata, F., Matsuda, K., Notsu, Y., Hattori, T., and del Carmine, R. (1984). Phosphorylation at a tyrosine residue of lipomodulin in mitogen-stimulated murine thymocytes. Proc. Natl. Acad. Sci. USA **81**:4717-4721.
Hirata, F., Schiffmann, E., Venkatasubramanian, K., Salomon, D., and Axelrod, J. (1980). A phospholipase A_2 inhibitory protein in rabbit neutrophils induced by glucocorticoids. Proc. Natl. Acad. Sci. USA **77**:2533-2536.
Hollenberg, M. D., Northup, J. K., Valentine-Braun, K. A., Johnson, L. K., and Severson, D. L. (1987). Human placental lipocortin-I: Dissociation of phospholipase-inhibitory from antiinflammatory activity. Clin. Res. 35, 639a.
Huang, K.-S., McGray, P., Mattaliano, R. J., Burne, C., Chow, E. P., Sinclair, L. K., and Pepinsky, R. B. (1987) Purification and characterization of proteolytic fragments of lipocortin I that inhibit phospholipase A_2. J. Biol. Chem. **262**:7639-7645.
Huang, K.-S., Wallner, B. P., Mattaliano, R. J., Tizard, R., Burne, C., Frey, A., Hession, C., McGray, P., Sinclair, L. K., Chow, E. P., Browning, J. L., Ramachandran, K. L., Tang, J., Smart, J. E., and Pepinsky, R. B. (1986). Two human 35-kd inhibitors of phospholipase A_2 are related to substrates of $pp60^{v-src}$ and of the epidermal growth factor receptor/kinase. Cell **46**:191-199.
Hunter, T., and Cooper, J. A. (1981). Epidermal growth factor induces rapid tyrosine phosphorylation of proteins in A431 human tumor cells. Cell **24**:741-752.

Isacke, C. M., Trowbridge, I. S., and Hunter, T. (1986). Modulation of p36 phosphorylation in human cells: studies using anti-p36 monoclonal antibodies. Mol. Cell. Biol. **6**:2745-2751.
Johnsson, N., Van, P. N., Söling, H.-D, and Weber, K. (1986a) Functionally distinct serine phosphorylation sites of p36, the cellular substrate of retroviral protein kinase; differential inhibition of reassociation with p11. EMBO J. **5**:3455-3460.
Johnsson, N., Vandekerchove, J., Van Damme, J., and Weber, K. (1986b). Binding sites for calcium, lipid and p11 on p36, the substrate of retroviral tyrosine-specific protein kinases. FEBS Letters **198**:361-364.
Khanna, N. C., Tokuda, M., and Waisman, D. M. (1986a). Phosphorylation of lipocortins in vitro by protein kinase C. Biochem. Biophys. Res. Commun. **141**:547-554.
Khanna, N. C., Tokuda, M., Chong, S. M., and Waisman, D. M. (1986b). Phosphorylation of p36 in vitro by protein kinase C. Biochem. Biophys. Res. Commun. **137**:397-403.
Kobayashi, N., and Kaji, A. (1980). Phosphoprotein associated with activation of the *src* gene product in myogenic cells. Biochem. Biophys. Res. Commun. **93**:278-284.
Kristensen, T., Saris, C. J. M., Hunter, T., Hicks, L. J., Noonan, D. J., Glenney, J. R., and Tack, B. F. (1986). Primary sequence of bovine calpactin I heavy chain (p36), a major cellular substrate for retroviral protein-tyrosine kinases: homology with the human phospholipase A_2 inhibitor, lipocortin. Biochemistry **25**:4497-4503.
Lehto, V-.P., Virtanen, I., Passivuo, R., Ralston, R., and Alitalo, K. (1983). The p36 substrate of tyrosine-specific protein kinases co-localizes with non-erythrocyte-spectrin, p230, in surface lamina or cultured fibroblasts. EMBO J. **2**:1701-1705.
Magee, A. I., and Courtneidge, S. A. (1985). Two classes of fatty acylated proteins exist in eukaryotic cells. EMBO J. **4**:1137-1144.
Mangeat, P. H., and Burridge, K. (1984). Immunoprecipitation of nonerythrocyte spectrin within live cells following microinjection of specific antibodies: relation to cytoskeletal structures. J. Cell Biol. **98**:1363-1377.
Martin, F., Derancourt, J., Capony, J.-P., Colote, S., and Cavadore, J.-C. (1987). Sequence homologies between p36, the substrate of $pp60^{src}$ tyrosine kinase and a 67 kDa protein isolated from bovine aorta. Biochem. Biophys. Res. Commun. **145**:961-968.
Michener, M. L., Dawson, W. B., and Creutz, C. E. (1986). Phosphorylation of a chromaffin granule-binding protein in stimulated chromaffin cells. J. Biol. Chem. **261**:6548-6555.
Mollenhauer, J., and von der Mark, K. (1983). Isolation and characterization of a collagen-binding glycoprotein from chondrocyte membranes. EMBO J. **2**:45-50.
Mollenhauer, J., and von der Mark, K. (1986). Anchorins: mediators of cell-matrix interaction. Front. Matrix Biol. **11**:110-122.
Mollenhauer, J., Bee, J. A., Lizarbe, M. A., and von der Mark, K. (1984). Role of anchorin CII, a 31,000-mol-wt membrane protein, in the interaction of chondrocytes with type II collagen. J. Cell Biol. **98**:1572-1578.
Moore, P. B. (1986). 67 kDa calcimedin, a new Ca^{2+}-binding protein. Biochem. J. **238**:49-54.
Moore, P. B., and Dedman, J. R. (1982). Calcium-dependent protein binding to phenothiazine columns. J. Biol. Chem. **257**:9663-9667.
Nakamura, K. D., and Weber, M. J. (1982). Phosphorylation of a 36,000 M_r cellular protein in cells infected with partial transformation mutants of Rous sarcoma virus. Mol. Cell. Biol. **2**:147-153.
Nigg, E. A., Cooper, J. A., and Hunter, T. (1983). Immunofluorescent localization of a 39,000-dalton substrate of tyrosine protein kinases to the cytoplasmic surface of the plasma membrane. J. Cell Biol. **96**:1601-1609.

Owens, R. J., and Crumpton, M. J. (1984). Isolation and characterization of a novel 68,000-M_r, Ca^{2+}-binding protein of lymphocyte plasma membrane. Biochem. J. **219**:309-316.
Owens, R. J., Gallagher, C. J., and Crumpton, M. J. (1984). Cellular distribution of p68, a new calcium-binding protein from lymphocytes. EMBO J. **3**:945-952.
Pepinsky, R. B., and Sinclair, L. K. (1986). Epidermal growth factor dependent phosphorylation of lipocortin. Nature **321**:81-84.
Pepinsky, R. B., Sinclair, L. K., Browning, J. L., Mattaliano, R. J., Smart, J. E., Chow, E. P., Falbel, T., Ribolini, A., Garwin, J. L., and Wallner, B. P. (1986). Purification and partial sequence analysis of a 37-kDa protein that inhibits phospholipase A_2 activity from rat peritoneal exudates. J. Biol. Chem. **261**:4239-4246.
Pollard, H. B., and Scott, J. H. (1982). Synhibin: a new calcium-dependent membrane-binding protein that inhibits synexin-induced chromaffin granule aggregation and fusion. FEBS Letters **150**:201-206.
Powell, M. A., and Glenney, J. R. (1987). Regulation of calpactin I phospholipid binding by calpactin I light chain binding and phosphorylation by $p60^{v-src}$. Biochem. J. **247**:321-328.
Radke, K., and Martin, G. S. (1979). Transformation by Rous sarcoma virus: Effects of *src* gene expression on the synthesis and phosphorylation of cellular polypeptides. Proc. Natl. Acad. Sci. USA **76**:5212-5216.
Radke, K., Gilmore, T., and Martin, G. S. (1980). Transformation by Rous sarcoma virus: a cellular substrate for transformation-specific protein phosphorylation contains phosphotyrosine. Cell **21**:821-828.
Radke, K., Carter, V. C., Moss, P., Dehazya, P., Schliwa, M., and Martin, G. S. (1983). Membrane association of a 36,000 dalton substrate for tyrosine phosphorylation in chicken embryo fibroblasts transformed by avian sarcoma viruses. J. Cell Biol. **97**:1601-1611.
Rocha, V., Hom, Y. K., and Marinkovich, M. P. (1986). Basal lamina inhibition suppresses synthesis of calcium-dependent proteins associated with mammary epithelial cell spreading. Exp. Cell Res. **165**:450-460.
Saris, C. J. M., Tack, B. F., Kristensen, T., Glenney, J. R., and Hunter, T. (1986) The cDNA sequence for the protein-tyrosine kinase substrate p36 (calpactin I heavy chain) reveals a multidomain protein with internal repeats. Cell **46**:201-212.
Saris, C. J. M., Kristensen, T., D'Eustachio, P., Hicks, L. J., Noonan, D. J., Hunter, T., and Tack, B. F. (1987). cDNA sequence and tissue distribution of the mRNA for bovine and murine p11, the S100-related light chain of the protein-tyrosine kinase substrate p36 (calpactin I). J. Biol. Chem. **262**:10663-10671.
Sawyer, S. T., and Cohen, S. (1985). Epidermal growth factor stimulates the phosphorylation of the calcium-dependent 35,000-dalton substrate in intact A-431 cells. J. Biol. Chem. **260**:8233-8236.
Schlaepfer, D. D., and Haigler, H. T. (1987). Characterization of Ca^{2+}-dependent phospholipid binding and phosphorylation of lipocortin I. J. Biol. Chem. **262**:6931-6937.
Sefton, B. M., Hunter, T., and Cooper, J. (1983). Some lymphoid cell lines transformed by Abelson murine leukemia virus lack a major 36,000-dalton tyrosine protein kinase substrate. Mol. Cell. Biol. **3**:56-63.
Sen-Majumdar, A., Murthy, U., and Das, M. (1986). A new trophoblast-derived growth factor from human placenta: purification and receptor identification. Biochemistry **25**:627-634.
Shadle, P. J., and Weber, K. (1987). Calcium binding protein from porcine intestine binds to phosphatidylserine vesicles in the presence of calcium. Biochim. Biophys. Acta **897**:502-506.
Shadle, P. J., Gerke, V., and Weber, K. (1985). Three Ca^{2+}-binding proteins from porcine liver and intestine differ immunologically and physicochemically and are distinct in Ca^{2+} affinities. J. Biol. Chem. **260**:16354-16360.

Sheets, E. E., Giugni, T. D., Coates, G. G., Schlaepfer, D. D., and Haigler, H. T. (1987). Epidermal growth factor dependent phosphorylation of a 35-kilodalton protein in placental membranes. Biochemistry **26**:1164-1172.

Silva, F. G., Sherrill, K., Spurgeon, S., Südhof, T. C., and Stone, D. K. (1986). High-level expression of the 32.5-kilodalton calelectrin in ductal epithelia as revealed by immunocytochemistry. Differentiation **33**:175-183.

Smith, V. L., and Dedman, J. R. (1986). An immunological comparison of several novel calcium-binding proteins. J. Biol. Chem. **261**:15815-15818.

Soric, J., and Gordon, J. A. (1985). The 36kd substrate of $pp60^{v\text{-}src}$ is myristylated in a transformation-sensitive manner. Science **230**:563-566.

Soric, J., and Gordon, J. A. (1986). Calcium-dependent isolation of the 36-kilodalton substrate of $pp60^{src}$-kinase. J. Biol. Chem. **261**:14490-14495.

Südhof, T. C. (1984a). Calelectrins are a ubiquitous family of Ca^{2+}-binding proteins purified by Ca^{2+}-dependent hydrophobic affinity chromatography by a mechanism distinct from that of calmodulin. Biochem. Biophys. Res. Commun. **123**:100-107.

Südhof, T. C. (1984b). The calelectrins: new and ubiquitous Ca^{2+}-regulated proteins with a possible role in cytotic processes. Biochem. Soc. Trans. **12**:972-974.

Südhof, T. C., Ebbecke, M., Walker, J. H., Fritsche, U., and Boustead, C. (1984). Isolation of mammalian calelectrins: a new class of ubiquitous Ca^{2}-regulated proteins. Biochemistry **23**:1103-1109.

Südhof, T. C., Slaughter, C. A., Leznicki, I., Barjon, P., and Reynolds, G. A. (1988) Human 67 kDa calelectrin contains a duplication of four repeats found in 35 kDa lipocortins. Proc. Natl. Acad. Sci. USA, in press.

Südhof, T. C., Walker, J. H., and Obrocki, J. (1982). Calelectrin self-aggregates and promotes membrane aggregation in the presence of calcium. EMBO J. **1**:1167-1170.

Summers, T. A., and Creutz, C. E. (1985). Phosphorylation of a chromaffin granule-binding protein by protein kinase C. J. Biol. Chem. **260**:2437-2443.

Taylor, W.R., and Geiso, M. J. (1987). Prdicted structure for the calcium-dependent membrane-binding proteins p35, p36 and p32. Protein Engineering **1**:183-187.

Touqui, L., Rothhut, B., Shaw, A. M., Fradin, A., Vargaftig, B. B., and Russo-Marie, F. (1986). Platelet activation--a role for a 40K anti-phospholipase A_2 protein indistinguishable from lipocortin. Nature **321**:177-180.

Tufty, R. M., and Kretsinger, R. H. (1975). Troponin and parvalbumin calcium binding regions predicted in myosin light chain and T4 lysozyme. Science **187**:167-169.

Valentine-Braun, K. A., Northup, J. K., and Hollenberg, M. D. (1986). Epidermal growth factor (urogastrone)-mediated phosphorylation of a 35-kDa substrate in human placental membranes: relationship to the β subunit of the guanine nucleotide regulatory complex. Proc. Natl. Acad. Sci. USA **83**:236-240.

Walker, J. H. (1982). Isolation from cholinergic synapses of a protein that binds to membranes in a calcium-dependent manner. J. Neurochem. **39**:815-823.

Wallner, B. P., Mattaliano, R. J., Hession, C., Cate, R. L., Tizard, R., Sinclair, L. K., Foeller, C., Chow, E. P., Browning, J. L., Ramachandran K. L., and Pepinsky, R. B. (1986). Cloning and expression of human lipocortin, a phospholipase A_2 inhibitor with potential anti-inflammatory activity. Nature **320**:77-81.

Weber, K., and Johnsson, N. (1986). Repeating sequence homologies in the p36 target protein of retroviral protein kinases and lipocortin, the p37 inhibitor of phospholipase A_2. FEBS Letters **203**:95-98.

Weber, K., Johnsson, N., Plessmann, U., Van, P. N., Söling, H.-D., Ampe, C., and Vandekerckhove, J. (1987). The amino acid sequence of protein II and its phosphorylation site for protein kinase C; the domain structure Ca^{2+}-modulated lipid binding proteins. EMBO J. **6**:1599-1604.
Zokas, L., and Glenney, J. R., Jr. (1987). The calpactin light chain is tightly linked to the cytoskeletal form of calpactin I: studies using monoclonal antibodies to calpactin subunits. J. Cell Biol. **105**:2111-2121.

FUNCTION OF THE ANTIGEN RECEPTOR IN T CELL ACTIVATION

Mark A. Goldsmith and Arthur Weiss

Howard Hughes Medical Institute, Departments of Medicine
and of Microbiology and Immunology
University of California, San Francisco
San Francisco, CA 94143

I. INTRODUCTION

The T lymphocyte plays a central role in both regulatory and effector aspects of the immune response to foreign antigens. In its regulatory role, represented for example by the T helper cell, the T lymphocyte is involved in complex interactions with antigen presenting cells, B cells, and other T cells. Communication in these interactions rely on both cell-to-cell contacts as well as the release of soluble growth and differentiation factors (lymphokines and monokines). In its effector role, represented by the cytotoxic T cell, direct engagement of infected cells results in destruction of the target.

In general, these functions in a given immune response are confined to a small subset of the entire T lymphocyte pool, the cells of which express antigen receptors of appropriate specificity. Thus, unlike hormone receptors that are non-polymorphic, the T cell antigen receptor is variable from cell to cell, and it is the T cell population in aggregate that recognizes the universe of antigens. Moreover, the receptors expressed by all T cell clones in the repertoire are coupled to the same effector programs, including those that lead to gene expression and production of lymphokines, and those that lead to expression of cytolytic function. This paper presents studies performed to characterize the structure and function of the T lymphocyte antigen receptor complex, emphasizing its role in transmembrane signalling and how these signal transduction events may be coupled to cellular activation programs and regulation of gene expression.

II. THE MOLECULES OF THE T CELL ANTIGEN RECEPTOR COMPLEX

A. Antigen-Specificity Derived from Rearranging Genes

Both the immunoglobulin and T cell receptor genes encode for large families of related, but unique, antigen-binding proteins. Thus, each antigen receptor is expressed from a gene that represents the concatenation of three or four coding segments drawn from a pool of such segments, with highly variable segments lying at the 5' end of the newly assembled gene (N-terminal end of the encoded protein) and constant segments at the 3' end (C-terminal end of the resulting protein).[1,2] By this construction, the protein product has a variable, clone-specific antigen-binding domain at its N-terminus and a relatively constant anchoring domain at its C-terminus. Presumably this organization accounts for the paradox of extremely variable antigen-reactivity coupled to conserved effector functions.

Most T cells express T cell antigen receptor proteins derived from two genes (Ti-α and Ti-ß) having the organization described above.[3,4] Peptide mapping studies have indicated the presence of variable and invariant domains within these proteins, consistent with the genomic studies. These two proteins together form an 80-90 kd disulfide-linked heterodimer which reduces to an acidic α chain of 43-54 kd and a more basic ß chain of 38-44 kd, each an integral membrane protein bearing multiple N-linked glycosylation sites. Peculiarly, the putative transmembrane domain of each chain contains one (ß chain) or two (α chain) net positively charged amino acid residues in the midst of an otherwise hydrophobic sequence. The cytoplasmic tails consist of only 5 highly charged amino acid residues.

A number of studies have tested the presumption that these proteins together represent the complete antigen-reactive complex. Among the earliest experiments was the somatic fusion by Yague et al.[5] of two antigen-specific clones, which yielded a somatic hybrid expressing precisely the combined antigen reactivity of the two donor cells. More definitively, recent gene transfection experiments by Dembic et al.[6] have demonstrated that introduction of genes encoding the α and ß chains of one murine antigen receptor into another T cell was sufficient to confer the antigen reactivity of the donor cell on the recipient murine T cell. Recently Saito et al.[7], in collaboration with our laboratory, accomplished the introduction of chains of a defined murine T cell antigen receptor into a human T cell line and observed that this receptor both reacts with, and mediates response to, its original antigen. Thus, it is now clear that the Ti heterodimer itself defines fully the antigen responsiveness of the cell bearing it, and that both the antigen reactivity and the function of these receptors are transferrable across species boundaries.

B. Molecules Associated with the Antigen Receptor

It has been appreciated for several years that the variable Ti chains are non-covalently associated with a number of other non-polymorphic proteins on the cell surface.[3,4] The function of these proteins, collectively referred to as the CD3 (formerly T3) complex, remains a mystery. However, under appropriate conditions, they may be co-immunoprecipited with, and even cross-linked to, the Ti heterodimer. Moreover, while engagement of the receptor itself by monoclonal antibodies (mAb) directed against the Ti chains results in the loss of surface expression of the receptor due to internalization (modulation),[8] it also results in a concomitant loss of surface expression of CD3 proteins (co-modulation). These observations therefore suggest that the Ti and CD3 proteins are both functionally and structurally associated.

In a series of experiments designed to develop T cells stably expressing either Ti or CD3 proteins alone,[9] we used immunoselection with monoclonal antibodies and complement to isolate cells that had lost a Ti or CD3 epitope. In all cases analyzed, selection against either Ti or CD3 resulted in the concomitant loss of both proteins from the cell surface. Furthermore, in all cases this phenotype appeared to result from a mutation impairing the transcription of either the Ti-α or -ß chain, and leaving apparently normal CD3 proteins inside the cell. Introduction of a cDNA clone encoding the appropriate Ti chain (either mouse or human) was sufficient to reconstitute expression of both the Ti and CD3 structures.[7,10] This study, then, provided evidence that surface expression of CD3 proteins required expression of both Ti-α and -ß chains. Likewise, surface expression of Ti chains transfected into L cells in the absence of CD3 proteins has proven unsuccessful (unpublished observation). These studies confirm the principle that the Ti and CD3 proteins exist together as a multi-subunit complex, and this structural association may have functional implications.

The CD3 complex consists of five to seven, presumably non-polymorphic, integral membrane proteins detected only on T lymphocytes.[3,4] Initial studies described three proteins, including two glycoproteins (the 26-28 kd γ chain and the 22 kd δ chain) and one non-glycosylated protein (the 21 kd ε chain). Isolation of the complete cDNA clones of the genes encoding these molecules revealed that the γ and δ chains have significant homology with each other, and that all three proteins have cytoplasmic tails containing 44-80 amino acid residues.[11-13] Interestingly, each chain contains a net negatively charged residue in the hydrophobic, putative transmembrane region, implying potential complementarity with the Ti chain transmembrane domains. Homologous murine molecules have also been identified.[3]

Recently additional chains that may be part of the CD3 complex have been identified. Biochemical studies have revealed a cell surface, 32 kd disulfide-linked homo- or heterodimer (the ζ chain) expressed on murine and human T cells that appears to co-modulate with the CD3/Ti complex.[14,15] Antibodies against this protein also co-immunoprecipitate from mouse T cells a 42 kd disulfide-linked heterodimer (p21).[16] Therefore, it appears that the CD3/Ti complex may consist of as many as nine proteins, the complexity of which is undoubtedly a function of its role in responding to receptor ligands (Fig. 1).

III. T CELL ACTIVATION VIA THE ANTIGEN RECEPTOR

A. General Considerations

In order to explore the functional characteristics of the T cell antigen receptor complex it has been useful to study extensively one T cell model, the CD3/Ti-bearing leukemic cell line Jurkat. In attempting to uncover the events that culminate in cellular activation, we searched for stimuli that result in expression of lymphokine genes in Jurkat, particularly of the interleukin-2 (IL-2) and interferon-γ (IFN-γ) genes. As described more fully in Section IV, we found that monoclonal antibodies against either the Jurkat Ti or CD3 molecules would, in the presence of the phorbol ester PMA, stimulate transcription from these genes.[4] Moreover, we observed that the Ca^{2+} ionophores ionomycin and A23187 could substitute for the anti-receptor antibodies in synergizing with PMA. The same combinations of signals trigger normal peripheral blood lymphocytes (PBL) to proliferate, to produce IL-2, and to express receptors for IL-2, implying that the Jurkat model bears resemblance to its normal T cell counterpart.

These observations initially suggested that an elevation of intracellular free Ca^{2+} ($[Ca^{2+}]_i$) synergizes with a PMA-mediated pathway to promote gene expression. Moreover, they implied that anti-receptor antibodies may act by somehow causing an elevation of $[Ca^{2+}]_i$. This model formed a basis for making certain predictions about the biochemical events that follow receptor-ligand interaction.

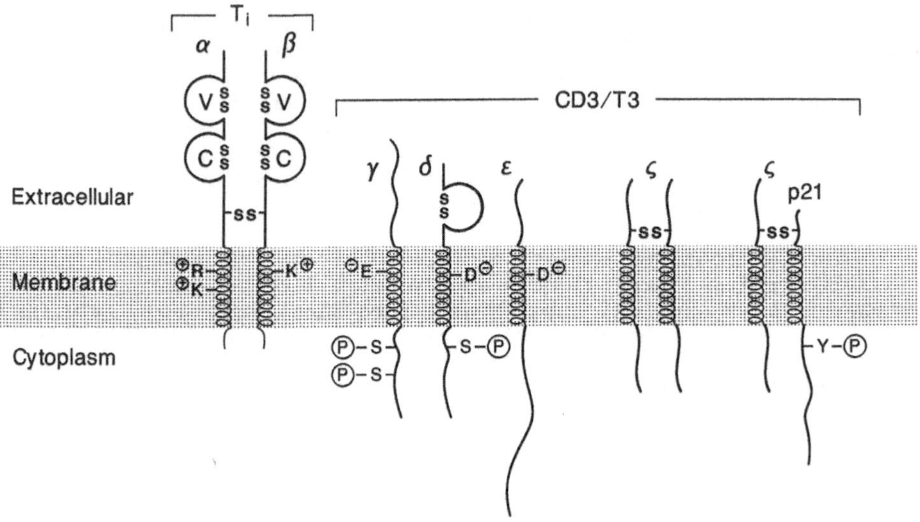

Fig. 1. The Human CD3/Ti Complex

B. Coupling of the Antigen Receptor to Phosphatidylinositol Hydrolysis

The first prediction was that activating ligands (antibodies and lectins) would cause an elevation of $[Ca^{2+}]_i$ in Jurkat and other T cells. Indeed, experiments with the intracellular fluorescent indicators Quin2 and Indo-1 in T cell lines, PBL, and T cell clones reveal that these ligands do trigger significant, prompt, and sustained increases in $[Ca^{2+}]_i$.[17-21] That these responses are receptor-mediated is implied by the lack of responsiveness in CD3/Ti-negative mutants of Jurkat (described above), and the failure of other non-receptor ligands to trigger this response.[4]

Careful examination revealed that the initial phase of $[Ca^{2+}]_i$ increase is not dependent on the presence of extracellular Ca^{2+}, while the sustained elevation requires extracellular Ca^{2+}.[22] This finding is consistent with the an initial release from intracellular stores of Ca^{2+}, followed by repletion of these stores with Ca^{2+} derived from the extracellular milieu. This pattern of Ca^{2+} processing parallels that of the well-known phosphatidylinositol (PI) second messenger pathway.[23] Receptors coupled to this second messenger system trigger hydrolysis of phosphatidylinositol 4,5-bisphosphate (PIP_2), releasing inositol 1,4,5-trisphosphate (IP_3) and diacylglycerol (DG), putative second messenger molecules that promote release of Ca^{2+} from internal stores and activation of protein kinase C (pkC), respectively. The studies of Imboden and Stobo[22] demonstrated convincingly that the CD3/Ti complex is functionally coupled to this second messenger cascade. Others have confirmed that activation of pkC is a detectable consequence of receptor triggering[24]. Receptor-mediated activation of the PI pathway requires continual receptor occupancy as demonstrated by the immediate fall in $[Ca^{2+}]_i$ when the agonist lectin concanavilin A (ConA) is competitively removed from stimulated cells by the antagonist α-methyl mannoside (αMM) (Fig. 2).[25] Thus, one of the major consequences of antigen receptor-ligand interaction is the generation of the PI second messengers, which subsequently contribute to events culminating in nuclear activation (Fig. 3).

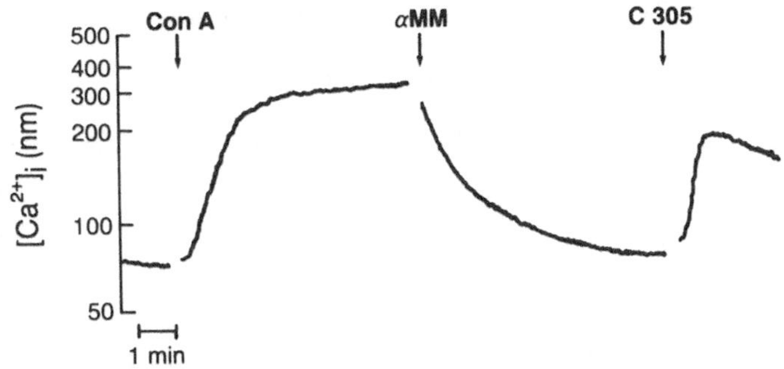

Fig. 2. Reversible Stimulation of the PI Pathway via the Antigen Receptor

With regard to structure/function relationships within the CD3/Ti complex, we have no detailed molecular understanding of how binding to the receptor actually transmits a signal intracellularly. The favored model is that binding to Ti results in a conformational change that is transmitted to CD3, and that the C-terminal domains of the CD3 molecules allosterically trigger the next molecule in the signal transduction cascade. To date, only one piece of evidence supports such an "effector" function for the CD3 complex: antibodies directed against invariant CD3

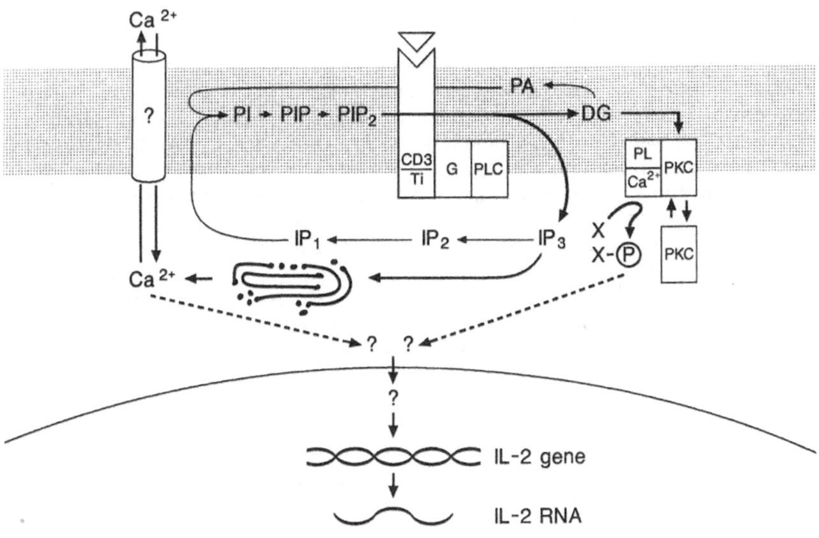

Fig. 3. Signal Transduction Events Leading to Nuclear Activation

Table 1. AlF_4^- Induces PIP_2 Hydrolysis in Jurkat

% of Basal:

Stimulus	Total Inositol Phosphates	Phosphatidic Acid
None	100	100
Anti-CD3 mAb	1000	260
AlF_4^-	400	212

epitopes function as agonists like those directed against Ti epitopes.[3,4] Of course, this fact may be an artifactual consequence of the close physical association of CD3 and Ti, and not an indication of intrinsic functional capacity of CD3. Furthermore, there has been no direct demonstration of an essential conformational change upon ligand binding. Compelling indirect evidence of such an allosteric mechanism comes from two experiments. First, it has been shown that the T cell antigen receptor, unlike the immunglobulin receptor on B cells, is responsive to monovalent antibody Fab fragments,[20] implying that aggregation of receptors (i.e., crosslinking) is not essential to receptor signalling. Second, in collaboration with Lewis Lanier, we have described a panel of anti-Ti antibodies reactive with spatially distinct determinants on the T cell line HPB-ALL, which have equal crosslinking potential to their agonist counterparts but which are dramatically less potent as stimulators of Ca^{2+} metabolism.[26] Further evidence for a requisite conformational shift has been obtained recently in experiments with signal transduction mutants (see Section IV.E.).

C. Role of GTP-Binding Proteins

It is well documented that many cell surface receptors are coupled to the relevant second messenger system via a GTP-binding protein (G protein) that shuttles between the ligand-binding structure and the second messenger enzyme.[27] Best characterized are the ß-adrenergic and rhodopsin receptor systems. Evidence has been presented that several PI-coupled receptors likewise may be dependent on G proteins.

The evidence in T cells has been relatively indirect, in large part due to the absence of an *in vitro* assay of CD3/Ti function. The first evidence came from studies by Imboden et al.,[28] who found that pre-treatment of whole cells with cholera toxin (a known disrupter of G protein function) significantly inhibited the receptor-mediated signalling function in these cells through a cAMP-independent process. While this observation contrasts with some conventional paradigms of ADP-ribosylating toxins and G proteins (i.e., cholera toxin is an activator of the adenylate cyclase-linked G_s), it strongly suggests a critical role for a G protein proximally in the signal transduction process. We have extended these studies with the use of AlF_4^-, a known direct activator of some G proteins.[27,29] Based on the studies of Exton et al. in hepatocytes,[30] we tested whether AlF_4^- would stimulate PI turnover and Ca^{2+} mobilization in Jurkat. As shown in Table 1, we found that this agent did indeed trigger these events, consistent with the notion of a G protein residing proximally in the transmembrane signalling process. More direct experiments await a broken-cell assay.

D. Phosphorylation of Receptor Structures

Protein phosphorylation has long been thought to play an important role in the regulation of enzymatic cascades. Many receptor proteins have been demonstrated to contain potential or actual sites of phosphorylation.[31] For receptors coupled to the PI second messenger system, phosphorylation of serine and threonine residues of the receptor coincidentally with activation of pkC has been presumed to indicate a feedback mechanism within the receptor/second messenger communication. Both T cell antigen receptor ligands[32] (e.g., lectins) and direct activators of pkC[33] (e.g., PMA) have been demonstrated to promote serine phosphorylation of the γ and δ chains of the human CD3 complex. At the same time, these stimuli promote loss of the receptor complex from the cell surface, leading to the hypothesis that receptor-mediated activation of pkC normally results in phosphorylation of CD3, thereby providing a trigger for modulation. However, no direct requirement for phosphorylation in CD3/Ti modulation or desensitization has been proven. Furthermore, we have observed essentially complete modulation in mutant T cells that demonstrate no detectable PI metabolism (see Section IV.D.). Therefore, although some regulatory role for pkC-mediated activation has not been excluded, further demonstration of its relevance to T cell activation is needed.

The identification of numerous receptors and receptor-like proteins with intrinsic tyrosine kinase activities has prompted a search for such an activity associated with the antigen receptor. The putative p21 chain of the CD3/Ti complex was initially identified by its tyrosine phosphorylation upon receptor activation, but not upon treatment with PMA.[14] Currently it is unknown whether p21 is a substrate of an unknown tyrosine kinase, or whether its phosphorylation represents autophosphorylation resulting from intrinsic tyrosine kinase activity. Other proteins also become tyrosine phosphorylated with receptor activation,[34] suggesting that both the PI second messenger system and a tyrosine kinase system may act in parallel to execute the complete CD3/Ti signal transduction program.

IV. A NOVEL SYSTEM FOR INVESTIGATING TRANSMEMBRANE SIGNALLING BY THE ANTIGEN RECEPTOR COMPLEX

A. Development of the System

Although a number of signal transduction events are measurable following receptor stimulation, little is understood about how engagement of the receptor is translated into these events. In the case of T cells, no direct demonstration of a role for CD3 proteins in signal transduction has been provided. Likewise, the evidence for a G protein coupled to the PIP_2 phospholipase (phospholipase C) is at present indirect. We sought to develop a system for applying somatic cell genetics to this problem. Specifically, we assumed that mutants with functional deficits in transmembrane signalling would be useful reagents for identifying proteins that are relevant to signal transduction, for defining the "active" domains within these molecules, and for characterizing the functional relationships among the relevant proteins.

To accomplish these objectives, we began with the cell line Jurkat, and developed a multi-step selection protocol designed to enrich for, and subsequently isolate clonally, rare mutants that express wildtype levels of receptor, but in which the CD3/Ti complex is uncoupled from the known second messenger events. From our own previous studies with receptor-loss mutants, we knew that the lectin phytohemagglutinin (PHA) would stimulate Jurkat cells to mobilize $[Ca^{2+}]_i$ in a CD3/Ti-dependent manner, a process which would synergize with PMA to promote production of IL-2. We observed that Jurkat cells cultured continuously in the presence of PHA failed to proliferate (Fig. 4); this inhibition was in large part dependent on the presence of a CD3/Ti

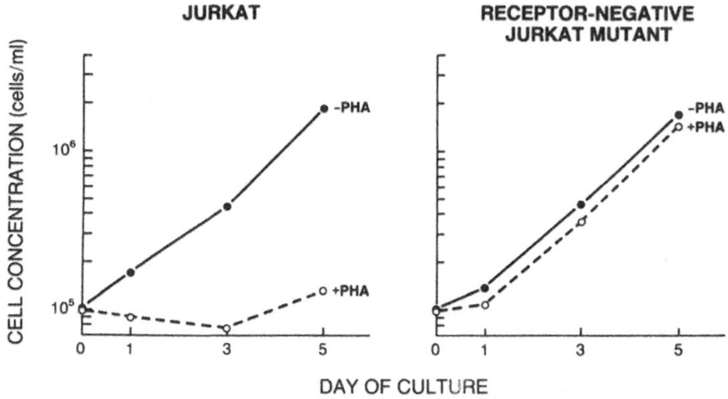

Fig. 4. Growth Inhibition by PHA

complex, as receptor-negative mutants were almost completely resistant to PHA. Since PHA binds to, but does not inhibit the growth of, CD3/Ti-deficient mutants, growth inhibition is likely to depend on an event associated with receptor-mediated transmembrane signalling. We reasoned that at least some of the cells that continued to proliferate in the presence of PHA would fail to express functional receptors. One class of signalling mutants would be expected to include cells which have lost surface expression of CD3/Ti. Consistent with this prediction, approximately half of the cells that grew out of this negative selection (applied to cells mutagenized first with ethane methane sulfonate, EMS) were found not to express receptors, as determined by fluorescent-activated flow cytometry (Fig. 5). We predicted that among the receptor-positive cells that survived PHA treatment (and which could be recovered readily by fluorescent-activated cell sorting (FACS)), some would have non-functional receptors and others would have mutations more distally in the metabolic pathway leading to growth arrest.

To enrich for more proximal mutations, we turned to an additional selection procedure that relies both on FACS and on the advent of

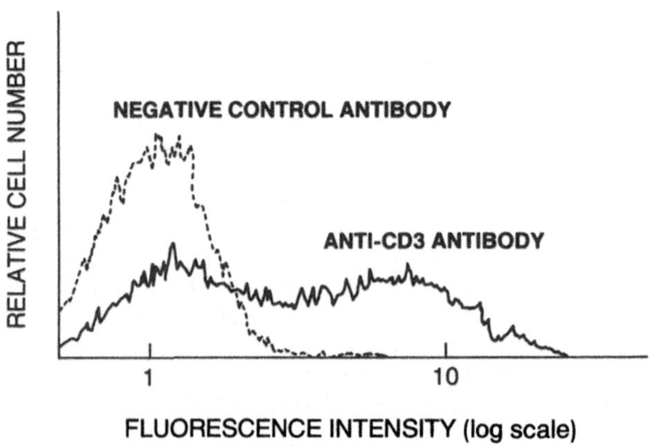

Fig. 5. Receptor-Positive and Receptor-Negative Cells Resistant to PHA

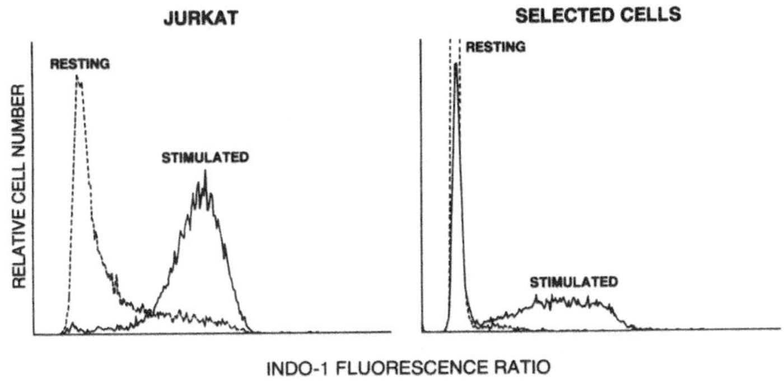

Fig. 6. Selection and Screening of Ca^{2+} Mutants

sensitive, fluorescent Ca^{2+}-indicators. The assumption here was that cells which failed to elevate $[Ca^{2+}]_i$ upon receptor binding with mAb directed against the receptor complex likely would have mutations relatively early in the triggering of second messenger production. Using the probe Indo-1,[35] it was possible to sort for cells that failed to increase their $[Ca^{2+}]_i$ when stimulated by a usually agonistic antibody. As shown in Fig. 6a, Jurkat cells with a resting $[Ca^{2+}]_i$ exhibit a low relative ratio of fluorescence emission at 404 nm and 486 nm following excitation by a 364 nm laser, since the predominant emission by Ca^{2+}-free Indo-1 is at 486 nm. Cells stimulated with anti-receptor mAb (C305, anti-clonotype of Jurkat) exhibit a relatively high ratio, since the predominant emission by Ca^{2+}-bound Indo-1 is at 404 nm.

Sorting for cells that failed to increase $[Ca^{2+}]_i$ in response to receptor stimuli simultaneously enriched for receptor-loss mutants, as would be predicted, and it became necessary to recover the receptor-bearing cells by conventional sorting. Eventually, after 4 pairs of alternating Ca^{2+} and receptor sorts, a large population of non-responding cells became evident (Fig. 6b). These cells were then cloned by limiting dilution, to yield individual clones for further analysis. The overall selection protocol is summarized in Fig. 7.[36]

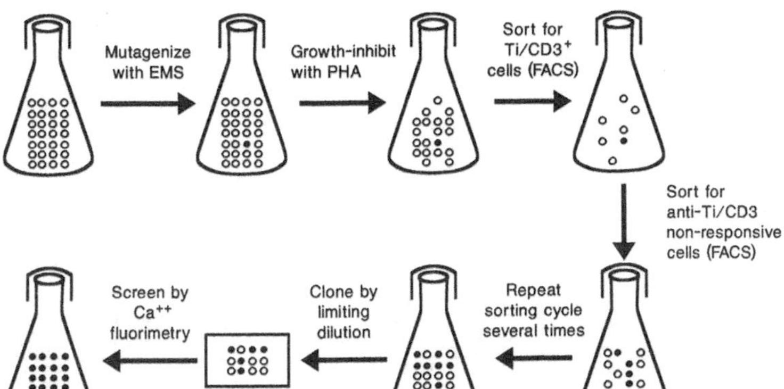

Fig. 7. Summary of Selection Protocol

Table 2. Ca^{2+} Non-Responsiveness of J.CaM1

Stimulus	Maximal $[Ca^{2+}]_i$ (nM): Jurkat	J.CaM1
None	114	130
C305	1975	146
OKT3	888	141

B. Signal Transduction Characteristics

Seven individual clones were assayed by conventional Ca^{2+} fluorimetry for responsiveness to C305, and four of the seven had the desired non-responsive phenotype. One of these, labelled J.CaM1 (Jurkat-derived Ca^{2+} mutant 1), was chosen for further analysis. Unlike parental Jurkat cells, J.CaM1 (Table 2) cells failed to increase $[Ca^{2+}]_i$ even when stimulated with saturating amounts of either C305 (anti-Ti) or OKT3 (anti-CD3) mAb.

To begin analyzing J.CaM1 further, it was essential to determine whether the block in Ca^{2+} mobilization occurred in the generation of PI second messengers (IP_3) that lead to mobilization of Ca^{2+}, or in the apparatus for responding to IP_3. Therefore, we assayed the synthesis of inositol phosphates following receptor stimulation, and found that J.CaM1 failed to produce IP_3 or its metabolites (Fig. 8). Likewise, these cells failed to generate phosphatidic acid, the primary metabolite of DG (not shown). These results indicated that the mutation in J.CaM1 prevented the usually efficient coupling of the receptor complex to the PI second messenger system.

To confirm this conclusion, we attempted to bypass the antigen receptor complex and to trigger PI metabolism independently of the receptor. To achieve receptor-independent stimulation of the PI pathway, we treated J.CaM1 cells with AlF_4^-, which is thought to act directly upon the putative G protein.[30] $[Ca^{2+}]_i$ in these cells rose from 212 nM to 682 nM after stimulation with AlF_4^-, and total IP's increased to 655 $\pm$ 13% of the basal levels after 20 mins. of

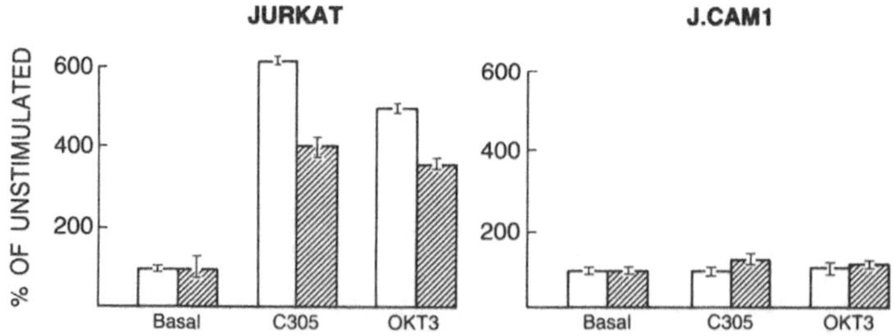

Fig. 8. Inositol Phosphate Production in J.CaM1. Stimulation for 10 mins. with designated antibody (C305 - anti-Ti; OKT3 - anti-CD3). Open bars - IP_1 + IP_2; hatched bars - IP_3.

Table 3. IL-2 Production by J.CaM1 (units/ml)

PMA (50 ng/ml) plus:

Cell Line	Ionomycin (1 uM)	C305 (1:1000)	OKT3 (1:1000)
J.CaM1	29 ± 15	1 ± 1	1 ± 1
Jurkat	36 ± 6	18 ± 1	19 ± 2

stimulation. These results suggest that a G protein/PI segment of the signal transduction system is intact, and are supportive of the conclusion that the mutation in these cells lies proximal to second messenger generation.

C. Receptor Structure

We used two approaches to analyze the receptor molecules. First, we applied immunoprecipitation and SDS-PAGE analysis to cell-surface radioiodinated cells for comparison with Jurkat. Immunoprecipitation with an anti-CD3 antibody under conditions that preserve the association with Ti molecules revealed the expected Ti and CD3-γ, δ, and -ε proteins, with no gross deletions or alterations (not shown).[36] Likewise, the ζ dimer was detectable using an anti-ζ antiserum for immunoprecipitation. Finally, two-dimensional isoelectric focusing revealed no differences in charge between the CD3/Ti molecules of the mutant and parental cells. Therefore, we concluded that the mutation either resided in another molecule not visualized, or was too subtle to be apparent by these studies.

A second approach was immunofluoresence as detected by FACS. Fig. 9 reveals that anti-Ti and anti-CD3 reagents stained J.CaM1 brightly, with slightly greater intensity than that of Jurkat (not shown). Moreover, four other anti-CD3/Ti antibodies demonstrated comparable binding to the two cells (not shown). Therefore, the mutation did not grossly alter the conformation of the receptor complex, as assessed with reagents reactive with several individual epitopes.

D. Functional Correlates of Impaired Transmembrane Signalling

Since up-regulation of expression of the IL-2 genes appears to result at least in part from the action of the PI second messengers, we determined the influence of defective PI signalling on receptor-mediated nuclear activation in J.CaM1 (Table 3). As predicted, the ionophore ionomycin was able to synergize with PMA to promote IL-2 production by J.CaM1, but receptor ligands were ineffective. Therefore, uncoupling from the PI pathway correlates with uncoupling from the receptor-mediated lymphokine gene regulation program.

Since receptor modulation or down-regulation has been thought to relate to activation of pkC, we next examined the modulation response of J.CaM1. As shown in Fig. 9, the receptors expressed by J.CaM1 were easily modulated from the cell surface. Like parental Jurkat cells, J.CaM1 cells expressed less than 4% of their resting receptor number after a 2.5 hour incubation with anti-Ti mAb. Therefore, receptor modulation in these cells can be dissociated from PI metabolism. This finding suggests that pkC-mediated phosphorylation of CD3 proteins is not critical to modulation, and that the PMA- and receptor-induced modulation events may have different mechanisms.

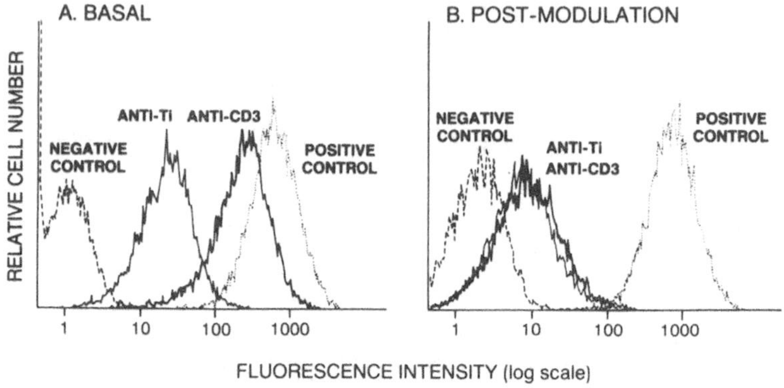

Fig. 9. Expression of CD3/Ti and Modulation Response

E. The Signalling Defect is Not Absolute

To assess whether the receptor expressed by J.CaM1 is entirely devoid of signal transduction potential, we applied combinations of anti-receptor mAb during Ca^{2+} fluorimetry experiments. As shown in Fig. 10A, while neither OKT3 nor C305 were agonists when used alone even at saturating doses (see Table 2), the two together elicited a substantial increase in $[Ca^{2+}]_i$. This Ca^{2+} response appears to result from PI turnover, since the two mAb together stimulated a 2- to 5-fold increase in IP levels relative to basal levels and a 5-fold increase in phosphatidic acid. Similar Ca^{2+} responses were seen with either mAb alone in the presence of a rabbit anti-mouse Ig antiserum (not shown). However, the restoration of agonist potential is not due entirely to an aggregation or cross-linking phenomenon since monovalent Fab fragments of the anti-CD3 mAb anti-Leu4 also synergized with C305 to yield a substantial, although submaximal, Ca^{2+} response.[36]

We were also surprised to observe that the anti-CD3 mAb 235 (IgM isotype) alone was capable of eliciting a Ca^{2+} response (Fig. 10b). The potency of this antibody likely is due to the particular protein or epitope to which it binds rather than to its isotype, since C305 (also an IgM mAb) is a non-agonist in J.CaM1. Therefore, while the receptor expressed by J.CaM1 is non-responsive to some stimuli, it remains responsive to others and to some combinations of non-agonist stimuli.

F. Implications

The results presented here suggest several features of CD3/Ti function. First, coupling to the PI pathway does indeed correlate with coupling to the subsequent nuclear activation events. Second, receptor down-regulation can occur independently of PI metabolism, and pkC-mediated CD3 phosphorylation may not be crucial to all modulation events. Third, conformational change rather than receptor aggregation may be critical to receptor function. The mutation in J.CaM1 appears to reside proximal to second messenger generation, and likely involves the

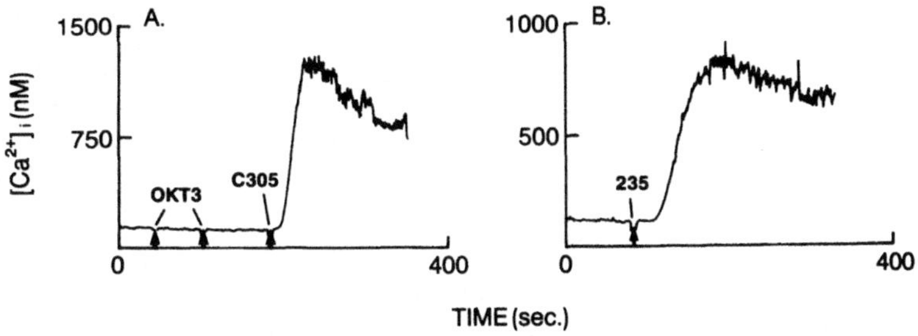

Fig. 10. Ca^{2+} Responses of J.CaM1 to Some Stimuli

receptor or receptor-associated molecules. This mutation appears to decrease the allosteric responsiveness of the receptor complex to engagement by mAb, but this inhibition may be overcome by some mAb directed against particular epitopes or by combinations of antibodies which together can force conformational changes. One hypothesis regarding normal receptor function is that the CD3 complex serves to amplify a receptor-mediated signal into a more pronounced intracellular second messenger event. In J.CaM1, then, the ligand binding function of the Ti structure and the signal amplification activity of the CD3 complex are dissociated.

Two further goals are being pursued with this somatic mutant system. First, we are attempting to identify the molecule and domain mutated in J.CaM1 by a series of complementation and gene transfer studies. Second, we are continuing to generate independent mutants to provide a family of such mutants with complementing and non-complementing mutations. We suggest that the strategy outlined here could form the basis for performing similar genetic analyses in other receptor systems linked to Ca^{2+} mobilization and PI metabolism.

V. EXPRESSION OF THE INTERLEUKIN-2 GENE DURING T CELL ACTIVATION

The immediate consequence of the binding of agonists to the CD3/Ti complex is the generation of a transmembrane signal with resultant appearance of the second messengers IP_3 and DG, followed by an increase in $[Ca^{2+}]_i$ and activation of pkC. These cytoplasmic biochemical events have been causally linked to various manifestations of T cell function including the activation of the cytolytic mechanism in effector T cells and the secretion of an array of lymphokines by helper T cells. An understanding of the mechanisms by which an increase in $[Ca^{2+}]_i$ and activation of pkC elicit such responses is not yet at hand. The most detailed studies aimed at linking such proximal events occurring at the plasma membrane with nuclear events have examined the activation of the interleukin-2 (IL-2) gene.

Utilizing Jurkat, we have previously demonstrated that ligands which induce transmembrane signalling events by the CD3/Ti complex, together with PMA, stimulate Jurkat cells to secrete IL-2.[4] The ability of CD3/Ti agonists to increase $[Ca^{2+}]_i$, and of calcium ionophores to substitute for CD3/Ti ligands, demonstrates a role for an increase in $[Ca^{2+}]_i$ in synergizing with activation of pkC to induce IL-2 production by Jurkat cells. Similar results have been observed with normal resting T cells and T cell clones.[37,38] The production of IL-2 by stimulated Jurkat or normal T cells involves the transcriptional activation of the IL-2 gene.[4] Moreover, studies have revealed that both the increase in $[Ca^{2+}]_i$ and activation of pkC are required for transcriptional activation of the IL-2 gene and accumulation of IL-2 transcripts (Fig. 11). Moreover, both stimuli are required simultaneously for 2-4 hours for cellular commitment to IL-2 production and this commitment coincides with the accumulation of IL-2 transcripts[35] (Fig. 12A).

Since the increase in $[Ca^{2+}]_i$ and activation of pkC initiated by ligands acting at the plasma membrane are evident within seconds to minutes after stimulation, the delayed appearance of IL-2 transcripts suggests the possibility that expression of the IL-2 gene might not represent a primary gene activation event. Indeed, this notion was reinforced by the observation that transcripts of the cellular nuclear proto-oncogene c-fos were inducible by CD3/Ti agonists, calcium ionophores, or PMA within 15 minutes and peaked by 30 minutes after stimulation (Fig. 12B). This observation suggested that the product of an early activation gene, typified by c-fos, might be necessary for the activation of the IL-2 gene. To examine this possibility, Jurkat cells were stimulated with ConA and PMA in the presence or absence of the protein synthesis inhibitor cycloheximide. Complete inhibition of the appearance of IL-2 transcripts was observed in the presence of cycloheximide (Fig. 11) whereas no change in the level of transcripts of the constitutively active Tiß-chain gene was observed. Interestingly, cycloheximide did not inhibit the appearance of c-fos transcripts and instead, as reported by others, superinduced the appearance of c-fos mRNA in stimulated cells (not shown). The resistance of c-fos transcription to cycloheximide strongly argues that the protein synthesis inhibitor is not interfering with signal transduction pathways, and that c-fos activation is regulated differently without the requirement for new protein synthesis. The putative newly induced protein required for IL-2 transcription has not been identified, nor is it clear whether a single protein, multiple proteins, or post-translational modification of such proteins are involved.

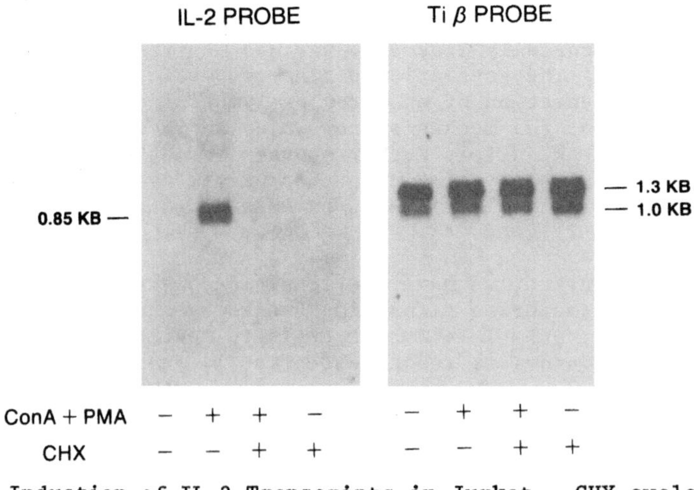

Fig. 11. Induction of IL-2 Transcripts in Jurkat. CHX-cycloheximide.

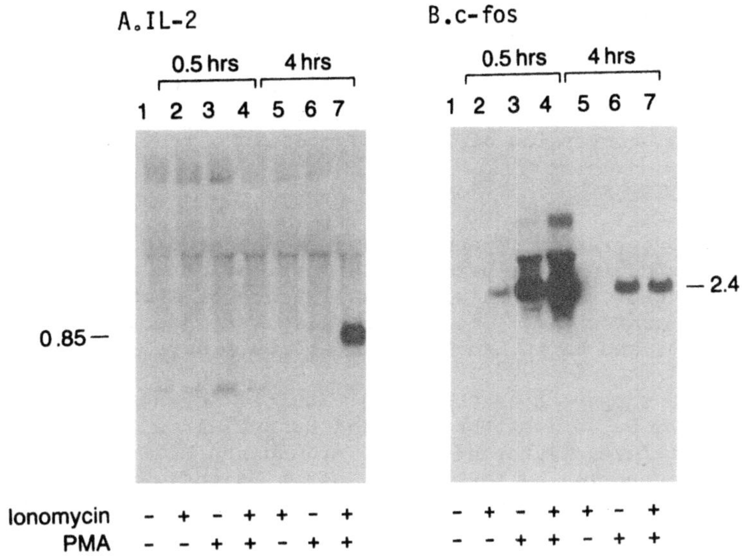

Fig. 12. Requirements for Induction of IL-2 and c-fos Transcripts

Regions of the IL-2 gene must be responsive to inducible regulatory elements since the gene is transcriptionally silent in unstimulated cells. Durand et al.[39] in collaboration with our laboratory have demonstrated that a 275 base pair (bp) segment upstream of the IL-2 gene may represent such a responsive region. This DNA segment, together with a heterologous promoter, can confer activation responsiveness to the chloramphenicol acetyl transferase (CAT) gene. Thus, this sequence, when transfected into Jurkat cells, induces the transcription of CAT only when such cells are stimulated with either CD3/Ti ligands plus PMA or ionophore plus PMA. The identification of the regulatory proteins influencing this region and how these proteins are regulated by the transmembrane signalling events discussed is of great interest. Such studies should provide lessons that are relevant to regulation of other sets of responsive genes which become transcriptionally active in stimulated T cells.

VI. SUMMARY

The studies described here, including ours and those of others, were aimed at understanding the mechanism(s) by which the T cell antigen receptor transmits signals that result in expression of lymphokine genes. They have focused both on the proximal molecules and biochemical events that account for transmembrane signalling, as well as on the more distal nuclear regulatory events. It is now clear that the receptor is coupled to the PI pathway, apparently via conformational changes in Ti, possibly amplification by CD3, and likely a G protein. PI second messengers appear to contribute to subsequent gene expression events, but may not be required for receptor down-regulation. How these biochemical events are coupled to gene expression still remains a mystery. However, studies of the regulation of the IL-2 gene indicate that the proximal receptor-mediated events must occur continuously for 2-4 hours to support transcriptional activation of the IL-2 gene, and that protein synthesis during this time is required. Other genes, such as the proto-oncogene c-fos, become transcriptionally active within 15 minutes of receptor triggering even in the presence of protein synthesis inhibitors. Thus, coupling of the proximal biochemical events to expression of the IL-2 gene may be dependent on the induction of a regulatory protein(s). This protein(s) may act to activate the IL-2 gene via a 275 bp segment that lies upstream of the IL-2 coding region. Studies are now in progress to understand better the molecular requirements for competence in signal transduction and conversion of these signals into gene expression events.

VII. ACKNOWLEDGEMENTS

The authors gratefully acknowledge the superb technical assistance of Marianne Newton and Paul Dazin, FACS Operator of the Howard Hughes Medical Institute, and the excellent assistance of Michael Armanini in preparing this manuscript. The authors also are indebted to Drs. William Lee, Shu Man Fu, and Lewis Lanier for gratiously supplying reagents.

M.A.G. is supported by the N.I.H. Medical Scientist Training Program of U.C.S.F., and by the Rosalind Russell Arthritis Center. A.W. is an Assistant Investigator of the Howard Hughes Medical Institute. This work was supported by the Howard Hughes Medical Institute, and by an Arthritis Investigator Award to A.W. from the Arthritis Foundation.

VIII. REFERENCES

1. T. Honjo, Ann. Rev. Immunol. 1:499 (1983).
2. B. Toyonaga, and T. Mak, Ann. Rev. Immunol. 5:585 (1987).
3. J. P. Allison, and L. L. Lanier, Ann. Rev. Immunol. 5:503 (1987).
4. A. Weiss, J. Imboden, K. Hardy, B. Manger, C. Terhorst, and J. Stobo, Ann. Rev. Immunol. 4:593 (1986).
5. J. Yague, J. White, C. Coleclough, J. Kappler, E. Palmer, and P. Cell 42:81 (1985).
6. Z. Dembic, W. Haas, S. Weiss, J. McCubrey, H. Kiefer, H. Von Boehmer, and M. Steinmetz, Nature 320:232 (1986).
7. T. Saito, A. Weiss, J. Miller, M. A. Norcross, and R. N. Germain, Nature 325:125 (1987).
8. E. L. Reinherz, S. Meuer, K. A. Fitzgerald, R. E. Hussey, H. Levine, and S. F. Schlossman, Cell 30:735 (1982).
9. A. Weiss, and J. Stobo, J. Exp. Med. 160:1284 (1984).
10. P. S. Ohashi, T. W. Mak, P. Van den Elsen, Y. Yanagi, Y. Yoshikai, A. F. Calman, C. Terhorst, J. D. Stobo, and A. Weiss, Nature 316:606 (1985).
11. P. Van den Elsen, B. A. Shepley, J. Borst, J. E. Coligan, A. F. Markham, S. Orkin, and C. Terhorst, Nature, 312:413 (1984).
12. D. P. Gold, J. M. Puck, C. L. Pettey, M. Cho, J. Coligan, J. N. Woody, and C. Terhorst, Nature 321:431 (1986).
13. G. W. Krissansen, M J. Owen, W. Verbi, and M. J. Crumpton, EMBO J. 5:1799 (1986).
14. L. E. Samelson, J. B. Harford, and R. D. Klausner, Cell 43:223 (1985).
15. A. M. Weissman, L. E. Samelson, and R.D. Klausner, Nature 324:480 (1986).
16. L. E. Samelson, M. D. Patel, A. M. Weissman, J. B. Harford, and R. D. Klausner, Cell 46:1083 (1986).
17. A. Weiss, J. Imboden, D. Shoback, and J. Stobo, Proc. Natl. Acad. Sci. USA 81:4169 (1984).
18. J. Imboden, A. Weiss, and J. D. Stobo, J. Immunol. 134:663 (1985).
19. K. O'Flynn, E. Zanders, J. Lamb, P. Beverly, D. Wallace, P. Tatham, W. Tax, and D. Linch, Eur. J. Immunol. 15:7 (1985).
20. J. Oettgen, C. Terhorst, L. Cantley, and P. Rosoff, Cell 40:583 (1985).
21. M. J. Weiss, J. F. Daley, J. C. Hodgdon, and E. L. Reinherz, Proc. Natl. Acad. Sci. USA 81:6836 (1985).

22. J. B. Imboden, and J. D. Stobo, J. Exp. Med. 161:446 (1985).
23. M. J. Berridge, and R. F. Irvine, Nature 312:315 (1984).
24. W. L. Farrar, and F. W. Ruscetti, J. Immunol. 136:1266 (1985).
25. A. Weiss, R. Shields, M. Newton, B. Manger, and J. Imboden, J. Immunol. 138:2169 (1987).
26. L. L. Lanier, J. J. Ruitenberg, J. P. Allison, and A. Weiss, J. Immunol. 137:2286 (1986).
27. L. Stryer, and H. R. Bourne, Ann. Rev. Cell. Biol. 2:391 (1986).
28. J. B. Imboden, D. M. Shoback, G. Pattison, and J. D. Stobo, Proc. Natl. Acad. Sci. USA 15:5673 (1986).
29. J. K. Northrup, M. D. Smigel, P. C. Sternweis, and A. G. Gilman, J. Biol. Chem. 258:11369 (1983).
30. P. F. Blackmore, S. B. Bocckino, L. E. Waynick, and J. H. Exton, J. Biol. Chem. 260:14477 (1985).
31. D. R. Sibley, J. L. Benovic, M. G. Caron, and R. J. Lefkowitz, Cell 48:913 (1987).
32. L. E. Samelson, J. Harford, R. H. Schwartz, and R. D. Klausner, Proc. Natl. Acad. Sci. USA 82:1969 (1985).
33. D. A. Cantrell, A. A. Davies, and M. J. Crumpton, Proc. Natl. Acad. Sci. USA 82:8158 (1985).
34. M. D. Patel, L. E. Samelson, and R. D. Klausner, J. Biol. Chem. 262:5831 (1987).
35. G. Grynkiewicz, M. Poenie, and R. Y. Tsien, J. Biol. Chem. 260:3440 (1985).
36. M. A. Goldsmith, and A. Weiss, Proc. Natl. Acad. Sci. USA In press.
37. D. K. Kim, G. Otten, R. L. Moldwin, D. E. Dunn, G. J. Nau, and F. W. Fitch, J. Immunol. 137:2755 (1986).
38. Y. Yamamoto, T. Ohmura, K. Fujimoto, and K. Onoue, Eur. J. Immunol. 15:1204 (1985).
39. D. B. Durand, M. R. Bush, J. B. Morgan, A. Weiss, and G. R. Crabtree, J. Exp. Med. 165:395 (1987).

THE ENDOSOMAL APPARATUS AND TRANSMEMBRANE SIGNALLING

J.J.M. Bergeron, W.H. Lai, D.G. Kay, J.J. Doherty, II, M.N. Khan and B.I. Posner

McGill University
Montreal, Quebec, H3A 2B2

INTRODUCTION

The binding of insulin, prolactin, and epidermal growth factor to their respective receptors on liver parenchymal cells rapidly leads to their internalization along with their associated receptors. As reviewed elsewhere, [1-5] high concentrations of these ligand-receptor complexes accumulate initially in a heterogeneous group of compartments collectively termed the endosomal apparatus. Conceptually, the endosomal apparatus bridges the plasmalemma with secondary lysosomes and involves at least indirectly the Golgi apparatus (Fig. 1 and [14-17]). In liver, components of the endosomal apparatus have been visualized and identified by electron microscope radioautography, [18] EM immunocytochemistry, [19-23] and subcellular fractionation [7-11,24]. These techniques have demonstrated that the endosomal components are not lysosomes, [10,12,18,25] that a sequential transport of ligands and receptors among endosomal components can be observed, [8,9,26,29] and that morphologically different microdomains can be observed in continuous structures segregating ligands (vesicular) from receptors (tubular)[19-21].

FUNCTIONS OF ENDOSOMES

Sorting

From such studies three main functions for the endosomal apparatus have been inferred. The most universally accepted is that of sorting. The receptors for insulin as well as EGF (at least in liver) recycle to the cell surface [27-30] while their corresponding ligands are degraded. As at least a subset of endosomal components are of low pH [1,31] mediated by an ATP-dependent proton translocase, [32,33] and as many ligand-receptor complexes dissociate at low pH, [34-36,57] then the proton ATPase and tubulovesicular microdomains have been causally linked to sorting [4,20]. Indeed, Geuze and colleagues [20,21] have coined the term CURL (Compartment for Uncoupling of Receptor from Ligand) for the sorting domains of the endosomal apparatus. Not only are several ligands sorted from their receptors but ligand-receptor complexes are sorted from each other. That various ligand-receptor complexes share the same endosomal components has been demonstrated for insulin, asialoglycoprotein, prolactin, epidermal growth factor, and polymeric IgA in the liver [37,38]. A spectacular demonstration of sorting of polymeric IgA away from other ligands and their receptors has been observed by Courtoy and colleagues, [37] who used

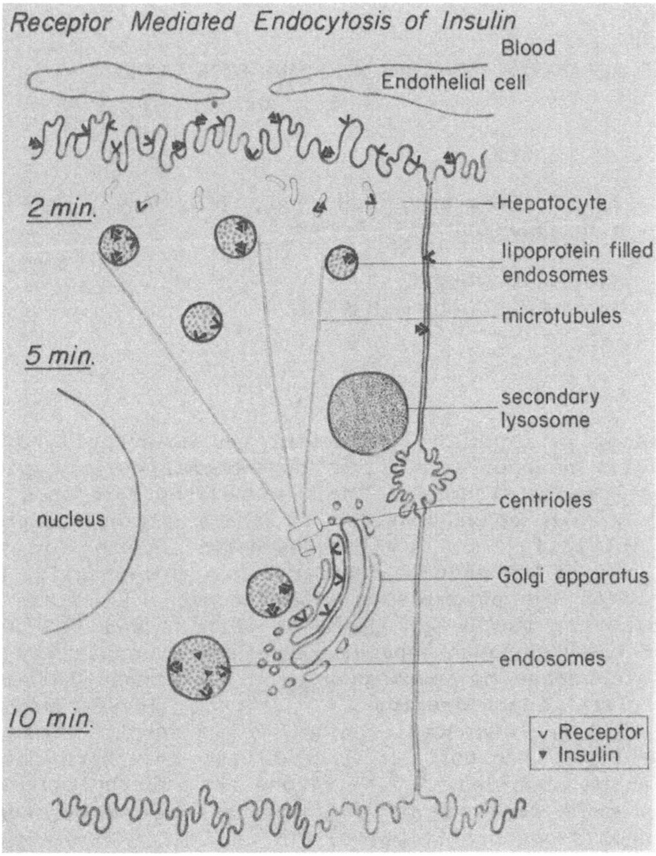

Figure 1. Endocytosis of insulin and its receptor by liver parenchyma. Shortly after binding, insulin and its receptor are internalized into a peripheral, small (100-150m) tubulo-vesicular compartment (2 min) which appears to be insensitive to the acidotropic agent chloroquine [6]. Lipoprotein-filled vesicles (early endosomes) of ca. 200-300 nm diameter then accumulate ligand (peak at 5 min) followed by accumulation into denser lipoprotein-filled vesicles (late endosomes, 300-400nm in diameter see [7-11]). Microtubules appear important in the transport and/or transformation of 'early' endosomes into 'late' endosomes [10,12]. Early endosomes accumulate little chloroquine and are concluded to be less acidic than 'late' endosomes and lysosomes [9,13].

the DAB shift protocol to demonstrate entry of polymeric IgA into common endosomal components followed by the segregation of IgA into distinct structures prior to transcytosis of polymeric IgA and a portion of its receptor into bile.

Selective Ligand Degradation

From detailed studies on the endosomal components internalizing prolactin and insulin and their receptors, we have concluded that insulin is selectively degraded early during its transit through the endosomal apparatus. This was previously inferred from a selective effect of the acidotropic agent chloroquine on the accumulation of insulin but not prolactin in endosome fractions, (Fig. 2 and [10,25]). More recently we have demonstrated directly that isolated endosome fractions will degrade in an ATP-dependent fashion intraluminal insulin but not another coincident ligand, i.e. epidermal growth factor (J.J. Doherty, II and J.J.M. Bergeron, unpublished observations). Furthermore, by the use of insulin molecules radiolabeled at position A^{14} and B^{26} we have concluded that the B chain is cleaved in early endosomes with the A chain degraded in late endosomes [40]. Peters and colleagues have also concluded that insulin is degraded at an endosomal locus [41] and Stahl, Costlow and coworkers [42,43] have identified the selective degradation of ligands bound to the mannose receptor of macrophages via a specific endosomal cathepsin D-like enzyme. As pointed out by Stahl, [43] selective ligand degradation within endosomes may be related to antigen processing in lymphocytes [44] and parathyroid hormone maturation by Kupffer cells.

Transmembrane Signalling

The receptors for insulin and epidermal growth factor are both transmembrane proteins with cytosolically exposed tyrosine specific protein kinases which are stimulated by ligand binding [45-47]. Current data strongly suggests that these kinases are causally linked to many if not all of the biological activities of the insulin and EGF receptors [48,51]. Following the uncovering of the internalization of insulin, prolactin, and other bioactive hormones in non-lysosomal components (endosomes), it was proposed that internalization could be related to the regulation of hormone action [9,18,52,53].

Both negative and positive regulation have been envisaged at the locus of the endosomal apparatus. It has been long known that internalization under conditions of receptor occupancy over prolonged times leads to down-regulation. This has been extensively described elsewhere [54-58] and will not be elaborated further. Evidence has also been accumulating for a positive role in transmembrane signalling in endosomes. Thus, during internalization active EGF receptor kinase activity has been detected by us (Fig. 3 and [24]) as well as by Cohen's group [59]. With regard to the insulin receptor, highly active insulin receptor kinase activity has been observed in endosomes (Fig. 4 & [11]). A noteworthy feature is that the biochemical properties of the insulin receptor in endosomes reveals that is has been covalently modified by phosphorylation to an 'activated' state. This activated receptor is no longer ligand dependent and shows increased Vmax and a lower Km for phosphorylation of exogenous substrates in vitro (M.N. Khan, J.J.M. Bergeron, and B.I. Posner, manuscript submitted).

Transmembrane signalling is expected to be a composite of ligand-receptor interactions in the endosome (stimulation) and modifications of the primary or quaternary structure of the receptor (activation).

Primary structure modifications, i.e. by phosphorylation, have been linked to the activated state of the insulin-receptor in endosomes (vide supra). Structural changes of receptors at the quaternary level have not

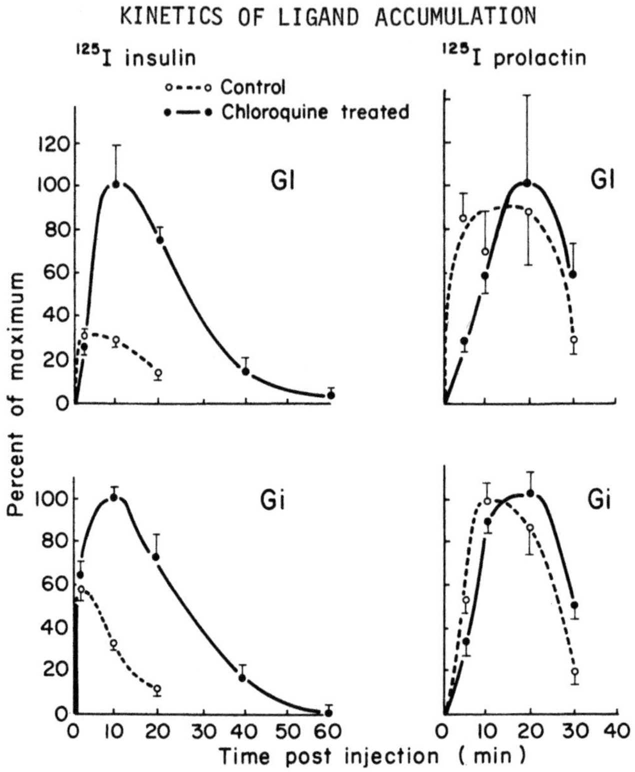

Figure 2. Effect of the acidotropic agent chloroquine (●–●) on the selective accumulation of ^{125}I-insulin but not ^{125}I-prolactin by endosomes sedimenting in the Golgi light (Gl) (top) or Golgi intermediate (Gi) fraction (lower). Results are expressed as percent of the radioactive content of the fractions at the peak time of uptake and are compared to animals not treated by chloroquine (o--o), (taken from [6,39] and unpublished data of the authors). Chloroquine acts by entering low pH compartments and may directly inhibit the low pH induced degradation of insulin within endosomes (J.J. Doherty, II and J.J.M. Bergeron, unpublished observations).

yet been studied within endosomes. This should however be a productive line of investigation as current studies by Schlesinger and colleagues [60-62] have linked EGF receptor dimerization with activation.

Although several groups have emphasized the endosome as a locus for low pH ligand-receptor dissociation (vide supra) initial studies from our group have shown a close association of ^{125}I-EGF and ^{125}I-insulin with the bounding membrane of endosomes (Fig. 5 and [62]). The concept of near neutral pH endosomes as suggested by the work of Fuchs et al. [63] and Kielian et al. [64], as well as the prior studies of Khan et al. [13,39] would implicate a subcompartment of endosomes in tight ligand-receptor interactions (see ref. 1 for a fuller discussion on this point).

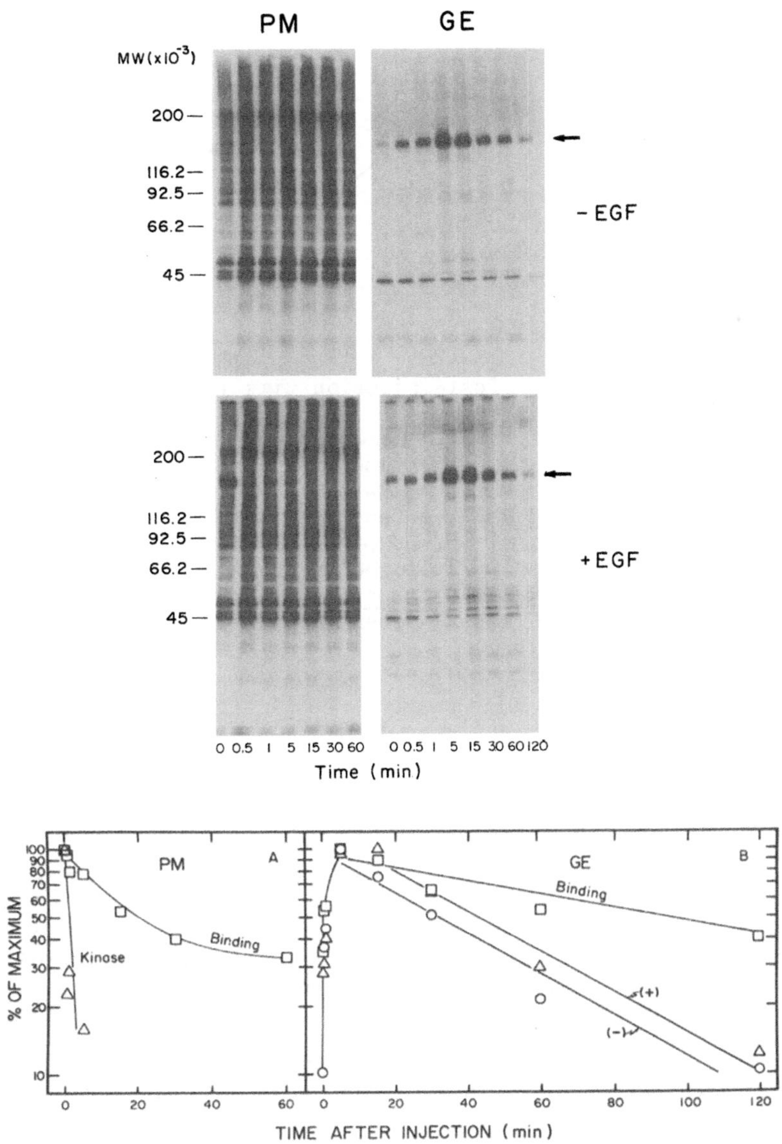

Figure 3. Ligand induced transport of EGF receptor from plasmalemma (PM) to endosomes (GE) as evaluated by direct binding (binding) and autophosphorylation activity (kinase), the latter assayed in the absence (-) or presence (+) of EGF in vitro. In the radioautographs, the band at Mr 170,000 (arrow) corresponds to the EGF receptor, (data taken from [24]).

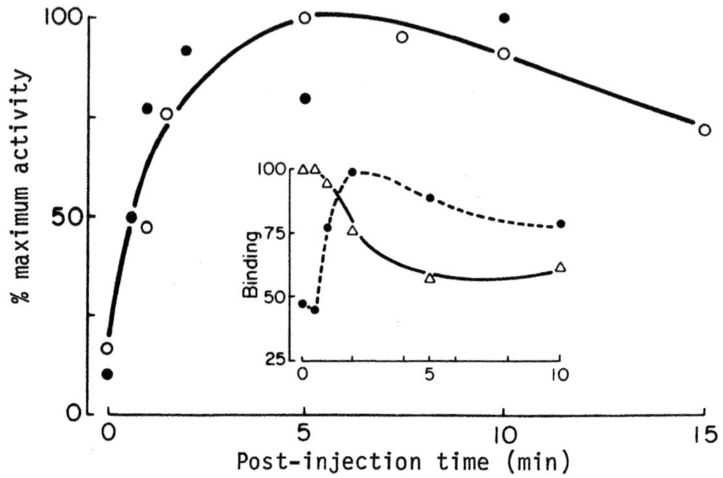

Figure 4. Time course of insulin receptor activation and translocation. At various times after the injection of insulin (150 ug/100 g body weight) endosomes[11] were isolated from liver homogenates. Autophosphorylation of solubilized receptors was carried out after lectin purification (WGA Sepharose) and subjected to autophosphorylation in the absence of in vitro insulin followed by receptor immunoprecipitation (endosomes •-•; microsomes o-o). Inset: kinetics of receptor change in PM (Δ —— Δ) and endosomes (•---•) after insulin injection.

CONCLUSION

The endosomal apparatus is a distinct accumulation of tubules and vesicles dispersed throughout the cytosol. The heterogeneous components interact in series as well as in parallel for three major functions. The first and best established is of sorting, most probably controlled by an ATP-dependent proton pump responsible for lowering the intraluminal pH of late endosomal components. Low pH appears to regulate ligand dissociation from receptors,[65,66] ligand degradation, (J.J. Doherty, II and J.J.M. Bergeron, unpublished observations and [10,42]) and receptor recycling [67,68] as these phenomena are prevented by acidotropic weak bases and proton ionophores. Not mentioned in this article is the functional association of endosomes with microtubules and the role of microtubules in the regulation of sorting [1]. As microtubules are responsible for transporting endosomes to centriolar regions of target cells near the Golgi apparatus and secondary lysosomes,[1,5,12,69,70] then an important role for microtubules in sorting seems apparent although as yet unproven.

Selective ligand degradation is another function of the endosomal apparatus and this seems clearest for the degradation of insulin in liver [10,25,41]. Finally, transmembrane signalling may be an important function of endosomes. As summarized in Fig. 6, ligand stimulated autophosphorylation of the insulin and EGF receptors of endosomes has been observed. Thus far no substrates of the insulin and EGF receptor kinases have been documented which mediate the physiological actions of these hormones. Particularly interesting will be studies which focus on the endosomal membrane for any such mediators of insulin and EGF actions.

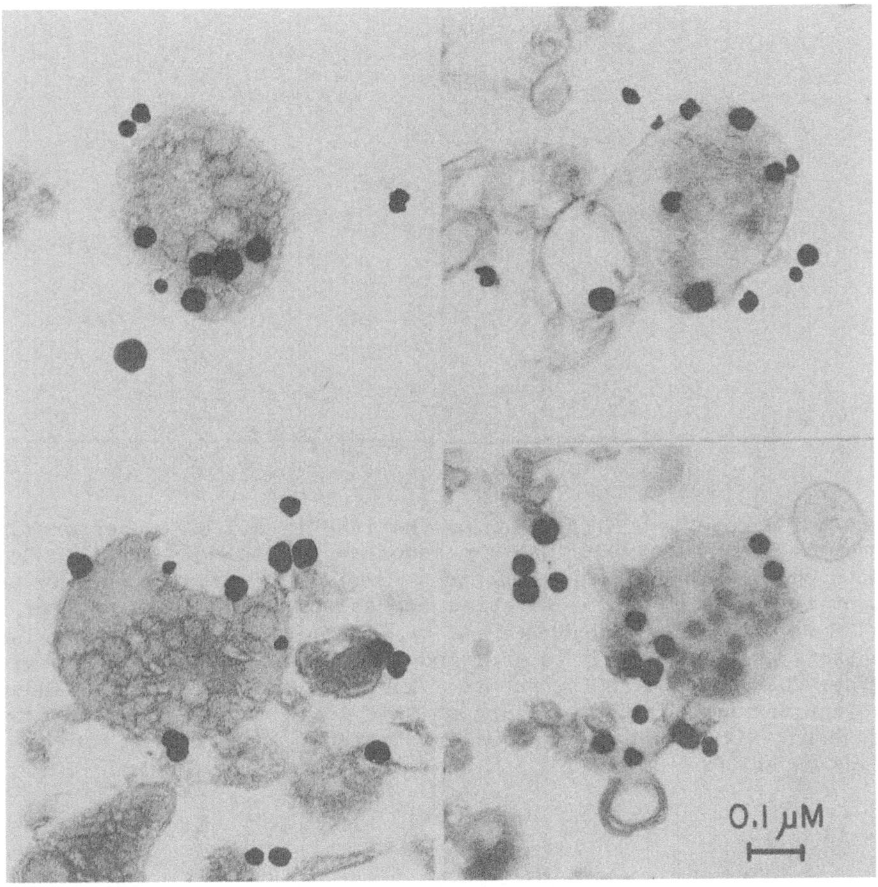

Figure 5. Electron microscope radioautography of lipoprotein-filled endosomes isolated at 15 min after the injection of ^{125}I-EGF. The small black dots correspond to radiation emanating from the ligand still associated with the endosomal membrane, presumably to the EGF receptor (J.J. Doherty, II and J.J.M. Bergeron, unpublished observations).

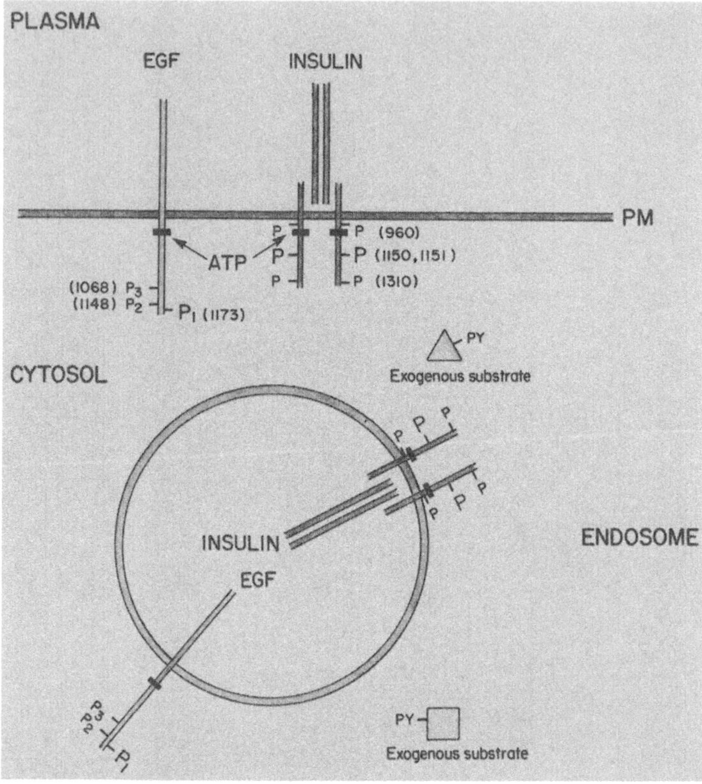

Figure 6. Schematic illustration of the insulin and EGF receptor kinases at the cell surface (PM) and in endosomes. The potential sites of autophosphorylation of the receptors are shown with the major sites thought to be related to activation emphasized with a larger size, i.e. P_1 1173 for the EGF receptor kinase [71,72] and 1150, 1151 for the insulin receptor kinase [51,73]. It is proposed that efficient autophosphorylation of the insulin and EGF receptor kinases occurs in the endosome. Efficient transmembrane signalling at this locus may be required in order to interact with unique exogenous substrates not available for the cell surface receptors.

ACKNOWLEDGEMENTS

This work was supported by the National Cancer Institute of Canada, The Medical Research Council of Canada, and The U.S. National Institutes of Health.

REFERENCES

1. J.J.M. Bergeron, J. Cruz, M.N. Khan, and B.I. Posner, Uptake of insulin and other ligands into receptor-rich endocytic components of target cells: the endosomal apparatus, Ann. Rev. Physiol. 47: 383-403 (1985).

2. T. Wileman, C. Harding, and P. Stahl, Receptor mediated endocytosis, Biochem. J. 232:1-14 (1985).
3. A. Helenius, I. Mellman, D. Wall, and A. Hubbard, Endosomes, Trends Biochem. Sci. 8:245-250 (1983).
4. I. Pastan and M.C. Willingham, Receptor-mediated endocytosis: coated pits, receptosomes, and the Golgi, Trends Biochem. Sci. 8:250-254 (1983).
5. C. Hopkins, Membrane boundaries involved in the uptake and intracellular processing of cell surface receptors, Trends Biochem. Sci. 11:473-477 (1986).
6. B.I. Posner, B.A. Patel, M.N. Khan, and J.J.M. Bergeron, Effects of chloroquine on the internalization of ^{125}I-insulin into subcellular fractions of rat liver. J. Biol. Chem. 257:5789-5799 (1982).
7. M.N. Khan, B.I. Posner, R.J. Khan, and J.J.M. Bergeron, Internalization of insulin into rat liver Golgi elements: evidence for vesicle heterogeneity and the path of intracellular processing, J. Biol. Chem. 257:5969-5976 (1982).
8. Z. Josefsberg, B.I. Posner, B. Patel, and J.J.M. Bergeron, The uptake of prolactin into female rat liver, J. Biol. Chem. 254:209-214 (1979).
9. B.I. Posner, B.A. Patel, A.K. Verma, and J.J.M. Bergeron, Uptake of insulin by plasmalemma and Golgi subcellular fractions of rat liver, J. Biol. Chem. 255:735-741 (1980).
10. J.J.M. Bergeron, N. Searle, M.N. Khan, and B.I. Posner, Differential and analytical subfractionation of rat liver components internalizing insulin and prolactin, Biochemistry, 25:1756-1764 (1986).
11. M.N. Khan, S. Savoie, J.J.M. Bergeron, and B.I. Posner, Characterization of rat liver endosomal fractions: in vivo activation of insulin-stimulable receptor kinase in these structures, J. Biol. Chem. 261:8462-8472 (1986).
12. J.J.M. Bergeron, L. Resch, R. Rachubinski, B.A. Patel, and B.I. Posner, Effect of colchicine on internalization of prolactin in female rat liver: an in vivo radioautographic study, J. Cell Biol. 96:875-876 (1983).
13. M.N. Khan, S. Savoie, R.J. Khan, J.J.M. Bergeron, and B.I. Posner. Insulin and insulin receptor uptake into rat liver, Diabetes 34:1025-1030 (1985).
14. D.J. Yamashiro, B. Tycko, S.R. Fluss, and F.R. Maxfield, Segregation of transferrin to a mildly acidic (pH 6.5) para-Golgi compartment in the recycling pathway, Cell 37:789-800 (1984).
15. H.J. Geuze, J.W. Slot, G.J. Strous, J.P. Luzio, and A.L. Schwartz, A cyclohexamide-resistant pool of receptors for asialoglycoproteins and mannose 6 phosphate resides in the Golgi complex of hepatocytes, EMBO J. 3:2677-2685 (1984).
16. J.J.M. Bergeron, D.G. Kay, W.H. Lai, J.J. Doherty, II, C.E. Smith, M.N. Khan, and B.I. Posner, Functional characteristics of the endosomal apparatus of rat liver parenchyma, in: "Cell Free Analysis of Membrane Traffic." J. Morre ed. Alan R. Liss Inc., N.Y., N.Y. (1988).
17. M.D. Snider and O.C. Rogers, Intracellular movement of cell surface receptors after endocytosis: resialylation of asialo-transferrin receptor in human erythroleukemia cells, J. Cell Biol. 100:826-834 (1985).
18. J.J.M. Bergeron, R. Sikstrom, A.R. Hand, and B.I. Posner, Binding and uptake of ^{125}I-insulin into rat liver hepatocytes and endothelium, J. Cell Biol. 80:427-443 (1979).
19. H.J. Geuze, J.W. Slot, G.J.A.M. Strous, H.F. Lodish, and A.L. Schwartz, Immunocytochemical localization of the receptor for asialoglycoprotein in rat liver cells, J. Cell Biol. 92:865-870 (1982).
20. H.J. Geuze, J.W. Slot, G.J.A.M. Strous, H.F. Lodish, and A.L.

Schwartz. Intracellular site of asialoglycoprotein receptor-ligand uncoupling: double-label immunoelectron microscopy during receptor-mediated endocytosis, Cell 32:277-287 (1983).
21. H.J. Geuze, J.W. Slot, G.J.A.M. Strous, J. Peppard, K. Von Figora, A. Hosilik, and A.L. Schwartz, Intracellular receptor sorting during endocytosis: comparative immunoelectron microscopy of multiple receptors in rat liver, Cell 37:195-204 (1984).
22. W.A. Dunn and A.L. Hubbard, Receptor mediated endocytosis of epidermal growth factor by hepatocytes in perfused rat liver: ligand and receptor dynamics, J. Cell Biol. 98:2148-2159 (1984).
23. W.A. Dunn, T.P. Connolly, and A.L. Hubbard, Receptor mediated endocytosis of epidermal growth factor by rat hepatocytes: receptor pathway, J. Cell Biol. 102:24-36 (1986).
24. D.G. Kay, W.H. Lai, M. Uchihashi, M.N. Khan, B.I. Posner and J.J.M. Bergeron, Epidermal growth factor receptor kinase translocation and activation in vitro, J. Biol. Chem. 261:8473-8480 (1986).
25. M.N. Khan, S. Savoie, J.J.M. Bergeron, and B.I. Posner, Differential kinetics and sensitivity to chloroquine of receptor-mediated insulin and prolactin endocytosis in liver parenchyma cells, Biochim. Biophys. Acta 888:100-106 (1986).
26. B.I. Posner, J.J.M. Bergeron, Z. Josefsberg, M.N. Khan, R.J. Khan, B.A. Patel, R.A. Sikstrom, and A.K. Verma, Polypeptide hormones: intracellular receptors and internalization, Rec. Prog. Horm. Res. 37:539-582 (1981).
27. M. Fehlmann, J.L. Carpentier, E. Van Obberghen, P. Freychet, P. Thamm, D. Saunders, D. Brandenburg, and L. Orci, Internalized insulin receptors are recycled to the cell surface in rat hepatocytes, Proc. Natl. Acad. Sci. U.S.A. 79:5921-5925 (1982).
28. J.L. Carpentier, H. Gazzano, E. Van Obberghen, M. Fehlmann, P. Freychet, and L. Orci, Intracellular pathway followed by the insulin receptor covalently coupled to ^{125}I-photoreactive insulin during internalization and recycling, J. Cell Biol. 102:989-996 (1986).
29. V. Murthy, M. Basu, A. Sen-Majumdar, and M. Das, Perinuclear location and recycling of epidermal growth factor kinase: immunofluorescent visualization using antibodies directed to kinase and extracellular domains, J. Cell Biol. 103:333-342 (1986).
30. W.H. Lai, D.G. Kay, M.N. Khan, B.I. Posner, and J.J.M. Bergeron, In vivo studies on the apparent recycling of EGF and insulin receptor kinase activities, J. Cell Biol. 103:450a (1986).
31. B. Tycko and F.R. Maxfield, Rapid acidification of endocytic vesicles containing 2 macroglobulin, Cell 28:643-651 (1982).
32. C.J. Galloway, C.E. Dean, M. Marsh, G. Rudnick, and I. Mellman, Acidification of macrophage and fibroblast endocytic vesicles in vitro, Proc. Natl. Acad. Sci. U.S.A. 80:3334-3338 (1983).
33. M. Merion, P. Schlesinger, R.M. Brooks, J.M. Moehring, T.J. Moehring and W.S. Sly, Defective acidification of endosomes in Chinese hamster ovary cell mutants "cross-resistant" to toxins and viruses, Proc. Natl. Acad. Sci. U.S.A. 80:5315-5319 (1983).
34. B.I. Posner, Z. Josefsberg, and J.J.M. Bergeron, Intracellular polypeptide hormone receptors: characterization of insulin binding sites in Golgi fractions from the liver of female rats, J. Biol. Chem. 253:4067-4073 (1978).
35. W.H. Lai, H.J. Guyda, C.L. Branchard, and C.G. Goodyear, Insulin induced receptor regulation in early gestation and term human placental cell cultures, Placenta 6:505-517 (1985).
36. T. Wileman, R. Boshans, and P. Stahl, Uptake and transport of mannosylated ligands by alveolar macrophages: studies on ATP-dependent receptor-ligand dissociation. J. Biol. Chem. 260:7387-7393 (1985).
37. J.N. Liment, J. Quintart, Y.J. Schneider, and P. Courtoy, Receptor mediated endocytosis of polymeric IgA and galactosylated serum

albumin in rat liver. Evidence for intracellular ligand sorting and identification of distinct endosomal compartments, Eur. J. Biochem. 146:539-548 (1985).
38. D.G. Kay, M.N. Khan, B.I. Posner, and J.J.M. Bergeron, In vivo uptake of insulin into hepatic Golgi fractions: application of the diaminobenzidine-shift protocol, Biochem. Biophys, Res. Commun., 123:1144-1148 (1984).
39. R.J. Khan, M.N. Khan, J.J.M. Bergeron, and B.I. Posner, Prolactin uptake into liver endocytic components: reduced sensitivity to chloroquine, Biochim. Biophys. Acta 838:77-83 (1985).
40. F.G. Hamel, B.I. Posner, M. Vanderwel, J.J.M. Bergeron, and W.C. Duckworth, Insulin degradation by endosomes in intact liver, Diabetes 36 (supplement 1): 56a (1987).
41. R.J. Pease, G.D. Smith, T.J. Peters, Degradation of endocytosed insulin in rat liver is mediated by low-density vesicles, Biochem. J. 228:137-146 (1985).
42. S. Diment and P. Stahl, Macrophage endosomes contain proteases which degrade endocytosed protein ligands, J. Biol. Chem. 260:15311-15317 (1985).
43. S. Diment, M.E. Costlow, and P. Stahl, Rabbit alveolar macrophage endosomes and lysosomes contain both membrane associated and soluble forms of cathepsin D-like protease, J. Cell Biol. 103:358a (1986).
44. A. Lanzavecehia, Antigen-specific interaction between T and B cells. Nature 314:537-539 (1985).
45. A. Ullrich, J.R. Bell, E.Y. Chen, R. Herrera, M. Petruzzelli, T.J. Dull, A. Gray, L. Cousseus, Y.C. Liao, M. Tsubokawa, A. Mason, P.H. Seeburg, C. Grunfeld, O.M. Rosen, J. Ramachandran, Human insulin receptor and its relationship to the tyrosine kinase family of oncogenes, Nature 313:756-761 (1985).
46. A. Ullrich, L. Cousseus, J.S. Hayflick, T.J. Dull, A. Gray, A.W. Tam, J. Lee, Y. Yarden, T.A. Libermann, J. Schlesinger, J. Downward, E.L.V. Mayes, N. Whittle, M.D. Waterfield, P.H. Seeburg, Human epidermal growth factor receptor cDNA sequence and aberrant expression of the amplified gene in A431 epidermoid carcinoma cells, Nature 309:418-425 (1984).
47. T. Hunter, At last the insulin receptor, Nature 313:740-761 (1985).
48. J. Schlesinger, Allosteric regulation of the epidermal growth factor receptor kinase, J. Cell Biol. 103:2067-2072 (1986).
49. D.O. Morgan and R.A. Roth, Acute insulin action requires insulin receptor kinase activity: introduction of an inhibitory monoclonal antibody into mammalian cells blocks the rapid effects of insulin, Proc. Natl. Acad. Sci. 84:41-45 (1987).
50. Y. Ebina, E. Araki, M. Taira, F. Shimada, M. Mori, C. Craik, K. Siddle, S.B. Peirce, R.A. Roth, W.J. Rutter, Replacement of lysine residue 1030 in the putative ATP-binding region of the insulin receptor abolishes insulin and antibody stimulated glucose uptake and receptor kinase activity, Proc. Natl. Acad. Sci. U.S.A. 84:704-708 (1987).
51. L. Ellis, E. Clauser, D.O. Morgan, M. Edery, R.A. Roth, and W.J. Rutter, Replacement of insulin receptor tyrosine residues 1162 and 1163 compromises insulin-stimulated kinase activity and uptake of 2-deoxyglucose, Cell 45:721-732 (1986).
52. B.I. Posner, M.N. Khan, J.J.M. Bergeron, Receptor-mediated uptake of peptide hormones and other ligands, in: "Polypeptide Hormone Receptors", ed. B.I. Posner, Marcel Dekker Inc., N.Y., N.Y. (1985).
53. B.I. Posner, M.N. Khan, J.J.M. Bergeron, Endocytosis of peptide hormones and other ligands, Endocr. Rev. 3:280-298 (1982).
54. F.C. Kosmakos and J. Roth, Insulin-induced loss of the insulin receptor in IM-9 lymphocytes: a biological process mediated through the insulin receptor, J. Biol. Chem. 255:9860-9869 (1980).
55. M.J. Prince, D. Baldwin, O. Tsai, J.M. Olefsky, Regulation of

insulin receptors in cultured human fibroblasts, Endocrinology 109:1754-1759 (1981).
56. A. Aharonov, R. Pruss, H.R. Herschman, Epidermal growth factor: relationship between receptor regulation and mitogenesis in 3T3 cells, J. Biol. Chem. 253:3970-3977 (1978).
57. W.H. Lai and H.J. Guyda, Characterization and regulation of epidermal growth factor receptors in human placental cell cultures, J. Clin. Endocrinol. Metab. 58:344-352 (1984).
58. S. Cohen and R.A. Fava, Internalization of functional epidermal growth factor: receptor/kinase complexes in A431 cells, J. Biol. Chem. 260:12351-12358 (1985).
59. Y. Yarden, and J. Schlesinger, Epidermal growth factor induces rapid, reversible aggregation of the purified epidermal growth factor receptor, Biochemistry 26:1443-1451 (1987).
60. Y. Yarden, and J. Schlesinger, Self-phosphorylation of epidermal growth factor receptor: evidence for a model of intermolecular allosteric activation, Biochemistry 26:1434-1442 (1987).
61. W.H. Lai, H.J. Guyda, J.J.M. Bergeron, Binding and internalization of epidermal growth factor in human term placental cells in culture, Endocrinology 118:413-423 (1986).
62. R. Fuchs, S. Schmid, P. Male, A. Helenius, and I. Mellman, Isolation and ATP-dependent acidification of endosome subcompartments from CHO cells, J. Cell Biol. 103:439a (1986).
63. M.C. Kielian, M. Marsh, and A. Helenius, Kinetics of endosome acidification detected by mutant and wild type Semliki Forest virus, EMBO. J. 12:3103-3109 (1986).
64. J. Harford, A.W. Walkoff, G. Ashwell, and R.D. Klausner, Monensin inhibits intracellular dissociation of asialoglycoproteins from their receptor, J. Cell Biol. 96:1824-1828 (1983).
65. J. Harford, K. Bridges, G. Ashwell, and R.D. Klausner, Intracellular dissociation of receptor bound asialoglycoproteins in cultured hepatocytes: a pH-mediated non-lysosomal event, J. Biol. Chem. 258:3191-3197 (1983).
66. G.J. Strous, A. DuMaine, J.E. Zijderhand-Bleekemolen, W.J. Slot, and A.L. Schwartz, Effect of lysosomotropic amines in the secretory pathway and on the recycling of the asialoglycoprotein receptor in human hepatoma cells, J. Cell Biol. 101:531-539 (1985).
67. A.L. Schwartz, A. Bolognesi, and S.E. Fridovich, Recycling of the asialoglycoprotein receptor and the effect of lysosomotropic amines in hepatoma cells, J. Cell Biol. 98:732-738 (1984).
68. M. De Brabander, R. Nuydens, H. Geerts, M. Moeremans, and C. Hopkins, Intracellular routing of receptors observed in living cells using Nanovid microscopy, J. Cell Biol. 103:61a (1986).
69. A. Robert, J.L. Carpentier, E. Van Obberghen, B. Canivet, P. Gorden, and L. Orci, The endosomal compartment of rat hepatocytes. Its characterization in the course of ^{125}I-insulin internalization, Exp. Cell Res. 159:113-126 (1985).
70. J. Downward, P. Parker, and M.D. Waterfield, Autophosphorylation sites in the EGF receptor, Nature 311:483-485 (1984).
71. P.J. Bertics and G.N. Gill, Self-phosphorylation enhances the protein-tyrosine kinase activity of the epidermal growth factor receptor, J. Biol. Chem. 260:14642-14647 (1986).
72. L. Stadtmaner and O.M. Rosen, Phosphorylation of synthetic insulin receptor peptides by the insulin receptor kinase and evidence that the preferred sequence containing Tyr-1150 is phosphorylated in vivo, J. Biol. Chem. 261:10000-10005 (1986).

CYTOSOLIC SUBSTRATES FOR THE INSULIN RECEPTOR KINASE

Yan C. Kwok and Cecil C. Yip

Banting and Best Department of Medical Research
University of Toronto, Toronto
Ontario, Canada, M5G 1L6

INTRODUCTION

Insulin receptor belongs to a family of cell surface receptors possessing ligand-activated tyrosine kinase activity. In the case of the insulin receptor, phosphorylation of tyrosine residues in its β subunit is stimulated by the binding of insulin to the α subunit (1-8), resulting in an increased receptor tyrosine phosphotransferase activity towards several artificial substrates (9-14). That receptor kinase activity is involved in insulin action is supported by the observation that the effect of insulin was abolished in intact cells when their insulin receptor kinase was inhibited (15) or in transfectant cells expressing insulin receptors deficient in kinase activity (16). As is the case with other receptor kinase, the exact relationship between receptor kinase activity and insulin action remains unresolved. It is possible that the activated insulin receptor may initially phosphorylate one or more membrane or cytosolic proteins which catalyze subsequent changes in the activities of key enzymes. The identification of such cellular substrate(s) for the insulin receptor kinase, or for the receptor kinase of other growth factors, is an important step to understanding the mechanism of action of insulin or other growth factors. Since the same tissue can possess more than one kind of receptor with kinase activity, for example, receptors for both EGF and insulin are present in hepatocyte, and since different tissues or target organs respond differently to the same hormone, for example the different actions of insulin on fat and liver tissue, it is reasonable to postulate that endogenous substrates would be both tissue and hormone specific. Thus an endogenous substrate S of the insulin receptor in the liver would not be phosphorylated by the EGF receptor, and substrate S may not be found in fat tissue. Several putative substrates of the insulin receptor kinase have been detected recently. These are listed in Table I. They were detected primarily as additional insulin-stimulated phosphorylated components in membrane preparations or in preparations of partially purified insulin receptors. The role of these proteins in the action of insulin is not known, except the 15-kDa protein detected in the 3T3-L1 adipocytes. This protein was linked circumstantially to the insulin-stimulated glucose oxidation by the 3T3-L1 adipocytes.

Table 1. Putative Endogenous Substrates of Insulin Receptor Kinase

M_r (kDa)	Tissue or Cell	Reference
180	FaO hepatoma cells	21
150	CHO insulin receptor transfectants	16
120	H-35 hepatoma cells	17
120	Liver membranes	20
110	Adipose and liver membranes	19
45	Adipocytes	22
35	CHO insulin receptor transfectants	16
15	3T3-L1 adipocytes	18

In our study we reason that the putative endogenous substrate of the insulin receptor kinase should also serve as a substrate in vitro. Our approach was to use solubilized insulin receptors immobilized on Sepharose beads to probe for the receptor substrate(s) in rat liver cytosolic fractions. A purified and concentrated preparation of receptor tyrosine kinase can be prepared utilizing the high affinity binding of the insulin receptor to insulin-Sepharose (11). The immobilized receptor kinase can be activated by autophosphorylation with millimolar concentrations of ATP and phosphorylation of substrates can be carried out subsequently in micromolar ATP. Thus we can regulate and optimize the conditions of phosphorylation and identify the cytosolic fractions that would contain the putative substrates.

MATERIALS AND METHODS

Our experimental protocol is outlined in Figure 1.

Preparation of immobilized insulin receptor

Rat livers (10g) were homogenized in 25 ml of 10 mM Tris-HCl, pH 7.4, containing 5 mM EDTA, 10% glycerol, 10 mM aminocaproic acid, 10 mM benzamidine, 0.25 M sucrose, 100 uM phenylmethylsulfonyl fluoride (PMSF), bacitracin (1 mg/ml), leupeptin (1 ug/ml), pepstatin (1 ug/ml) and aprotinin (1 ug/ml). The preparation of microsomal membranes, and the solubilization of the insulin receptor in Triton X-100 were carried out as described (11). The insulin-Sepharose was prepared by reacting insulin with succinyl diamino-dipropylamine-Sepharose (1 mg of insulin/ml of settled beads) according to the method of Cuatrecasas and Parikh (23). Binding of the solubilized insulin receptor to insulin-Sepharose was carried out at 4° C for 16 h. The receptor-insulin-Sepharose was washed with 50 ml of 50 mM Tris-HCl, pH 7.4, containing 0.2% Triton X-100 and 1 M NaCl, and then with 50 ml of the same buffer without NaCl. Where indicated, the immobilized insulin receptor was phosphorylated with 2 mM ATP in the presence of 10 mM $MnCl_2$ at 24° C for 30 min, washed and then used.

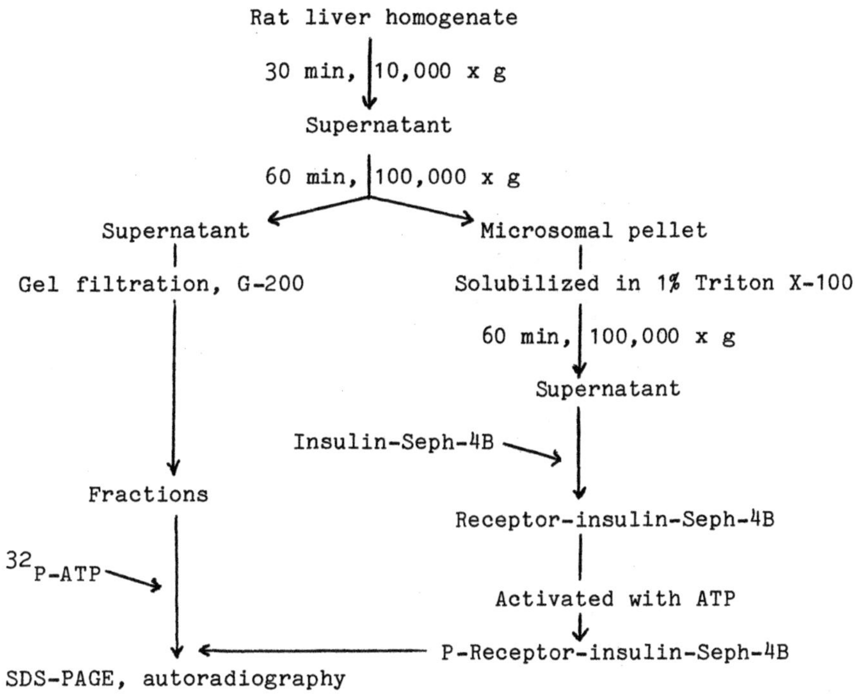

Figure 1. Experimental Protocol

<u>Preparation and phosphorylation of cytosolic substrates</u>

Two ml of the liver supernatant from the preparation of microsomal membranes, as described above, were fractionated at 4 C on a column (1.5 cm x 90 cm) of Sephadex G-200 equilibrated with 50 mM Tris-HCl, pH 7.4. Fractions of 3 ml were collected and stored at -76 C. Phosphorylation was carried out at 24 C for 15 min in a reaction mixture (80 ul) containing 20 ul of insulin receptor-Sepharose, 40 ul of each fraction, 10 mM $MnCl_2$ or $MgCl_2$, 5 uM ATP (containing 5 uCi of ^{32}P-ATP), 50 mM Tris-HCl, pH 7.4, and 0.5% Triton X-100 for 15 min. The reaction was terminated by the addition of 20 ul of a solution containing 0.45 M DTT, 60 mM EDTA, 6% SDS and 30% sucrose. The reaction mixture was subjected to sodium dodecyl sulfate-polyacrylamide gel electrophoresis (SDS-PAGE) in a 10 % gel. The gel was stained and dried. Autoradiography was carried out overnight on Kodak XAR-5 film enhanced by a Lightning Plus screen.

RESULTS AND DISCUSSION

The phosphorylation of liver cytosolic fractions by the insulin receptor kinase immobilized on insulin-Sepharose beads is shown in Figure 2. The receptor kinase catalyzed the phosphorylation of three protein bands of 95 kDa, 50 kDa and 35 kDa. The 95-kDa band, present in all the fractions incubated with the insulin receptor beads, was the β subunit of the insulin receptor. A 50-kDa band in fractions 23 to 27 and a 35-kDa band in fractions 31 to 39 were also phosphorylated by the insulin receptor beads. In fractions 35 to 39, a doublet of 35 kDa was evident. The 50-kDa and the 35-kDa band evidently were not linked by disulfide bonds to other components since their electrophoretic mobilities were not

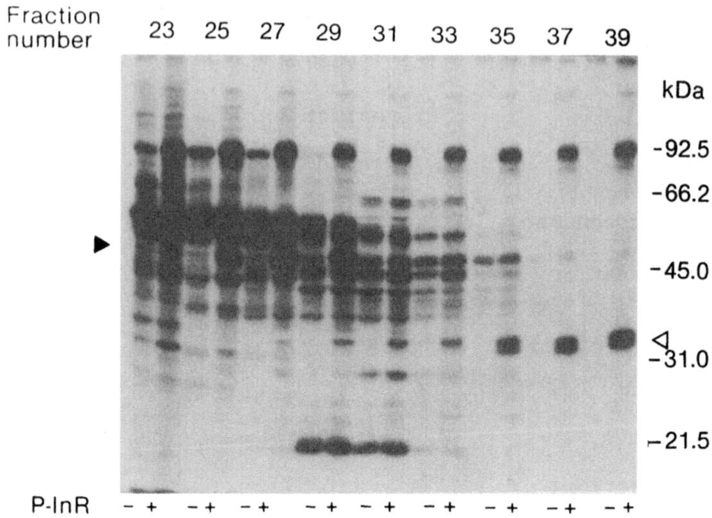

Fig. 2. Phosphorylation of the hepatic cytosolic proteins by insulin receptor-Sepharose. Two ml of rat liver supernatant were fractionated on a column of Sephadex G-200 as described in the text. An aliquot of 40 ul from alternate fractions as indicated was incubated with Mn-^{32}P-ATP in the absence (-) or presence (+) of activated receptor-Sepharose (P-InR). The reaction mixture was analyzed by SDS-PAGE and autoradiography. The phosphorylated 50-kDa and 35-kDa bands are identified, respectively, by solid and open triangles. Positions of molecular mass markers in kDa are indicated. (Reprint from Ref. 29 with permission from the British Biochemical Society).

altered by reduction with dithiothreitol. On the other hand, the 50-kDa band was collected at an elution volume slightly ahead of bovine serum albumin suggesting that under non-denaturing conditions the 50-kDa protein either self-aggregated or associated with other cytosolic components. It is interesting to note that when unfractionated cytosol was used, no receptor-catalyzed phosphorylation could be demonstrated. Gel-filtration might have removed inhibitory or interferring substances from the cytosol.

As shown in Figure 3, compared to the non-activated insulin receptor (InR), insulin receptors (P-InR) which had been activated with ATP increased the rate of phosphorylation of these two bands. Phosphorylation of the 50-kDa band was evident after only 1 min at 24 C in the presence of the P-InR. Activated receptors required a lower concentration of ATP (5 uM) than the non-activated receptors (more than 10 uM) in the incubation mixture to achieve a maximal phosphorylation of the bands. Both the 50-kDa and the 35-kDa bands were tyrosine-phosphorylated because their radioactivity was not released by 85% KOH at 55 C for 2 h (24), and phosphoamino acid analysis showed mainly phosphotyrosine. As well, both proteins were precipitated by antibodies against phosphotyrosine (Fig. 3, lane b and f).

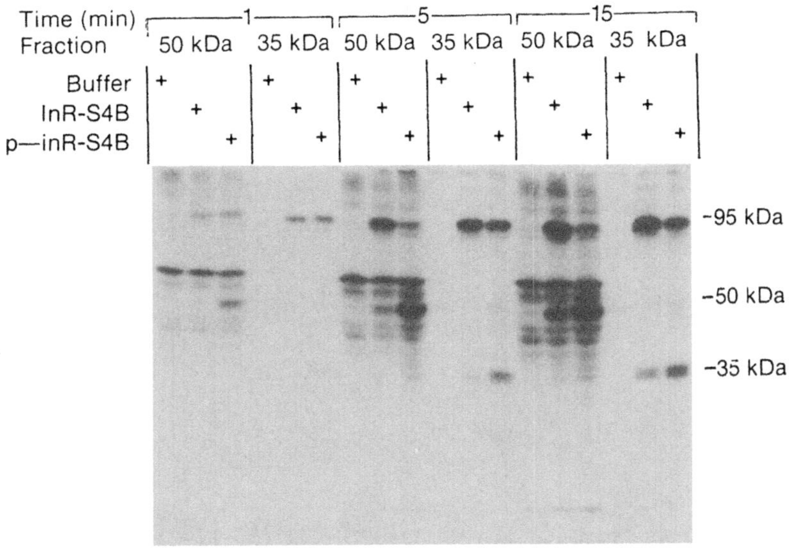

Fig. 3. Time course of the phosphorylation of the 50-kDa and 35-kDa bands by insulin receptor beads not previously activated with ATP (InR-S4B) or by ATP-activated insulin receptor-Sepharose (P-InR-S4B). Aliquots of 40 ul from the column fractions containing the 50-kDa or 35-kDa proteins were incubated with 5 uM ATP and 10 mM $MnCl_2$ at 24 C for the periods indicated. Positions of the phosphorylated bands of the receptor β subunit (95 kDa) and the 50-kDa and 35-kDa substrates are indicated. (Reprint from Ref. 29 with permission from the British Biochemical Society).

We have determined that the 50-kDa and 35-kDa proteins were not derived from the 95-kDa subunit of the insulin receptor since the peptide maps of these two bands and the receptor 95-kDa subunit were very different. Furthermore, antibodies that we have prepared against the 50-kDa protein were able to precipitate the phosphorylated 50-kDa protein but not the receptor β subunit (Fig. 4, lane c) nor the phosphorylated 35-kDa protein (Fig. 4, lane g). Similarly, antibodies prepared against the 35-kDa protein were specific for the phosphorylated 35-kDa protein (Fig. 4, lanes d and h). Taken together, these results demonstrate that the 95-kDa, 50-kDa and the 35-kDa proteins are unrelated to each other, and they are not associated with each other as an immunoprecipitable complex.

The phosphorylated 50-kDa band and the 35-kDa doublet did not correspond to any major silver-stained or Coomassie Blue-stained bands on the gel. Assuming the molar incorporation of phosphate to be one, we estimate that there are about 100,000 molecules of each substrate per cell or 1 nanomole of each substrate per 15 g of rat liver.

These two hepatic proteins appear to be specific substrates for insulin receptor kinase, since epidermal growth factor (EGF) receptors prepared from A431 cell lines (a gift from Dr. J. Kudlow, Department of Clinical Biochemistry, University of Toronto, Toronto) did not stimulate the phosphorylation of the two proteins under conditions optimal for EGF receptor kinase activity at 4 C (Fig. 5) or at 24 C. Insulin receptor-

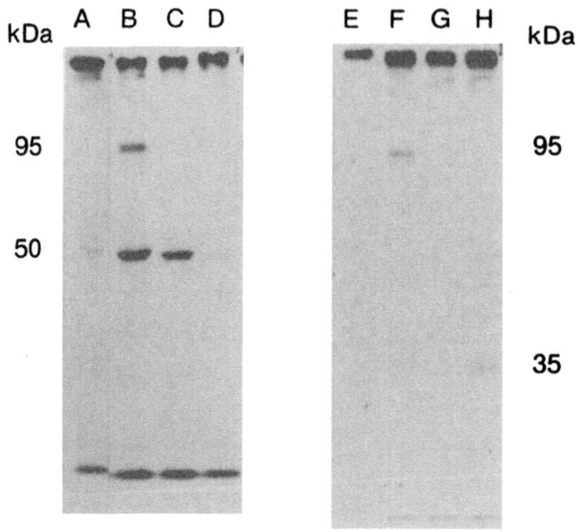

Fig. 4. Immunoprecipitation of the 95-kDa, 50-kDa and 35-kDa phosphoproteins by polyclonal antibodies against phosphotyrosine, 50-kDa protein and 35-kDa protein. After phosphorylation of the 50-kDa and 35-kDa proteins by the insulin receptor-Sepharose, the reaction mixtures were frozen and thawed several times to dissociate the phosphorylated insulin receptor from insulin-Sepharose. The supernatants containing the phosphorylated 50-kDa protein (lanes a, b, c and d), or the phosphorylated 35-kDa protein (lanes e, f, g and h) were precipitated with normal sera (a, e), polyclonal anti-phosphotyrosine antibodies (b, f), anti-50-kDa antibodies (c, g) and anti-35-kDa antibodies (d, h). The phosphorylated receptor β subunit of 95 kDa present in the supernatant was also precipitated by the anti-phosphotyrosine antibodies (b and f). (Reprint from Ref. 29 with permission from the British Biochemical Society).

catalyzed phosphorylation of these proteins seems to be unique to the liver, since we have not been able to demonstrate the phosphorylation of the two proteins in fractionated cytosol from brain, heart and adipose tissues of the rat. Under similiar conditions, activated insulin receptors catalyzed the phosphorylation of a 50-kDa and a 35-kDa protein in the fractionated cytosols from livers of rabbit and mouse.

The roles of these phosphoproteins in the action of insulin remain to be established. Since the liver 50-kDa and 35-kDa proteins were cytosolic and were not phosphorylated by EGF receptor kinase it is unlikely that they are related to lipocortin which is associated with the cell membrane cytoskeleton (25). It is significant that the insulin receptor kinase was found not to phosphorylate EGF receptor substrate. It was reported that EGF receptor kinase and insulin receptor kinase have similar but not identical specificities for peptide substrates and divalent metal ion requirements (10, 26). Insulin receptor and EGF receptor are both growth factor receptors and their kinase domains are similar (27, 28). However, insulin is different from EGF in its

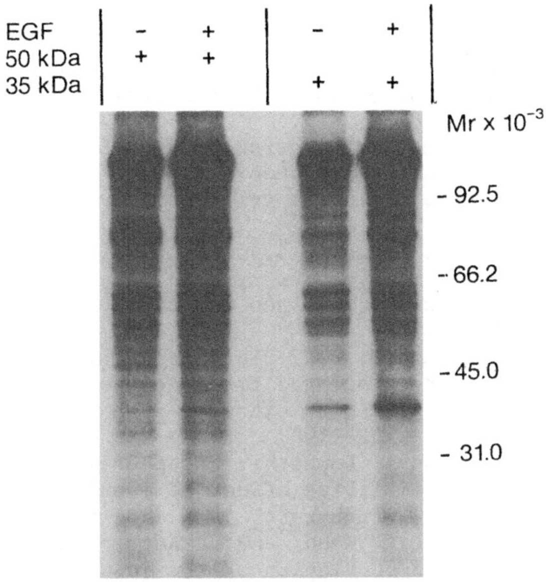

Fig. 5. Inability of epidermal growth factor (EGF) receptor tyrosine kinase to phosphorylate the rat liver cytosol 50-kDa and 35-kDa proteins. Plasma membranes prepared from A431 cells were incubated with (+) or without (-) EGF at 10^{-7} M for 10 min at 4° C in 20 mM HEPES buffer, pH 7.4. The phosphorylation reaction was carried out in 20 mM $MgCl_2$, 1 mM $CaCl_2$, 10 uM sodium orthovanadate, and 0.5 uM ATP for 10 min at 4° C. Fractions containing the 50-kDa or 35-kDa protein were added as indicated. The phosphorylated 170-kDa EGF receptor band is indicated by the triangle. The phosphorylation of the endogenous substrate for the EFG receptor kinase is also indicated (o). (Reprint from Ref. 29 with permission of the British Biochemical Society).

metabolic effects on the liver. It is therefore reasonable to suggest that transduction of hormonal signals may involve the phosphorylation of substrates that are receptor specific.

In conclusion, we have detected two liver cytosolic proteins which appear to be unique substrates for the insulin receptor kinase. The two proteins seem to be present in very low abundance since they were not identified with a stainable band on SDS-PAGE. If we assume a one-to-one molar ratio of phosphorylation, we estimate that the liver contains about 1 nmole of each protein per 15 g of tissue. This would be expected of a putative substrate that might play a key role in the phosphorylation cascade leading to the appropriate cellular response.

ACKNOWLEDGEMENT

This study was supported by a grant from the Medical Research Council of Canada. YCK was a postdoctoral fellow of the C.H. Best Foundation and is now a recipient of a Juvenile Diabetes Foundation International Postdoctoral Fellowship.

REFERENCES

1. M. Kasuga, F.A. Karlsson, and C.R. Kahn, Insulin stimulates the phosphorylation of the 95,000-Dalton subunit of its own receptor, Science 215: 185 (1982).
2. D.T.Pang, B.R. Sharma, J.A. Shafer, M.F. White, and C.R.Kahn, Predominance of tyrosine phosphorylation of insulin receptors during the initial response of intact cells to insulin, J. Biol. Chem. 260: 7131 (1985).
3. L.M. Petruzzelli, S. Ganguly, C.J. Smith, M.H. Cobb, C.S. Rubin, and O.M. Rosen, Insulin activates a tyrosine-specific protein kinase in extracts of 3T3-L1 adipocytes and human placenta, Proc. Natl. Acad. Sci. USA 79: 6792 (1982).
4. J. Avruch, R.A. Nemenoff, P.J. Blackshear, M.W. Pierce, and R. Osathanondh, Insulin-stimulated tyrosine phosphorylation of the insulin receptor in detergent extracts of human placental membranes, J. Biol. Chem. 257: 15162 (1982).
5. E. Van Obberghen, and A. Kowalski, Phosphorylation of the hepatic insulin receptor: Stimulating effect of insulin on intact cells and in a cell-free system, FEBS Lett. 143: 179 (1982).
6. Y. Zick, M. Kasuga, C.R. Kahn, and J. Roth, Characterization of insulin-mediated phosphorylation of the insulin receptor in a cell-free system, J. Biol. Chem. 258: 75 (1983).
7. Y. Zick, R.W. Rees-Jones, G. Grunberger, S.I. Taylor, V. Moncada, P. Gorden, and J. Roth, The insulin-stimulated receptor kinase is a tyrosine-specific casein kinase, Eur. J. Biochem. 137: 631 (1983).
8. L.A. Stadtmauer, and O.M. Rosen, Phosphorylation of exogenous substrates by the insulin receptor-associated protein kinase, J. Biol. Chem. 258: 6682 (1983).
9. L. Petruzzelli, R. Herrera, and O.M. Rosen, Insulin receptor is an insulin-dependent tyrosine protein kinase: Copurification of insulin-binding activity and protein kinase activity to homogeneity from human placenta, Proc. Natl. Acad. Sci. USA. 81: 3327 (1984).
10. L.J. Pike, E.A. Kuenzel, J.E. Casnellie, and E.G. Krebs, A comparison of the insulin- and epidermal growth factor-stimulated protein kinases from human placenta, J. Biol. Chem. 259: 9913 (1984).
11. R.A. Nemenoff, Y.C. Kwok, G.I. Shulman, P.J. Blackshear, R. Osathanondh, and J. Avruch, Insulin-stimulated tyrosine protein kinase: Characterization and relation to the insulin receptor, J. Biol. Chem. 259: 5058 (1984).
12. K.T. Yu, and M.P. Czech, Tyrosine phosphorylation of the insulin receptor subunit activates the receptor-associated tyrosine kinase activity, J. Biol. Chem. 259: 5277 (1984).
13. K.T. Yu, and M.P. Czech, Tyrosine phosphorylation of insulin receptor subunit activates the receptor tyrosine kinase in intact H-35 hepatoma cells, J. Biol. Chem. 261: 4715 (1986).
14. Y.C. Kwok, R.A. Nemenoff, A.C. Powers, and J. Avruch, Kinetic properties of the insulin receptor tyrosine protein kinase: Activation through an insulin-stimulated tyrosine-specific, intramolecular autophosphorylation, Arch. Biochem. Biophys. 244: 102 (1986).
15. D.O. Morgan, and R.A.Roth, Acute insulin action requires insulin receptor kinase activiTy: Introduction of an inhibitory monoclonal antibody into mammalian cells blocks the rapid effects of insulin, Proc. Natl. Acad. Sci. USA. 84: 41 (1987).
16. C.K. Chou, T.J. Dull, D.S. Russell, R. Gherzi, D. Lebwohl, A. Ullrich, and O.M. Rosen, Human insulin receptors mutated at the ATP-binding site lack protein tyrosine kinase activity and fail to mediate postreceptor effects of insulin, J. Biol. Chem. 262: 1842 (1987).

17. N. Perrotti, D. Accili, B. Marcus-Samuels, R.W. Rees-Jones, and S.I. Taylor, Insulin stimulates phosphorylation of a 120-kDa glycoprotein substrate (pp120) for the receptor-associated protein kinase in intact H-35 hepatoma cells, Proc. Natl. Acad. Sci. USA. 84: 3137 (1987).
18. M. Bernier, D.M. Laird, and M.D. Lane, Insulin-activated tyrosine phosphorylation of a 15-kilodalton protein in intact 3T3-L1 adipocytes, Proc. Natl. Acad. Sci. USA. 84: 1844 (1987).
19. J.L. Sadoul, J.F. Peyron, R. Ballotti, A. Debant, M. Fehlmann, and E. Van Obberghen, Identification of a cellular 110000-Da protein substrate for the insulin-receptor kinase, Biochem. J. 227: 887 (1985).
20. R.W. Rees-Jones, and S.I. Taylor, An endogenous substrate for the insulin receptor-associated tyrosine kinase, J. Biol. Chem. 260: 4461 (1985).
21. M.F. White, R. Maron, and C.R. Kahn, Insulin rapidly stimulates tyrosine phosphorylation of a M_r-185,000 protein in intact cells, Nature (London) 318: 183 (1985).
22. H.U. Haring, M.F. White, F. Machicao, B. Ermel, E. Schleicher, and B. Obermaier, Insulin rapidly stimulates phosphorylation of a 46-kDa membrane protein on tyrosine residues as well as phosphorylation of several soluble proteins in intact fat cells, Proc. Natl. Acad. Sci. USA. 84: 113 (1987).
23. P. Cuatrecasas, and I. Parikh, Adsorbants for affinity chromatography, use of N-hydroxysuccinimide esters of agarose, Biochemistry 11: 2291 (1972).
24. J.A. Cooper, B.M. Sefton, and T. Hunter, Detection and quantitation of phosphotyrosine in proteins, in Methods in Enzymology 99: 387 (1983).
25. R.B. Pepinsky, and L.K. Sinclair, Epidermal growth factor dependent phosphorylation of lipocortin, Nature (London) 321: 81 (1986).
26. H.H. Klein, G.R. Freidenberg, R. Cordera, and J.M. Olefsky, Substrate specificities of insulin and epidermal growth factor receptor kinases, Biochem. Biophys. Res. Commun. 127: 254 (1985).
27. A. Ullrich, J.R. Bell, E.Y. Chen, R. Herrera, L.M. Petruzzelli, T.J. Dull, A. Gray, L. Coussens, Y.C. Liao, M. Tsubokawa, A. Mason, P.H. Seeburg, C. Grunfeld, O.M. Rosen, and J. Ramachandran, Human insulin receptor and its relationship to the tyrosine kinase family of oncogenes, Nature (London) 313: 756 (1985).
28. A. Ullrich, L. Coussens, J.S. Hayflick, T.J. Dull. A. Gray, A.W. Tam, Lee, Y. Yarden, T.A. Libermann, J. Schlessinger, J. Downward, E.L.V. Mayes, N. Whittle, M.D. Waterfield, and P.H. Seeburg, Human epidermal growth factor receptor cDNA sequence and aberrant expression of The amplified gene in A431 epidermoid carcinoma cells, Nature (London) 309: 418 (1984).
29. Y.C. Kwok, and C.C. Yip, Tyrosine phosphorylation of two cytosolic proteins of 50 kDa and 35 kDa in rat liver by insulin-receptor kinase in vitro, Biochem. J. 248: 27 (1987).

INDEX

GPSR Compliance

The European Union's (EU) General Product Safety Regulation (GPSR) is a set of rules that requires consumer products to be safe and our obligations to ensure this.

If you have any concerns about our products, you can contact us on ProductSafety@springernature.com

In case Publisher is established outside the EU, the EU authorized representative is:

Springer Nature Customer Service Center GmbH
Europaplatz 3
69115 Heidelberg, Germany

Batch number: 10615115

Printed by Printforce, the Netherlands